The
Hamlyn Guide to
Minerals, Rocks and Fossils

W.R. Hamilton
A.R. Woolley
A.C. Bishop

Hamlyn
London · New York · Sydney · Toronto

Preface

In preparing this book we have been fortunate in being able to use for the colour illustrations material in the collections of the British Museum (Natural History), and we have selected typical, rather than the most spectacular, specimens. The authors and publishers are grateful to the Director, Dr G F Claringbull, for permission to photograph the specimens.

Many of our colleagues have helped us by reading and correcting the manuscript, and by making suggestions for its improvement. In particular we should like to thank Miss Valerie Jones, who has drawn most of the diagrams illustrating the sections on minerals and rocks. We are grateful to our editor, Ian Jackson, whose care and patience during the progress of the work made our task so much easier.

A C Bishop, A R Woolley (minerals and rocks) and W R Hamilton (fossils)

The size of the specimens shown in the photographs is indicated by a rule which is positioned, wherever possible, at the foot of each plate. The length of the rule varies from page to page, depending on the scale of the specimens on each plate (the longer the rule, the smaller the specimens), but it always represents 5 cm and therefore allows quick comparison of the sizes of specimens from anywhere in the book.

The photographs in this book were taken by Peter J Green/Imitor and the line drawings prepared by Valerie Jones and Oxford Illustrators Limited.

Published by
The Hamlyn Publishing Group Limited
London · New York · Sydney · Toronto
Astronaut House, Feltham, Middlesex, England

First published in paperback 1974
Ninth Impression 1982
Cased edition published 1980

ISBN 0 600 34398 7 (paperback)
ISBN 0 600 35353 2 (cased edition)

Text set in Univers by London Filmsetters Limited

**Printed in Spain by Printer industria gráfica sa
Sant Vicenç dels Horts Barcelona D.L.B. 14388-1980**

Contents

Introduction

This field guide is divided into three sections, namely minerals, rocks (including meteorites and tektites) and fossils. Each section comprises an introductory part, which is illustrated by line drawings, and a descriptive part, which is illustrated by line drawings and colour photographs. The introductory sections include the minimum basic information required to follow the descriptive sections adequately, while the descriptive sections, for ease of reference, are always arranged so that photographs and accompanying text are opposite one another.

To make the best use of the book the contents page and index should be used freely. The contents list will enable you to turn quickly to the appropriate section of the book, whereas if a tentative identification has been made, then reference to the index will immediately direct you to the relevant page. The index includes not only the names of specific minerals, rocks and fossils, but also technical terms which are used in describing them. By consulting the index you will be referred to the page on which the term is defined, and possibly illustrated.

The stratigraphical column is given on page 310, and will be a particularly valuable reference for collectors of fossils.

How to collect

The basic equipment required is a hammer, chisel, notebook and pencil, felt-tipped pen, wrapping materials and a bag. The usual geological hammer has a square head and a chisel edge, which is particularly useful for splitting rocks when looking for fossils. Do not be tempted to use any other kind of hammer. Geological hammers are specially tempered and others are likely to splinter when hammering, and metal splinters could damage the eyes. A steel chisel is sometimes required to prize open rocks which resist hammering, or for carefully breaking specimens which might be damaged by blows from a hammer. **When hammering be very careful indeed of flying splinters of rock.** Protective goggles can be obtained. Specimens should be carefully numbered; use either a felt-tipped pen or sticky tape on which a number can be written. The exact locality from which the specimens were collected should be recorded in the notebook. Specimens should always be wrapped in plenty of newspaper in order to prevent chipping or scratching, and small or delicate specimens are best carried in a small box, such as a match or cigar box. If a large collection is to be made, or if long distances are to be walked, then a stout rucksack is the most suitable kind of bag to have.

The best places to collect minerals, rocks and fossils are usually quarries, cliffs, road cuttings and mine dumps, but any outcrop of rock may prove fruitful. Particular care is needed, however, when collecting near quarry faces or from the foot of cliffs, and permission must always be sought if it is intended to collect from outcrops on private land. Remember to take care and precautions if you intend to do field work on your own, and always tell someone of your intended route before setting out.

Geological maps, sometimes on a large scale, are

available for most parts of the world, and they show the distribution and geological ages of the different rock types. This information should indicate where fossils are likely to be found, and where it is probably best to look for minerals, or for interesting rock types. If there is a museum in your area a visit may well be worth while. Many museums not only have exhibits illustrating the geology of their vicinity, but they also usually have displays of minerals, fossils, and sometimes rocks, which will help you to 'get your eye in', and give you some idea of what is to be found in your neighbourhood.

Housing a collection

A collection is best kept in a cabinet of shallow drawers, with the specimens placed in individual cardboard trays. Under no circumstances should specimens be placed one on top of the other. Each specimen should have its own label giving details of what it is and where and when it was collected. A number should also be firmly glued to each specimen, and a corresponding entry made in a notebook or a card index giving details such as name and locality, and any other relevant information. This entry is a safety precaution against accidental loss of or damage to the label attached to the specimen.

The system followed in this book will prove a useful guide in arranging specimens, though there are, of course, other systems which you may prefer to follow.

Mineral specimens, in particular, look their best when they are clean. To remove loose dust and dirt first take off the label, then immerse the specimen in clean water to which a little detergent has been added, and lightly scrub it with a soft brush. This should not be done, of course, with specimens which are soluble in water, or with very delicate material.

Further reading

Although in the introductory sections of this guide outlines of the subjects of mineralogy, petrology (the study of rocks), and palaeontology (the study of fossils) are given, it is obviously not possible in a single volume to do justice to these subjects. Further, although something like 600 specific types of mineral, rock and fossil are described in the following pages, there are many other types which, for reasons of space, cannot be included. To help the reader who would like to widen his knowledge, a list of recommended books is given on page 311. It would also be useful to include a list of the available geological maps and guides of particular areas, but such a list, if it is to be comprehensive, would need to be very long indeed. To find such maps and guides we suggest that you enquire at your local library.

For the real enthusiast there is no substitute for joining a geological society. Most countries have such societies organized on a national basis, but there are also local societies which cater mainly for the enthusiastic amateur, and which often have geological libraries, and organize field excursions to good collecting localities.

Minerals

The rocks which form the Earth, the Moon and the planets are made up of minerals. Minerals are solid substances composed of atoms having an orderly and regular arrangement. This orderly atomic arrangement is the criterion of the crystalline state and it means also that it is possible to express the composition of a mineral as a chemical formula.

Crystals When minerals are free to grow without constraint, they are bounded by crystal faces which are invariably disposed in a regular way such that there is a particular relationship between them in any one mineral species. A crystal is bounded by naturally formed plane faces, and its regular outward shape is an expression of its regular atomic arrangement.

The structure of minerals The internal structure of minerals has been determined only during this century, by the use of X-rays, although for about 200 years it had been appreciated that crystals are almost incredibly regular. This is not at once apparent for crystals of the same substance, such as quartz, have faces that seem almost infinitely variable in their size and shape; it is only when the angles between corresponding pairs of faces are measured that the regularity becomes apparent. The angle between the same two faces in all crystals of the same mineral species is constant (Fig. 1). It is now known that this is

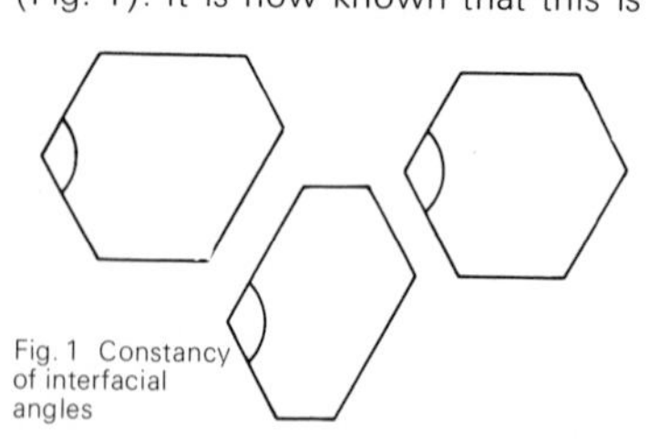

Fig. 1 Constancy of interfacial angles

because the constituent atoms pack together in a definite and orderly way. Crystals were studied long before this was appreciated, however, and from a study of external shape alone it was deduced that crystals were symmetrical and could be grouped according to their symmetry.

Crystal symmetry We are familiar with symmetrical objects such as boxes, furniture and even ourselves. Close inspection of such objects will reveal that they can be symmetrical about a *plane* such that if the object were to be cut in half along the plane, one half would be the mirror image of the other (Fig. 2). The human body is symmetrical externally about a vertical plane arranged from front to back.

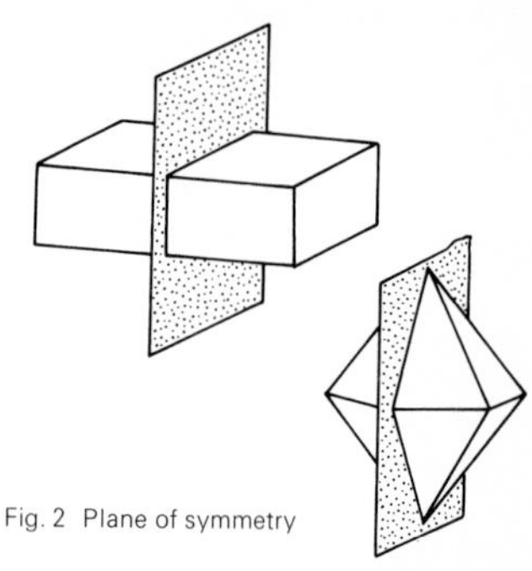

Fig. 2 Plane of symmetry

Objects can also be symmetrical about a *line* or *axis* which is considered to pass through their centre. When crystals are rotated about this axis they present the same appearance twice, three times, four times or six times during a complete revolution (Fig. 3). The axis is called an axis of two-fold, three-fold, four-fold or six-fold sym-

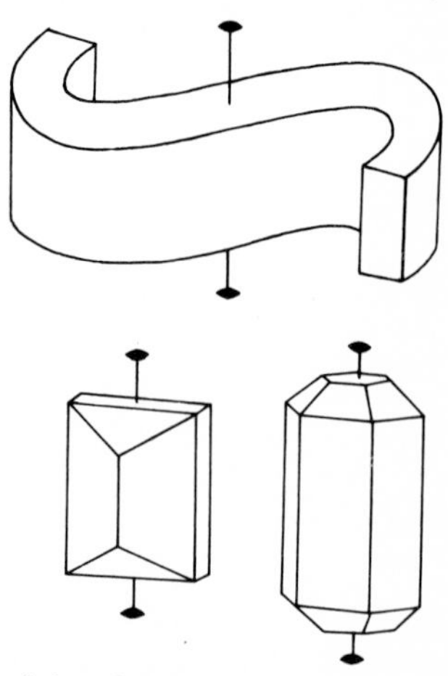

Fig. 3 Axes of symmetry

metry. Crystals never have an axis of five-fold symmetry. Finally, crystals can be said to be symmetrical about a *centre* if a face on one side of the crystal has a corresponding parallel face on the other (Fig. 4).

The crystal systems On the basis of their symmetry, crystals can be grouped into six crystal systems, and can be referred to imaginary reference axes, as shown in the diagrams (Fig. 5). A seventh crystal system, the trigonal, is recognized by many mineralogists. It has the same set of reference axes as the hexagonal system, but has a vertical three-fold axis of symmetry. These reference axes are chosen so as to be parallel to the edges of the unit cell (the repeat unit of pattern in a crystal structure), and hence they can be regarded as having length. The following table summarizes the seven crystal systems.

Fig. 4

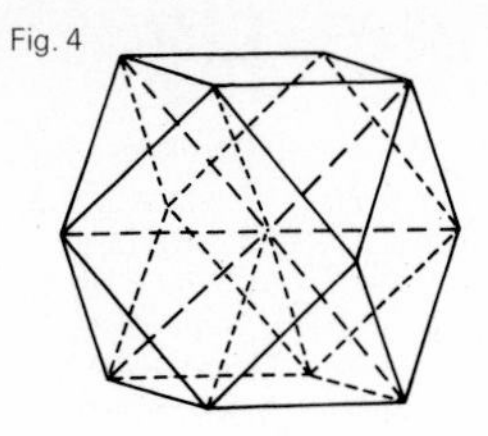

Centre of symmetry: cube and octahedron

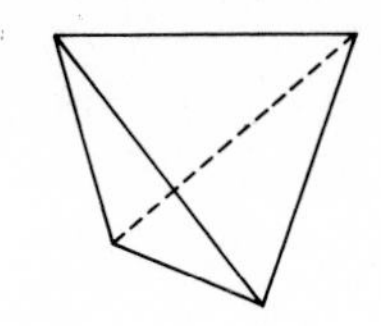

No centre of symmetry: tetrahedron

System	Symmetry	Reference axes
Cubic	4 three-fold axes	3 axes mutually at right-angles, and of equal length
Tetragonal	One vertical four-fold axis	3 axes mutually at right-angles; one axis, conventionally held vertically, differing in length from the other two
Orthorhombic	*Either* one two-fold axis at the intersection of two mutually perpendicular planes; *or* 3 mutually perpendicular two-fold axis	3 axes mutually at right-angles, all of different length
Monoclinic	One two-fold axis	3 axes of unequal length; two axes are not at right-angles; the third, the symmetry axis, is at right-angles to the plane containing the other two
Triclinic	*Either* a centre of symmetry; *or* no symmetry	3 axes, all of unequal length, none at right-angles to the others
Hexagonal	One vertical six-fold axis	4 axes, three of equal length arranged in a horizontal plane; the fourth perpendicular to this plane and of different length from the other three
Trigonal	One vertical three-fold axis	As for hexagonal

Crystal form It is useful when identifying minerals to determine to which crystal system they belong, but minerals that crystallize in the same crystal system, and even crystals of the same substance, can show remarkable differences in shape according to which crystal *form*, or combination of forms, is developed.

A crystal form comprises all the faces

Fig. 5 Reference axes of the crystal systems and some examples of crystals belonging to each

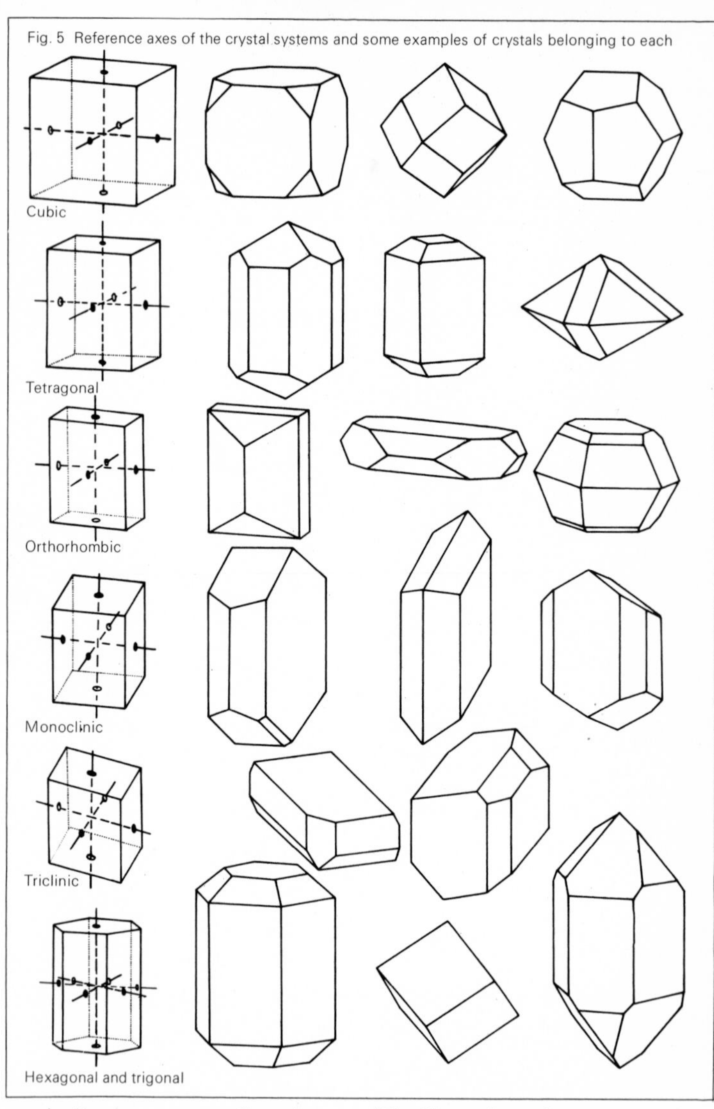

required by the symmetry. Some forms, such as the cube and octahedron (Fig. 6), totally enclose space and are called *closed forms* and can occur by themselves as crystals. Other forms, such as a pinacoid (a pair of parallel faces), or a prism (a form comprising three or more faces that meet in edges that are parallel) do not totally enclose space, and are called *open forms* (Fig. 7). Clearly they can occur only

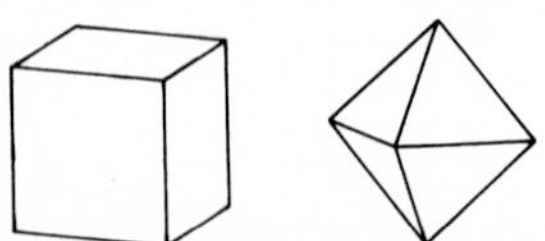

Fig. 6 Cube and octahedron

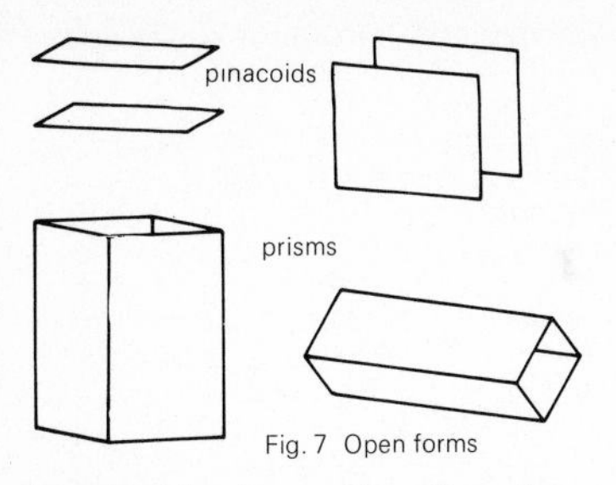

Fig. 7 Open forms

in *combination* with other forms, because crystals are solids. Forms are often used to describe the appearance of minerals, for example spinel is octahedral; hornblende occurs as

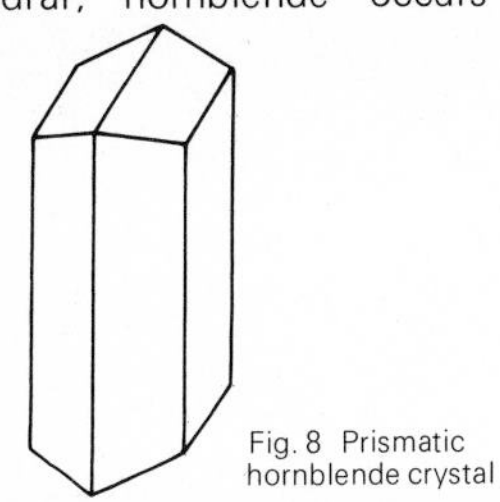
Fig. 8 Prismatic hornblende crystal

prismatic crystals (Fig. 8). A more detailed treatment of crystal shape will be found in the reference books.

The general aspect conferred on a mineral by the development of its faces is called the *habit*. Thus baryte commonly forms crystals of *tabular* habit

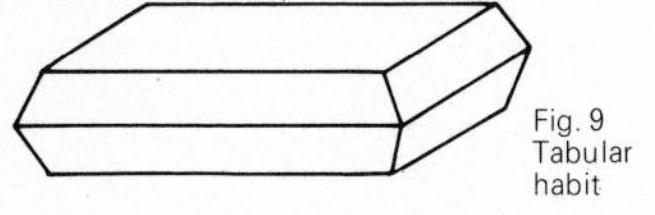
Fig. 9 Tabular habit

(Fig. 9); and zeolites such as natrolite frequently have *acicular* (needle-like) habit (Fig. 10).

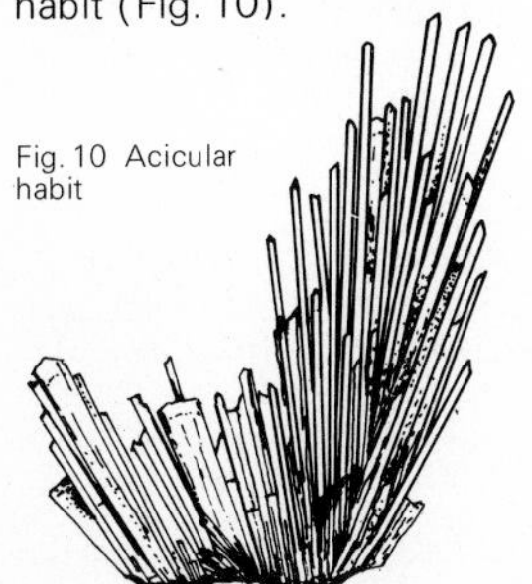
Fig. 10 Acicular habit

Mineral aggregates So far only single crystals have been discussed. Most minerals, however, occur as *aggregates* of crystals that rarely show perfect crystal shapes. The form of the aggregate, however, can be useful in identification. The *fibrous* zeolites have already been mentioned, and this adjective aptly describes their appearance. Sometimes crystals grow outwards from a centre, and the aggregate so formed is internally radiating, and outwardly may be rounded and nodular. The resulting form resembles a bunch of grapes and is called *botryoidal* (Fig. 11). Larger and more gently rounded shapes are said to be

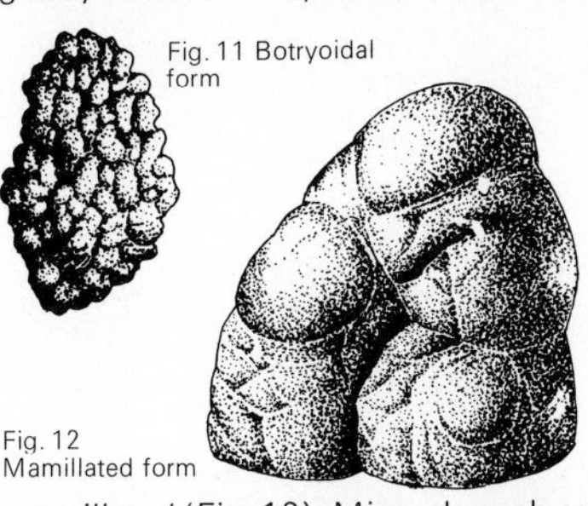
Fig. 11 Botryoidal form

Fig. 12 Mamillated form

mamillated (Fig. 12). Minerals such as native copper often form distinctive branching and divergent forms to which the term *dendritic* is applied (Fig. 13), and crystals forming

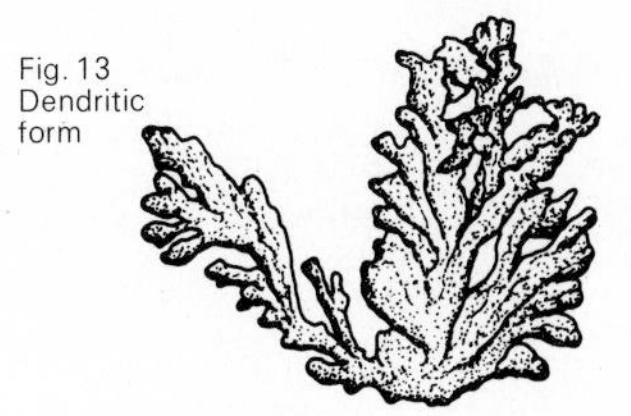
Fig. 13 Dendritic form

distinctly flat sheets are said to be *lamellar*. If the lamellae are very thin and can be readily separated, like the pages of a book, they are said to be *foliated* (Fig. 14). These and other examples are given in the mineral descriptions.

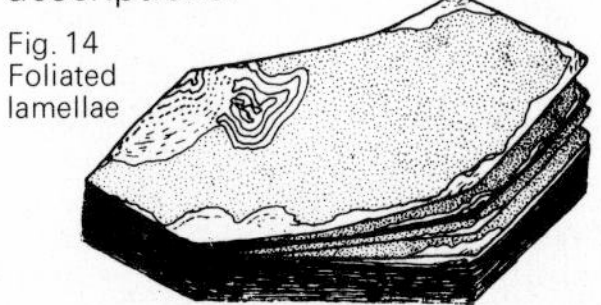
Fig. 14 Foliated lamellae

Physical properties There is a close link between the structure of a mineral and its physical properties which are, accordingly, of considerable value in identification. Some of the more useful physical properties are described below.

Density Defined strictly, density is mass per unit volume and is expressed in appropriate units, for example grams per cubic centimetre. It is often used synonymously, though not strictly correctly, with *specific gravity*, which is the weight of the substance compared to the weight of an equal volume of water. Density depends on several factors including the kind of atoms in the structure and how closely they pack together. Other things being equal, the heavier the atoms or the more closely packed they are, the greater the density. Tridymite and quartz are both silica (SiO_2) but quartz, the closely packed form, has a specific gravity of 2·65 at room temperature, whereas tridymite, with a more open structure, has a specific gravity of 2·26 under the same conditions. Similarly, celestine and anglesite (sulphates of strontium and lead respectively) have the same structure, but the presence of the heavier lead atoms gives anglesite a specific gravity of 6·32 compared with 3·97 for celestine. With practice, specific gravity can be roughly estimated by hand; methods of measurement are described in the books listed as 'further reading'.

Hardness Hardness is the resistance of a mineral to scratching or abrasion. F Mohs, in 1812, arranged ten minerals in order of hardness, so that each will scratch those lower in the scale, thus:

1	Talc (softest)	6	Orthoclase
2	Gypsum	7	Quartz
3	Calcite	8	Topaz
4	Fluorite	9	Corundum
5	Apatite	10	Diamond (hardest)

It says much for Mohs' careful selection that this scale is still used as the standard for hardness. Hardness is tested either by observing whether or not the minerals of Mohs' scale scratch the unknown mineral, or by observing whether objects of known hardness such as a knife blade or the fingernail will scratch the unknown. Minerals of hardness 1 feel soapy or greasy; the fingernail has a hardness of about $2\frac{1}{2}$; a steel pocket-knife blade has a hardness of about $5\frac{1}{2}$; and minerals of hardness 6 and over will scratch glass. Hardness is related to structure and to the strength of the chemical bonding; it is greater the smaller the atoms in the structure or the closer their packing. It should be appreciated that hardness is not the same as difficulty of breaking. A hard mineral may be brittle.

Cleavage and fracture If roughly handled, crystals will break. If the broken surface is irregular, the crystal possesses *fracture*, but if it breaks along a plane surface that is related to the structure, and parallel to a possible crystal face, then it has *cleavage*. Cleavage and fracture are expressions of the internal structure of the mineral. Cleavage occurs because of the variation in the strength of the bonds between different atoms, or planes of atoms. This is best illustrated by the layer silicates, of which mica is a familiar example. Chemical bonds are very strong within the silicon-oxygen layers, but the bonds between layers are weak, and so very little effort is needed to break them. Mica splits easily (cleaves) into thin sheets. Bond strength varies and so the degree of perfection of cleavage varies also. Mica, for example, has a *perfect* cleavage; less perfect cleavages are described as *good, poor* or *indistinct*. Cleavages may develop in several directions within a crystal and their quality and direction may well be of diagnostic value. Fracture has no such structural control but may still be of use in mineral identification. Thus glass, which has no orderly arrangement of its atoms, breaks with a characteristic *conchoidal* fracture (Fig. 15), so called because the overall appearance of the fractured surface, with its concentric ridges, resembles a shell with its growth lines. Quartz, though crystalline, has such a uniformly bonded structure that it breaks with a conchoidal fracture similar to that of glass. Another important type of fracture gives a broken surface that

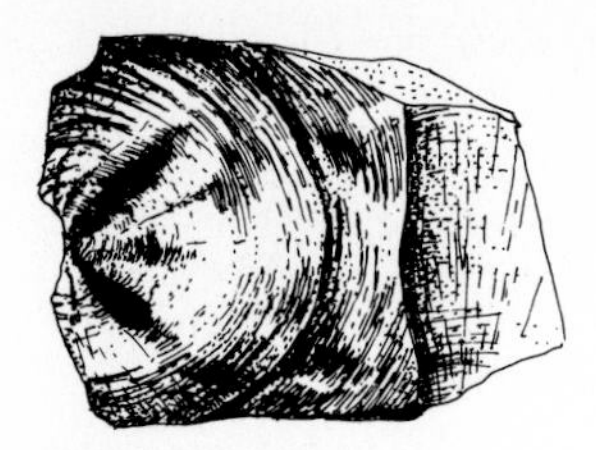

Fig. 15 Conchoidal fracture

resembles that of fractured wrought iron and is called *hackly* fracture.

Optical properties Optical properties depend on the interaction of light with minerals.

Transparency A most obvious property is whether a mineral in hand specimen is transparent, translucent or opaque. This is a function of the structure of the mineral and the kind of bonds that bind atom to atom. It is a measure of the amount of light absorbed by the mineral, and the subject is treated fully in some of the recommended reading. Many minerals which in the mass are opaque become translucent in very thin fragments.

Reflection and refraction When light meets a translucent mineral at an oblique angle, part is *reflected* from its surface and part enters the crystal or is *refracted* into it. Refraction is of little diagnostic value in the field, but it is a most useful property of minerals when they are investigated in the laboratory.

Lustre Lustre is a property of the surface of a mineral. The nature of the reflecting surface gives rise to the different kinds of lustre, and the amount of light reflected produces different intensities of lustre. Lustre, it should be remembered, is assessed independently of colour. The main kinds of lustre are described below.

Metallic lustre is the lustre of metals. It is produced by minerals which, like metals, absorb light strongly and are opaque even in the thinnest slices. In addition to the native metals themselves, most sulphides have a metallic lustre. Imperfect metallic lustre is called *submetallic*. There are various kinds of *non-metallic* lustre. *Adamantine* lustre is the lustre of diamond. *Resinous* lustre is the lustre of resin. This occurs in certain minerals with a yellow to brown colour. *Vitreous* lustre is the lustre of broken glass. This is the lustre most commonly displayed by minerals. Certain kinds of lustre are caused by the quality of the reflecting surface. A *greasy* lustre is often caused by minute irregularities in the surface which, if perfectly smooth, would give an adamantine or resinous lustre. *Pearly* lustre results from the reflection of light from a succession of parallel surfaces, such as cleavage planes, within a crystal. *Silky* lustre is due to the presence of small parallel fibres, as in asbestos and some varieties of gypsum. *Earthy* lustre is in effect a lack of lustre produced by surfaces that scatter the light. It is worth remembering that lustre may vary in different faces of a crystal. Heulandite, for example, shows a pearly lustre on one pair of faces and vitreous lustre on all the others.

Colour Colour in minerals is the result of the selective absorption of parts of the spectrum of white light, the observed colour being due to those wavelengths of light that are least absorbed. There is no single cause of colour in minerals. Sometimes it is a direct result of the presence in the structure of certain chemical elements; for example many copper minerals are blue or green. There are other more subtle reasons, however, and the reader is referred to the references for more information. It should be emphasized that, with experience, colour is one of the most valuable of the diagnostic properties of minerals.

Streak Streak is the colour of the powdered mineral. The most usual means of determining streak is to draw the mineral across a piece of white, unglazed porcelain, called a streak plate. Whereas the colour of the mineral in the mass can often be very variable, the colour of its streak is much less so. Streak is particularly valuable in the determination of opaque and coloured minerals. It is of little diagnostic value in the silicates, most of which have a white streak, and are too hard to powder readily.

Fluorescence When certain minerals are irradiated with ultraviolet light, they emit light in the visible part of the spectrum and are said to fluoresce. Fluorite—from which the name of the phenomenon is derived—and many other minerals show this property. Although interesting and sometimes spectacular, fluorescence only occasionally ranks as an important diagnostic property, because its effects are so variable. Different specimens of a single mineral may fluoresce with several different colours, and even specimens from the same locality may vary considerably in their fluorescence.

Other properties Some minerals have distinctive magnetic, electrical and radioactive properties of use in identification. They are mentioned where appropriate in the mineral descriptions, and full accounts are given in some of the books listed as further reading.

Chemistry of minerals It is possible to write a chemical formula to express the composition of a mineral, and such formulae are used as a short way of expressing mineral chemistry. Atoms can conveniently be regarded as electrically neutral because the positive charge on the nucleus is balanced by the negative charges of the surrounding electrons. Atoms can, however, gain or lose one or more electrons and so become either negatively or positively charged, when they are called *ions*. Negatively charged ions are called *anions* and positive ions are called *cations*. A chemical compound can be regarded as being made up of two parts, a positively charged or cationic part and a negatively charged or anionic part. The resulting compound is electrically neutral because the two sets of charges are in balance. The positive part is usually a metal, and is always the first part of a written chemical formula. The negative or anionic part of the formula can be either a non-metallic ion such as oxygen or sulphur or else a combination of several elements to form a negatively charged group such as carbonate (CO_3) or sulphate (SO_4). The following table lists the chemical symbols of the elements referred to in this book.

Ag Silver
Al Aluminium
As Arsenic
Au Gold
B Boron
Ba Barium
Be Beryllium
Bi Bismuth
C Carbon
Ca Calcium
Cd Cadmium
Ce Cerium
Cl Chlorine
Co Cobalt
Cr Chromium
Cu Copper
F Fluorine
Fe Iron
H Hydrogen
Hg Mercury
K Potassium
La Lanthanum
Li Lithium
Mg Magnesium
Mn Manganese
Mo Molybdenum
N Nitrogen
Na Sodium
Nb Niobium
Ni Nickel
O Oxygen
P Phosphorus
Pb Lead
S Sulphur
Sb Antimony
Si Silicon
Sn Tin
Sr Strontium
Ta Tantalum
Th Thorium
Ti Titanium
U Uranium
V Vanadium
W Tungsten
Y Yttrium
Zn Zinc
Zr Zirconium

Some common anionic groups and their names are given below.

Al_2O_4 etc	Aluminate
As, As_2 etc	Arsenide
AsO_4 etc	Arsenate
BO_3, B_3O_4 etc	Borate
Cl, Cl_2 etc	Chloride
CO_3	Carbonate
CrO_4 etc	Chromate
F, F_2 etc	Fluoride
MoO_4 etc	Molybdate
N, N_2 etc	Nitride
NO_3	Nitrate
NbO_3 etc.	Niobate
O, O_2 etc	Oxide
OH, $(OH)_2$ etc	Hydroxide
PO_4 etc	Phosphate
S, S_2 etc	Sulphide
SiO_4, Si_2O_7 etc	Silicate
SO_4	Sulphate
TaO_3 etc	Tantalate
TiO_3 etc	Titanate
UO_2 etc	Uranate
VO_4 etc	Vanadate
WO_4 etc	Tungstate

In chemical formulae the subscript numerals denote the numbers of atoms of the preceding element that are present in the formula unit. When referring to a chemical compound by name it is simply necessary to state, in turn, the cationic and then the anionic part that follows; for example $CaCO_3$ is calcium carbonate, FeS_2 is iron sulphide, CaF_2 is calcium fluoride, $(Mg,Fe)SiO_4$ is magnesium iron silicate, and so on. By contrast, $KAlSi_3O_8$ is potassium aluminium silicate, or better, potassium alumino-silicate; here there are two parts of the cationic group and they are emphasized in the way shown. Another example is $K_2(UO_2)_2(VO_4)_2.3H_2O$ which is called hydrated potassium uranylvanadate. Notice that water of crystallization (H_2O) is referred to by the adjective 'hydrated'. Atoms which can substitute the one for the other in a mineral are written so (Mg,Fe).

Field occurrence Nearly all rocks are composed of minerals, but fine specimens are rare and tend to occur in fissures and other cavities where the crystals have been unobstructed during their growth.

Many good specimens are obtained from mineral veins (Fig. 16). High-

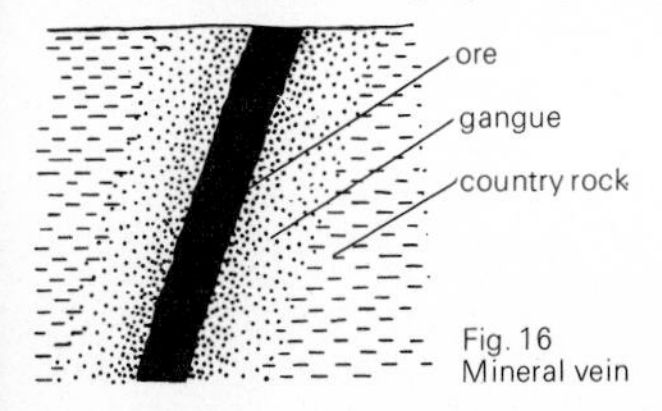

Fig. 16 Mineral vein

temperature fluids deposit minerals in cracks and fissures in rocks and many of these veins, often called hydrothermal veins, are worked as sources of ore. They frequently contain colourful specimens and good crystals, not only of the commercially valuable ore minerals, but also of the accompanying and economically valueless *gangue* minerals as well. It is not always necessary to examine or collect from the veins themselves—in many instances it is dangerous or impossible to do so—for mining activity usually results in dumps of discarded material which, if carefully searched, will often yield good specimens. Good crystals can often be found lining cavities in rocks of virtually every kind, though particular minerals tend to occur in certain environments. Sometimes weathered-out cavity linings, called *geodes*, are lined with well shaped crystals, and many fine specimens of amethyst occur in such associations

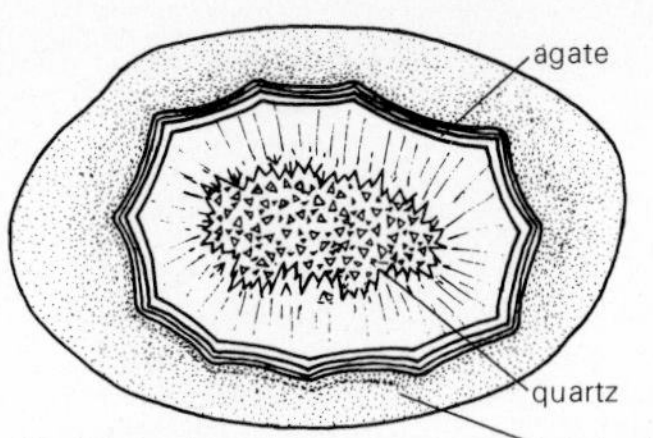

Fig. 17 Geode

(Fig. 17). Pegmatites, which crystallize from relatively low-temperature, volatile-rich magma, are another source of good crystals and rare minerals that frequently grow to large sizes.

The largest specimens, however, are not always the most spectacular, and there is a growing interest in *micromounts* in which small crystals, or groups of crystals, are carefully mounted in a transparent plastic or glass-topped box in which they can be examined by using a lens or a microscope. These small crystals have a beauty of their own, and have the advantages of occupying a minimum of space and of being more perfectly formed than larger crystals.

The collector will invariably find some specimens that are difficult to identify. He is urged to become acquainted with minerals that are displayed in many national and other museums. Time spent in this way will be amply repaid, not only in terms of identification of his specimens, but also in becoming more deeply involved in the study of natural history.

Organization of the mineral descriptions in this book The groups are described in the following order.

Native elements
Sulphides
Oxides and hydroxides
Halides
Carbonates
Nitrates and borates
Sulphates and chromates
Molybdates and tungstates
Phosphates
Arsenates and vanadates
Silicates

The silicates are such a large group that, although the primary classification is based on a chemical criterion, they are subdivided on a structural basis. This system has the advantage of grouping minerals with similar properties.

Gold Au **Crystal system** Cubic. **Habit** Usually as disseminated grains, or dendritic forms: crystals rare but octahedral; occasionally as cubes or rhombdodecahedra. Irregular rounded masses are called nuggets. **Twinning** Common, on octahedron. **SG** 19·3 (less if alloyed with other metals) **Hardness** $2\frac{1}{2}$–3 **Cleavage** None. **Fracture** Hackly. **Colour and transparency** Characteristic gold-yellow; lighter yellow when alloyed with silver: opaque except in thinnest sheets. **Streak** Gold-yellow. **Lustre** Metallic. **Distinguishing features** Colour, low hardness, insoluble in single acids. Gold may be mistaken for pyrite or chalcopyrite (fool's gold) but the colour, low hardness and ductility of gold contrast with the greater hardness and brittle nature of the other two. **Alteration** None. **Occurrence** In small amounts in hydrothermal veins, often in association with quartz; and in alluvial deposits in which gold, by reason of its density, is separated from other minerals during weathering and transport to become concentrated in stream or other sediments, which may be loose and unconsolidated, or hardened into rock. Tiny grains of gold are often carried long distances by streams and can be recovered from gravel by panning, which entails washing away all but the heavy minerals, and searching these for flecks of gold. South African gold is obtained from consolidated alluvial deposits, notably gold-bearing quartz conglomerates.

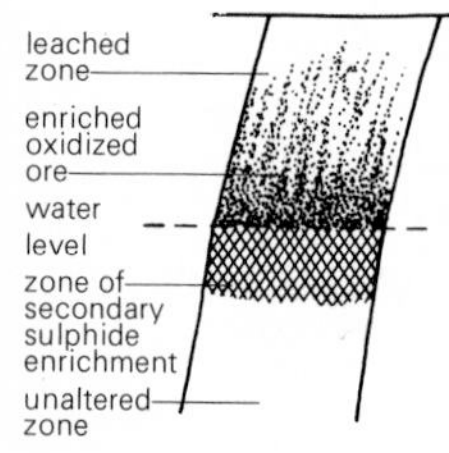

Mineral vein showing secondary sulphide enrichment

Silver Ag **Crystal system** Cubic. **Habit** Commonly as wiry or scaly forms; crystals rare. **SG** 10–11 **Hardness** $2\frac{1}{2}$–3 **Cleavage** None. **Fracture** Hackly. **Colour and transparency** Silver-white, tarnishes quickly to a black colour: opaque. **Streak** Silver-white. **Lustre** Metallic. **Distinguishing features** Colour, black tarnish, malleability, soluble in nitric acid. **Occurrence** In hydrothermal veins, or in small amounts in the oxidized zone of silver-bearing ore deposits.

Copper Cu **Crystal system** Cubic. **Habit** Dendritic, branching forms; crystals usually cubic or rhombdodecahedral. **SG** 8·9 **Hardness** $2\frac{1}{2}$–3 **Cleavage** None. **Fracture** Hackly. **Colour and transparency** Copper-red, deepening to dull brown with tarnish: opaque. **Streak** Metallic copper-red. **Lustre** Metallic. **Distinguishing features** Colour and ductility, readily soluble in nitric acid. **Occurrence** In basaltic lavas and in sandstones and conglomerates, in which it is secondary, having formed by reaction between copper-bearing solutions and other minerals, notably those of iron. Native copper, though widely distributed, occurs only in small amounts.

Gold nuggets
Gold on quartz

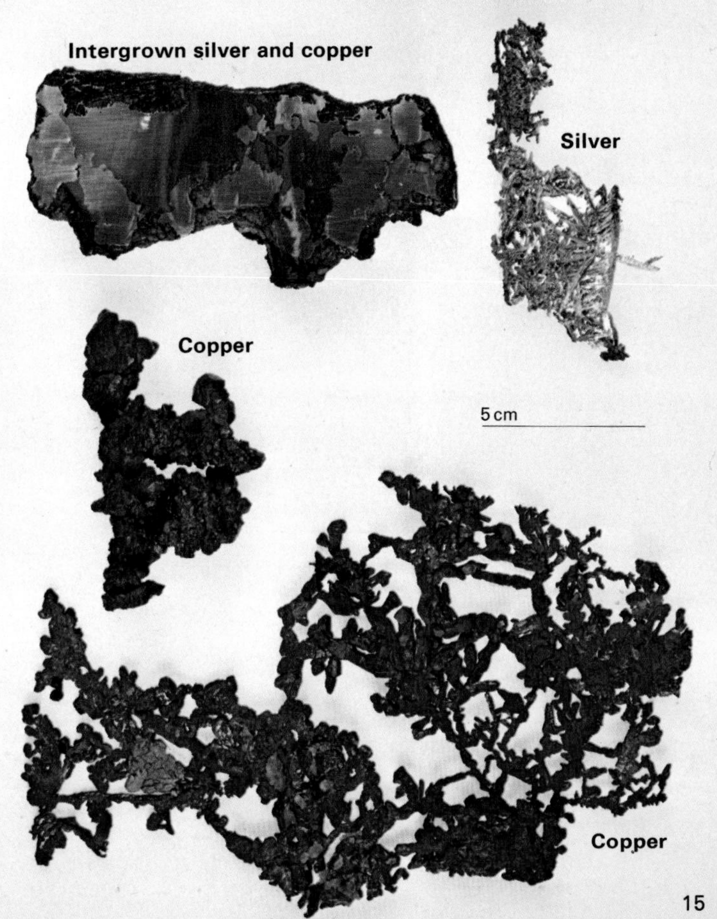
Intergrown silver and copper
Silver
Copper
5 cm
Copper

Arsenic As **Crystal system** Trigonal. **Habit** Crystals rare; usually massive as granular, botryoidal or stalactitic masses. **SG** 5·6–5·8 **Hardness** 3½ **Cleavage** Basal, perfect. **Colour and transparency** Light grey, tarnishes rapidly to dark grey: opaque. **Streak** Light grey. **Lustre** Metallic. **Distinguishing features** Smells like garlic when heated or struck with a hammer. **Occurrence** In hydrothermal veins, usually in igneous and metamorphic rocks, and associated with silver, cobalt or nickel ores. The name arsenic is derived from a Greek word for 'masculine', and dates from the time when metals were thought to be of different sexes.

Antimony Sb **Crystal system** Trigonal. **Habit** Usually massive and reniform (kidney-shaped); sometimes lamellar; crystals rare. **Twinning** Common. **SG** 6·6–6·7 **Hardness** 3–3½ **Cleavage** Basal, perfect; rhombohedral good. **Colour and transparency** Very light grey: opaque. **Streak** Grey. **Lustre** Metallic. **Occurrence** In hydrothermal veins, often associated with silver or arsenic. Accompanying minerals are stibnite, sphalerite, galena and pyrite.

Bismuth Bi **Crystal system** Trigonal. **Habit** Massive, granular or arborescent (tree-like or moss-like); crystals rare. **Twinning** Fairly common. **SG** 9·7–9·8 **Hardness** 2–2½ **Cleavage** Basal, perfect. **Colour and transparency** Silver-white, becoming reddish with tarnish: opaque. **Streak** Silver-white, shiny. **Lustre** Metallic. **Distinguishing features** Reddish silver colour, perfect cleavage, melts readily at 270°C. **Occurrence** In hydrothermal veins, often in association with ores of gold, silver, tin, nickel, cobalt and lead.

Iron Fe **Nickel-iron** NiFe **Crystal system** Cubic. **Habit** In grains and masses in terrestrial rocks. Nickel-iron is the major native metallic constituent of meteorites in the form of kamacite and taenite (see under meteorites). **SG** 7·3–7·9 **Hardness** 4½ **Cleavage** Poor. **Fracture** Hackly. **Colour and transparency** Steel-grey to black: opaque. **Lustre** Metallic. **Distinguishing features** Strongly magnetic character, malleability. **Occurrence** Native iron is uncommon in terrestrial rocks, occurring mainly where volcanic rocks cut coal seams.

Iron
Arsenic
Antimony
Antimony
Bismuth
Bismuth
5 cm

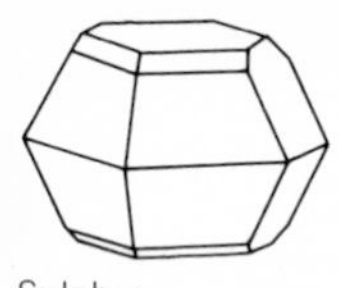

Sulphur

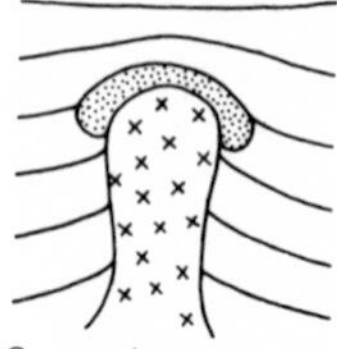

Cap rock containing sulphur above salt dome

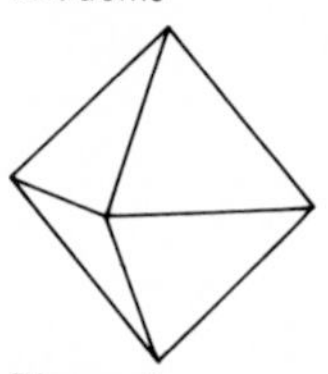

Diamond: octahedron

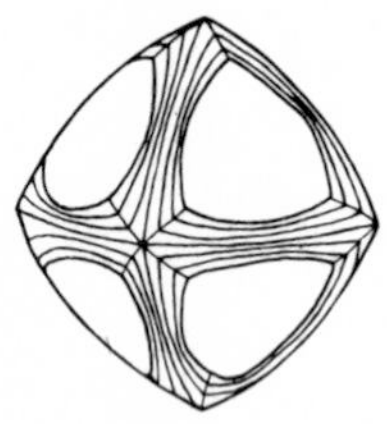

Diamond: octahedron with curved faces

Sulphur S **Crystal system** Orthorhombic. **Habit** Crystals tabular or bipyramidal; also occurs as stalactitic or encrusting masses. **SG** 2·0–2·1 **Hardness** $1\frac{1}{2}$–$2\frac{1}{2}$ **Cleavage** None. **Fracture** Uneven, sometimes conchoidal. **Colour and transparency** Bright yellow, sometimes brownish: transparent to translucent. **Streak** White. **Lustre** Resinous. **Distinguishing features** Colour, low hardness, low melting point (113°C), insoluble in water and dilute hydrochloric acid, soluble in carbon disulphide. **Occurrence** As encrusting masses produced by sublimation around volcanic vents and fumaroles; in sedimentary rocks, particularly limestones and those containing gypsum. Sulphur often occurs in the cap rock of salt domes in association with anhydrite, gypsum and calcite.

Diamond C **Crystal system** Cubic. **Habit** Commonly occurs as octahedral crystals frequently of flattened habit; more rarely as cubes, often with curved faces. **Twinning** Sometimes twinned on octahedron. **SG** 3·5 **Hardness** 10 **Cleavage** Octahedral, perfect. **Fracture** Conchoidal. **Colour and transparency** Colourless: transparent. May be yellowish, brown, red and even black. Gem quality diamonds are clear; grey to black, opaque, finely granular diamond is called bort. **Streak** White. **Lustre** Adamantine; uncut crystals look greasy. **Distinguishing features** Extreme hardness, octahedral cleavage. **Occurrence** Sporadically distributed in kimberlite, a rock forming pipe-like intrusions that have risen from great depth; also in alluvial deposits (mainly river and beach gravels) in which diamond is concentrated. Most diamonds came from alluvial deposits until the discovery of kimberlite pipes in South Africa in the mid-nineteenth century. The name comes from the Greek word meaning 'invincible' and alludes to the hardness and durability of diamond.

Graphite C **Crystal system** Hexagonal. **Habit** Flat tabular crystals but more commonly massive, foliated or earthy. **SG** 2·1–2·3 **Hardness** 1–2 **Cleavage** Basal, perfect. **Colour and transparency** Black: opaque. **Streak** Black. **Lustre** Dull metallic. **Distinguishing features** Extreme softness, greasy feel, readily marks paper and soils the fingers. Distinguished from molybdenite by its black streak, lower specific gravity and colour; molybdenite being bluish grey with grey to grey-green streak. **Occurrence** As disseminated flakes in metamorphic rocks derived from rocks having an appreciable carbon content. Graphite schists and limestones are fairly widely distributed. It occurs also as veins in igneous rocks and pegmatites. The name comes from the Greek word meaning 'to write'. Diamond and graphite have the same chemical composition and yet have quite different structures and physical properties. This phenomenon, in which a chemical substance can exist in two or more distinct forms which differ in structure and physical properties, is called *polymorphism*. There could hardly be a greater contrast in hardness than that between diamond and graphite.

Sulphur
Sulphur
Graphite
5 cm
Diamonds
Graphite
Diamond in matrix
Bort

Argentite-acanthite (Silver glance) Ag_2S **Crystal system** Cubic (argentite); orthorhombic (acanthite). **Habit** Crystals commonly cubic or octahedral, frequently occur as groups of crystals in parallel alignment. Acanthite may crystallize at low temperatures as pointed crystals. Also arborescent, filiform (wiry), massive. **SG** 7·2–7·4 **Hardness** 2–2½ **Cleavage** Cubic, poor. **Fracture** Subconchoidal. **Colour and transparency** Black: opaque. **Streak** Black, shiny. **Lustre** Metallic. **Distinguishing features** Colour, sectility (can be cut with a knife, like lead). **Alteration** Argentite is stable only above 180°C. Below this temperature Ag_2S has an orthorhombic structure and is called acanthite. The cubic forms shown are thus paramorphs of acanthite after argentite. **Occurrence** In hydrothermal veins in association with pyrargyrite, proustite and native silver. It also occurs as a weathering product of primary silver sulphides.

Bornite (Peacock ore, Erubescite) Cu_5FeS_4 **Crystal system** Cubic. **Habit** Crystals rough, cubic and rhombdodecahedral; usually massive. **Twinning** On octahedron. **SG** 5·0–5·1 **Hardness** 3 **Cleavage** None visible. **Fracture** Subconchoidal, uneven. **Colour and transparency** Reddish brown on fresh surface; tarnishes to characteristic purplish iridescence: opaque. **Streak** Pale grey-black. **Lustre** Metallic. **Distinguishing features** Iridescent colours, hence 'peacock ore'. Soluble in nitric acid. **Alteration** To chalcosine, covelline, cuprite, chrysocolla, malachite and azurite. **Occurrence** A common copper mineral, found in hydrothermal veins in association with chalcopyrite and chalcosine. It occurs also as a primary mineral in some igneous rocks and pegmatite veins.

Covelline (Covellite) CuS **Crystal system** Hexagonal. **Habit** Crystals tabular or platy; rare. Usually as foliated masses or coatings. **SG** 4·6–4·8 **Hardness** 1½–2 **Cleavage** Basal, perfect. **Colour and transparency** Indigo blue, purplish iridescent tarnish: opaque. **Streak** Dark grey to black. **Lustre** Metallic. **Distinguishing features** Perfect cleavage distinguishes covelline from bornite. Colour distinguishes it from chalcosine. **Occurrence** In hydrothermal veins as a primary sulphide: more commonly in the zone of secondary enrichment in association with chalcosine, bornite, chalcopyrite.

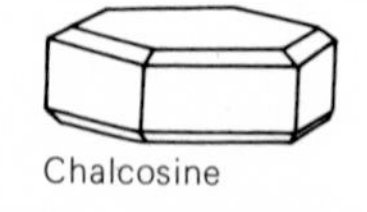
Chalcosine

leached zone
enriched oxidized ore
water level
zone of secondary sulphide enrichment
unaltered zone

Mineral vein showing secondary sulphide enrichment

Chalcosine (Copper glance, Chalcocite) Cu_2S **Crystal system** Orthorhombic. **Habit** Prismatic or tabular crystals, rare; usually massive, or as powdery coatings. **Twinning** Common, to give pseudo-hexagonal forms. **SG** 5·5–5·8 **Hardness** 2½–3 **Cleavage** Prismatic, indistinct. **Fracture** Conchoidal. **Colour and transparency** Dark lead-grey, tarnishing to black: opaque. **Streak** Black. **Lustre** Metallic. **Distinguishing features** Black colour, association with other copper minerals. Soluble in nitric acid. **Alteration** To covelline, malachite or azurite. **Occurrence** A widespread and valuable copper ore, often associated with native copper or cuprite. Most commonly found in the zone of secondary sulphide enrichment.

Argentite-acanthite
Bornite
Bornite
Covelline
Covelline
Chalcosine
5 cm

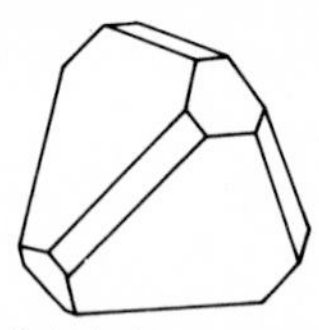

Sphalerite: combination of two tetrahedra and cube

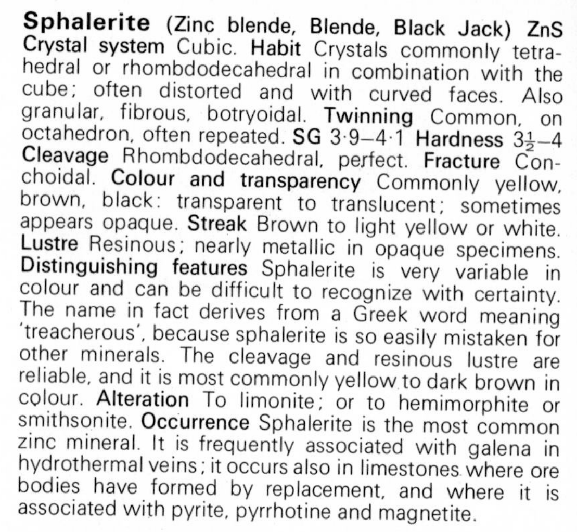

Sphalerite (Zinc blende, Blende, Black Jack) ZnS **Crystal system** Cubic. **Habit** Crystals commonly tetrahedral or rhombdodecahedral in combination with the cube; often distorted and with curved faces. Also granular, fibrous, botryoidal. **Twinning** Common, on octahedron, often repeated. **SG** 3·9–4·1 **Hardness** 3½–4 **Cleavage** Rhombdodecahedral, perfect. **Fracture** Conchoidal. **Colour and transparency** Commonly yellow, brown, black: transparent to translucent; sometimes appears opaque. **Streak** Brown to light yellow or white. **Lustre** Resinous; nearly metallic in opaque specimens. **Distinguishing features** Sphalerite is very variable in colour and can be difficult to recognize with certainty. The name in fact derives from a Greek word meaning 'treacherous', because sphalerite is so easily mistaken for other minerals. The cleavage and resinous lustre are reliable, and it is most commonly yellow to dark brown in colour. **Alteration** To limonite; or to hemimorphite or smithsonite. **Occurrence** Sphalerite is the most common zinc mineral. It is frequently associated with galena in hydrothermal veins; it occurs also in limestones where ore bodies have formed by replacement, and where it is associated with pyrite, pyrrhotine and magnetite.

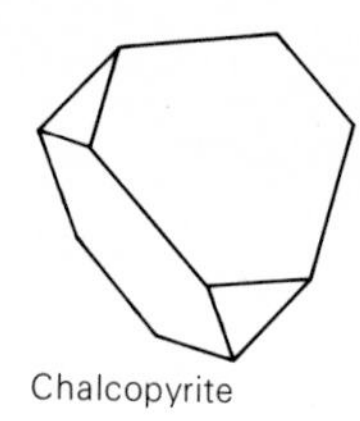

Chalcopyrite

Chalcopyrite (Copper pyrites) $CuFeS_2$ **Crystal system** Tetragonal. **Habit** Crystals appear tetrahedral; usually massive. **Twinning** Several kinds, giving rise to interpenetration twins and twins that resemble spinel twins. **SG** 4·1–4·3 **Hardness** 3½–4 **Cleavage** Very poor. **Fracture** Conchoidal to uneven. **Colour and transparency** Brass-yellow, often with slightly iridescent tarnish: opaque. **Streak** Greenish black. **Lustre** Metallic. **Distinguishing features** Distinguished from pyrite by its deeper yellow colour, tarnish and inferior hardness; and from gold by its brittle nature and greater hardness. Soluble in nitric acid. **Alteration** To chalcosine, covelline, chrysocolla and malachite. **Occurrence** Chalcopyrite is the most common copper mineral, and an important copper ore. It occurs as a primary mineral in igneous rocks and in hydrothermal vein deposits in association with pyrite, pyrrhotine, cassiterite, sphalerite, galena and gangue minerals such as quartz, calcite and dolomite. It is an important mineral in 'porphyry copper' deposits where it is disseminated with bornite and pyrite in veinlets in igneous intrusions of quartz diorite or diorite porphyry. It occurs also in pegmatites, crystalline schists, and in contact metamorphic deposits.

Wurtzite ZnS **Crystal system** Hexagonal. **Habit** Pyramidal crystals; also radiating, fibrous, massive. **SG** 4·0–4·1 **Hardness** 3½–4 **Cleavage** Prismatic, distinct; basal, imperfect. **Colour and transparency** Brownish black. **Streak** Brown. **Lustre** Resinous. **Occurrence** Rare; in sulphide ores. Wurtzite is a polymorph of ZnS, and is the form stable at high temperatures. It is named after A Wurtz, a French chemist.

Sphalerite
Sphalerite
Chalcopyrite
Chalcopyrite
Wurtzite
5 cm

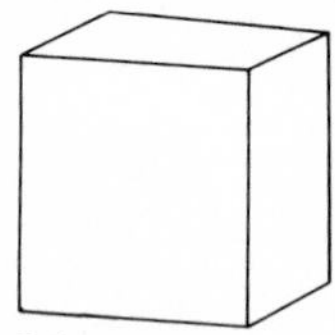
Galena: cube

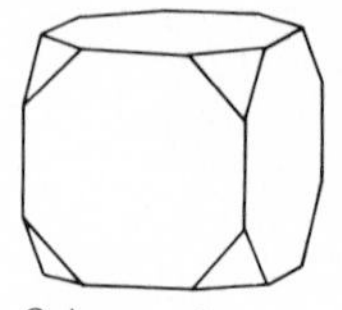
Galena: cube and octahedron

Galena PbS **Crystal system** Cubic. **Habit** Crystals often of cube/octahedral, or octahedral habit; sometimes as cubes. Also massive or granular. **Twinning** Penetration or contact twins on octahedron. **SG** 7·4–7·6 **Hardness** 2½ **Cleavage** Cubic, perfect. **Colour and transparency** Lead-grey; opaque. **Streak** Lead-grey. **Lustre** Metallic. **Distinguishing features** Colour, metallic lustre, perfect cubic cleavage, high specific gravity. **Alteration** Oxidizes readily to anglesite, cerussite, pyromorphite or mimetite. **Occurrence** Galena is very widely distributed and the most important lead ore. It occurs, following the bedding of sedimentary rocks, as hydrothermal veins and in pegmatites, and as replacement bodies in limestone and dolomitic rocks. In hydrothermal veins it is commonly associated with sphalerite, pyrite, chalcopyrite, tetrahedrite and bournonite, and with gangue minerals such as quartz, calcite, dolomite, baryte and fluorite. In high-temperature veins and replacement deposits it is associated with such minerals as garnet, feldspar, diopside, rhodonite and biotite. Replacement deposits occur in limestones which have sometimes been dolomitized. The name comes from the Latin word for 'lead ore'.

Pyrrhotine (Magnetic pyrites, Pyrrhotite) FeS (varies to $Fe_{0\cdot8}S$) **Crystal system** Hexagonal. **Habit** Usually massive, granular; crystals rare, platy or tabular. **Twinning** Rare. **SG** 4·6–4·7 **Hardness** 3½–4½ **Cleavage** Basal parting. **Fracture** Subconchoidal to uneven. **Colour and transparency** Bronze-yellow, darkening with exposure to more reddish bronze; opaque. **Streak** Greyish black. **Lustre** Metallic. **Distinguishing features** Reddish bronze colour, magnetism. Apart from magnetite it is the only common mineral that is noticeably magnetic. Distinguished from chalcopyrite by colour and magnetism; from pyrite by colour and inferior hardness. **Occurrence** Pyrrhotine occurs in igneous rocks such as gabbro or norite as disseminated grains, commonly in association with minerals such as chalcopyrite, pentlandite and pyrite. It is found also in contact metamorphic deposits, in veins and in pegmatites. It occurs (troilite) in iron meteorites. The name comes from a Greek word meaning 'reddish'.

Pyrrhotine

Pyrrhotine

Galena

Galena

Galena

Nickeline (Niccolite, Kupfernickel) NiAs **Crystal system** Hexagonal. **Habit** Usually massive in reniform or columnar aggregates; crystals rare. **SG** 7·8 **Hardness** 5–$5\frac{1}{2}$ **Cleavage** None. **Fracture** Uneven. **Colour and transparency** Pale copper-red: opaque. **Streak** Pale brownish black. **Lustre** Metallic. **Distinguishing features** Colour. **Alteration** Alters to pale green annabergite (nickel bloom). **Occurrence** It occurs in igneous rocks such as norite and gabbro with pyrrhotine, chalcopyrite and nickel sulphides, and in hydrothermal veins with silver, silver-arsenic and cobalt minerals.

Greenockite

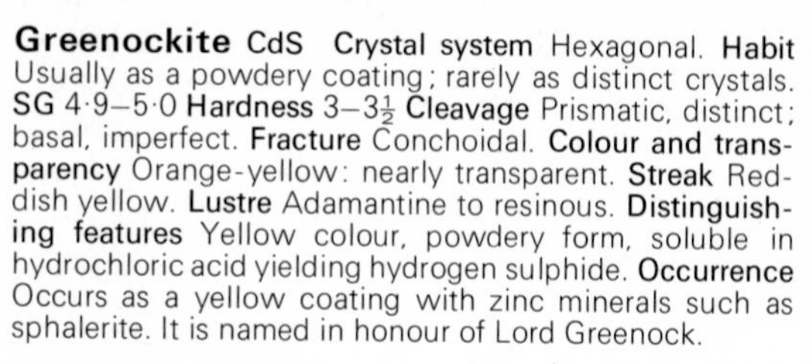

Greenockite CdS **Crystal system** Hexagonal. **Habit** Usually as a powdery coating; rarely as distinct crystals. **SG** 4·9–5·0 **Hardness** 3–$3\frac{1}{2}$ **Cleavage** Prismatic, distinct; basal, imperfect. **Fracture** Conchoidal. **Colour and transparency** Orange-yellow: nearly transparent. **Streak** Reddish yellow. **Lustre** Adamantine to resinous. **Distinguishing features** Yellow colour, powdery form, soluble in hydrochloric acid yielding hydrogen sulphide. **Occurrence** Occurs as a yellow coating with zinc minerals such as sphalerite. It is named in honour of Lord Greenock.

Cinnabar: thick tabular habit

Cinnabar HgS **Crystal system** Trigonal. **Habit** Crystals rhombohedral or thick tabular, sometimes short prismatic or acicular. Also granular, massive. **Twinning** Common, with basal pinacoid as twin plane. **SG** 8·0–8·2 **Hardness** 2–$2\frac{1}{2}$ **Cleavage** Prismatic, perfect. **Fracture** Uneven. **Colour and transparency** Scarlet-red to brownish red: transparent to translucent, occasionally nearly opaque. **Streak** Vermilion. Powdered cinnabar was used as the pigment. **Lustre** Adamantine; near metallic when opaque. **Distinguishing features** Red colour and streak, high specific gravity, perfect cleavage. **Alteration** Sometimes alters to calomel (mercurous chloride). **Occurrence** Cinnabar is the commonest of the mercury minerals, and is the only important ore of the metal. It occurs in fractures in sedimentary rocks in areas of recent volcanic activity and around hot springs. It is associated with pyrite, stibnite and realgar, and with gangue minerals such as chalcedony, quartz, calcite and baryte.

Millerite NiS **Crystal system** Trigonal. **Habit** Crystals usually slender and acicular, often in radiating groups. **SG** 5·2–5·6 **Hardness** 3–$3\frac{1}{2}$ **Cleavage** Rhombohedral, perfect. **Fracture** Uneven. **Colour and transparency** Brass-yellow: opaque. **Streak** Greenish black. **Lustre** Metallic. **Distinguishing features** Colour, slender acicular crystal form. Individual acicular crystals are elastic. **Occurrence** Commonly occurs as tufts of radiating fibres in cavities and as a replacement of other nickel minerals. It also occurs in veins carrying nickel minerals and other sulphides, and around some volcanoes as a sublimation product. It is named after WH Miller (1801–1880), a British mineralogist.

cm
Millerite
Greenockite
Nickeline
Cinnabar
Cinnabar

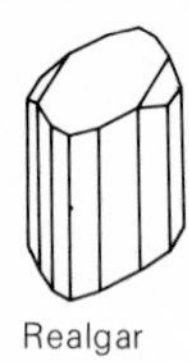

Realgar

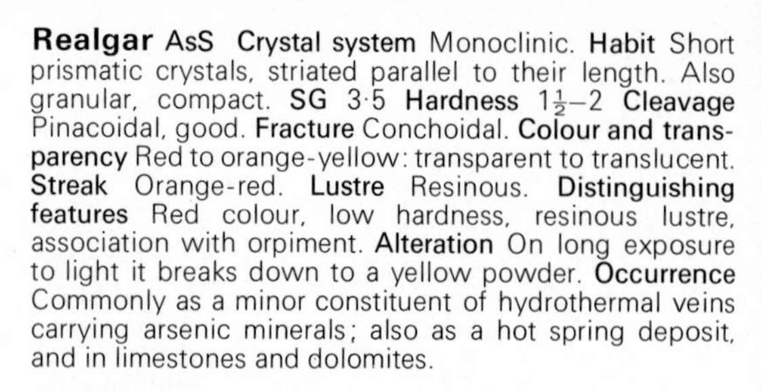

Realgar AsS **Crystal system** Monoclinic. **Habit** Short prismatic crystals, striated parallel to their length. Also granular, compact. **SG** 3·5 **Hardness** $1\frac{1}{2}$–2 **Cleavage** Pinacoidal, good. **Fracture** Conchoidal. **Colour and transparency** Red to orange-yellow: transparent to translucent. **Streak** Orange-red. **Lustre** Resinous. **Distinguishing features** Red colour, low hardness, resinous lustre, association with orpiment. **Alteration** On long exposure to light it breaks down to a yellow powder. **Occurrence** Commonly as a minor constituent of hydrothermal veins carrying arsenic minerals; also as a hot spring deposit, and in limestones and dolomites.

Orpiment As_2S_3 **Crystal system** Monoclinic. **Habit** Usually as foliated or columnar masses; crystals small and rare. **SG** 3·4–3·5 **Hardness** $1\frac{1}{2}$–2 **Cleavage** One perfect cleavage. **Colour and transparency** Lemon-yellow to brownish or reddish yellow: transparent to translucent. **Streak** Pale yellow. **Lustre** Pearly on cleavage surface: elsewhere resinous. **Distinguishing features** Yellow colour, perfect cleavage, pearly lustre on cleavage surface. **Occurrence** Orpiment often accompanies realgar as a low-temperature mineral in veins and hot spring deposits.

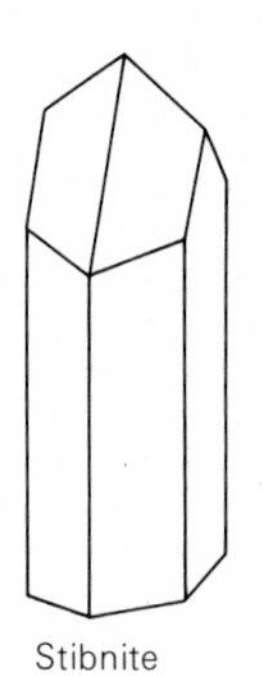

Stibnite

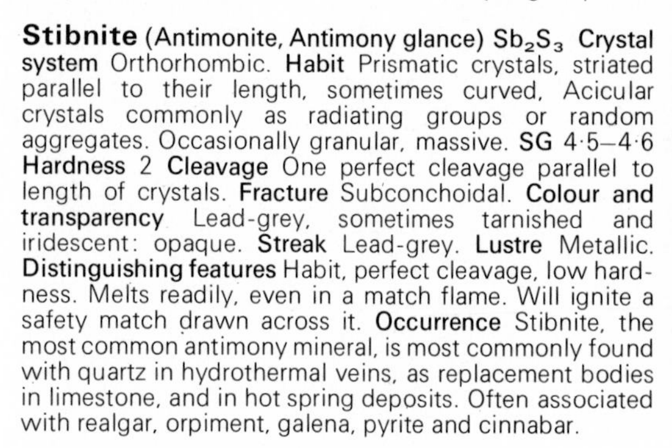

Stibnite (Antimonite, Antimony glance) Sb_2S_3 **Crystal system** Orthorhombic. **Habit** Prismatic crystals, striated parallel to their length, sometimes curved, Acicular crystals commonly as radiating groups or random aggregates. Occasionally granular, massive. **SG** 4·5–4·6 **Hardness** 2 **Cleavage** One perfect cleavage parallel to length of crystals. **Fracture** Subconchoidal. **Colour and transparency** Lead-grey, sometimes tarnished and iridescent: opaque. **Streak** Lead-grey. **Lustre** Metallic. **Distinguishing features** Habit, perfect cleavage, low hardness. Melts readily, even in a match flame. Will ignite a safety match drawn across it. **Occurrence** Stibnite, the most common antimony mineral, is most commonly found with quartz in hydrothermal veins, as replacement bodies in limestone, and in hot spring deposits. Often associated with realgar, orpiment, galena, pyrite and cinnabar.

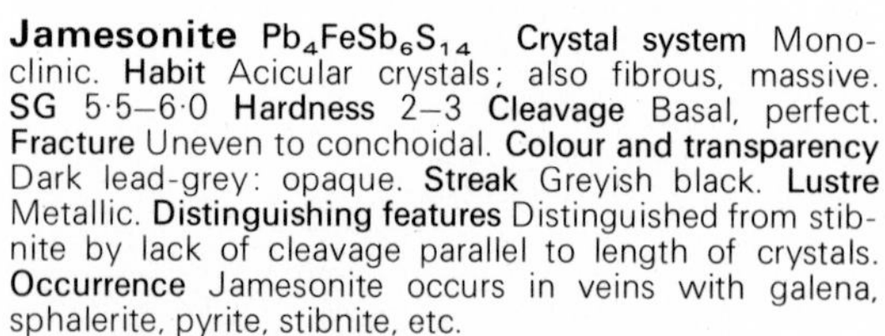

Jamesonite $Pb_4FeSb_6S_{14}$ **Crystal system** Monoclinic. **Habit** Acicular crystals; also fibrous, massive. **SG** 5·5–6·0 **Hardness** 2–3 **Cleavage** Basal, perfect. **Fracture** Uneven to conchoidal. **Colour and transparency** Dark lead-grey: opaque. **Streak** Greyish black. **Lustre** Metallic. **Distinguishing features** Distinguished from stibnite by lack of cleavage parallel to length of crystals. **Occurrence** Jamesonite occurs in veins with galena, sphalerite, pyrite, stibnite, etc.

Bismuthinite (Bismuth glance) Bi_2S_3 **Crystal system** Orthorhombic. **Habit** Usually massive, fibrous; rarely as acicular crystals. **SG** 6·8 **Hardness** 2 **Cleavage** One perfect cleavage. **Colour and transparency** Light lead-grey: opaque. **Streak** Light lead-grey. **Lustre** Metallic. **Distinguishing features** Similar to stibnite but sectile and less flexible. **Occurrence** In igneous rocks in association with such minerals as magnetite, pyrite, chalcopyrite, sphalerite and galena, and with tin and tungsten ores.

Jamesonite
Bismuthinite
Stibnite
Orpiment
Stibnite
Realgar
Realgar
5 cm

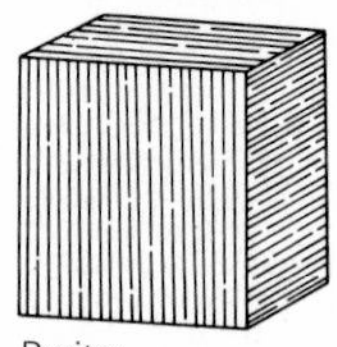

Pyrite: striated cube

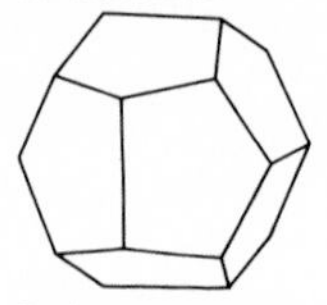

Pyrite: pyritohedron

Pyrite: 'iron-cross' twin

Pyrite (Iron pyrites) FeS_2 **Crystal system** Cubic. **Habit** Crystals usually cubes, pyritohedra or octahedra, or combinations of these forms. Cubes frequently show striations produced by oscillatory growth of cube and pyritohedron and which are perpendicular to each other on adjacent faces. Also massive, granular, stalactitic, spheroidal, radiating. **Twinning** Interpenetration 'iron cross' twins sometimes occur in forms showing pyritohedra. **SG** 4·9–5·2 **Hardness** 6–6½ **Cleavage** Cubic and octahedral, indistinct. **Fracture** Conchoidal to uneven. **Colour and transparency** Pale brass-yellow: opaque. **Streak** Greenish black. **Lustre** Metallic. **Distinguishing features** Colour, general lack of tarnish. Distinguished from chalcopyrite by lighter colour and greater hardness. It is difficult to distinguish from marcasite except by crystal form, though marcasite is paler in colour and has a lower specific gravity. **Alteration** Pyrite oxidizes either to iron sulphate or to the hydrated oxide, limonite. Limonite pseudomorphs after pyrite are not uncommon. **Occurrence** Pyrite is one of the most widely distributed of sulphide minerals, occurring in a variety of environments. It is present in igneous rocks as an accessory mineral and as segregations; in sedimentary rocks, particularly in black shales formed under stagnant, anaerobic conditions, and as nodules; and in metamorphic rocks, notably in slates when it frequently forms well shaped cubic crystals. It is a common mineral in hydrothermal sulphide veins, in replacement deposits and in contact metamorphic deposits. Fossils are often replaced by pyrite. The name comes from the Greek word for 'fire' and alludes to the sparks given off when the mineral is struck sharply.

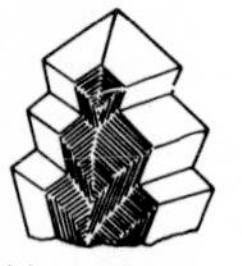

Marcasite: spear-shaped twin

Marcasite FeS_2 **Crystal system** Orthorhombic. **Habit** Crystals commonly tabular; also massive, stalactitic or as radiating fibres. **Twinning** Common; producing spear-shaped forms or 'cockscomb aggregates'. **SG** 4·8–4·9 **Hardness** 6–6½ **Cleavage** Prismatic, poor. **Fracture** Uneven. **Colour and transparency** Pale bronze-yellow: opaque. **Streak** Greyish black. **Lustre** Metallic. **Distinguishing features** Very similar to pyrite but has lower specific gravity, paler colour and distinctive spear-shaped forms. **Alteration** Like pyrite, it oxidizes easily to ferrous sulphate or limonite. It may also alter to pyrite. **Occurrence** Marcasite is deposited at low temperatures (less than 450°C) in hydrothermal veins containing zinc and lead ores. It is found in near-surface deposits, most commonly in sedimentary rocks such as limestone, especially chalk, and clay, as single crystals, concretions or as a replacement of fossils. The name comes from an Arabic word once used for pyrite.

Marcasite
Pyrite
Pyrite
Pyrite
Pyrite
5 cm

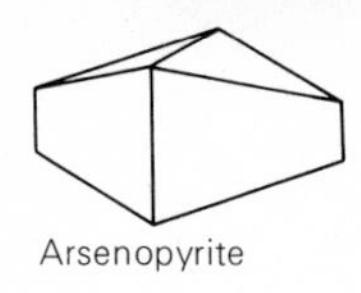
Arsenopyrite

Arsenopyrite (Mispickel) FeAsS **Crystal system** Monoclinic, pseudo-orthorhombic. **Habit** Prismatic crystals common, faces often striated. Columnar crystals have a rhombic cross-section. Also granular, columnar, massive. **Twinning** Common, on prism. **SG** 5·9–6·2 **Hardness** 5½–6 **Cleavage** Prismatic, indistinct. **Fracture** Uneven. **Colour and transparency** Silver grey-white often with a brownish tarnish: opaque. **Streak** Dark greyish black. **Lustre** Metallic. **Distinguishing features** Silver-white colour, crystal form. **Occurrence** Arsenopyrite forms under high to moderate temperature conditions and so it frequently accompanies gold, and ores of tin, tungsten and silver, together with sphalerite, pyrite, chalcopyrite, galena and quartz. Besides occurring in mineral veins, it is found disseminated in limestones, dolomites, gneisses and pegmatites.

Cobaltite CoAsS **Crystal system** Cubic. **Habit** Crystallizes as cubes, pyritohedra, or combinations of these forms. Also granular, massive. **SG** 6·0–6·3 **Hardness** 5½ **Cleavage** Cubic, perfect. **Fracture** Uneven. **Colour and transparency** Silver-white to grey with reddish tinge: opaque. **Streak** Grey-black. **Lustre** Metallic. **Distinguishing features** Cleavage, silver-white colour and inferior hardness distinguish it from pyrite. **Occurrence** Cobaltite occurs in high-temperature hydrothermal veins together with skutterudite, arsenopyrite and nickeline. It occurs as disseminated grains in metasomatic contact deposits.

Molybdenite MoS_2 **Crystal system** Hexagonal. **Habit** Crystals hexagonal, often tabular. Commonly occurs as foliated or scaly masses: also granular, massive. **SG** 4·6–4·8 **Hardness** 1–1½ **Cleavage** Basal, perfect; laminae flexible but not elastic. **Colour and transparency** Pale bluish lead-grey: opaque. **Streak** Greenish grey; bluish grey on paper. **Lustre** Metallic. **Distinguishing features** Similar hardness to graphite but greater specific gravity. The lighter, bluish tinge is distinctive, contrasting with the lead-grey of graphite. Greasy feel. **Occurrence** Molybdenite is widely distributed but never in large quantities. It is an accessory mineral in granites and occurs in associated pegmatites and quartz veins. It occurs also in contact metamorphic deposits with garnet, pyroxenes, scheelite, pyrite and tourmaline, and in veins together with scheelite, wolframite, cassiterite and fluorite. Molybdenite is an ore of molybdenum, and the name comes from the Greek word for 'lead'.

Arsenopyrite
Arsenopyrite
Cobaltite
Molybdenite with quartz
Molybdenite
5 cm

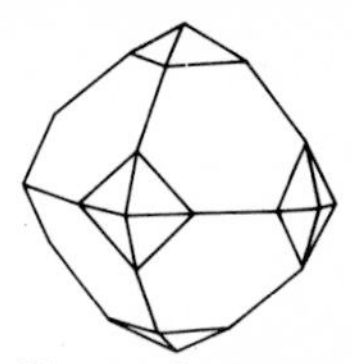
Skutterudite: octahedral habit

Smaltite - skutterudite - chloanthite series $(Co,Ni)As_{2-3}$ **Cr·stal system** Cubic. **Habit** Often massive, granular: crystals cubic, octahedral, or combinations of these forms; pyritohedra may also be present. **SG** 5·7–6·9 **Hardness** 5½–6 **Cleavage** Cubic and octahedral, indistinct. **Fracture** Uneven. **Colour and transparency** Tin-white, steel-grey when massive; iridescent or greyish tarnish: opaque. **Streak** Greyish black. **Lustre** Metallic. **Distinguishing features** The three names are applied to a series ranging from smaltite, which is essentially cobalt arsenide, through skutterudite, which approximates to $(Co,Ni)As_3$, to chloanthite, which is nickel arsenide. It is very difficult to distinguish the skutterudite minerals from arsenopyrite without chemical tests. **Occurrence** In veins, accompanying other cobalt and nickel minerals such as cobaltite and nickeline. Native silver, arsenopyrite and calcite are also associated minerals. Skutterudite, the name often applied to the series as a whole, comes from the Norwegian locality, Skutterude.

Pyrargyrite Ag_3SbS_3 **Crystal system** Trigonal. **Habit** Crystals prismatic; also as massive aggregates. **Twinning** Common. **SG** 5·8 **Hardness** 2½ **Cleavage** Rhombohedral, distinct. **Fracture** Uneven. **Colour and transparency** Black when opaque, but deep red by transmitted light. Darkens on exposure to light. Translucent to nearly opaque, transparent in thin fragments. **Streak** Purplish red. **Lustre** Adamantine; nearly metallic when opaque. **Distinguishing features** Deep red colour and streak; deeper in colour and less translucent than proustite. **Occurrence** Pyrargyrite and proustite are called ruby silver ores. They occur typically in low-temperature silver veins together with silver, argentite, tetrahedrite, galena, sphalerite, etc. The name comes from two Greek words meaning 'fire' and 'silver', in reference to its red colour and its composition.

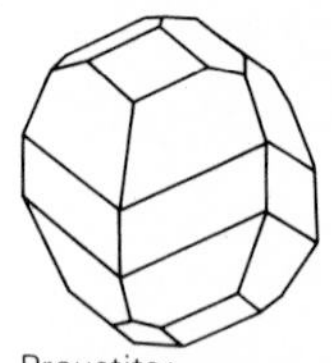
Proustite: combination of prism, scalenohedron and two rhombohedra

Proustite Ag_3AsS_3 **Crystal system** Trigonal. **Habit** Crystals prismatic, rhombohedral or scalenohedral. Also massive, compact. **Twinning** Common. **SG** 5·6 **Hardness** 2–2½ **Cleavage** Rhombohedral, distinct. **Fracture** Uneven. **Colour and transparency** Scarlet, darkens on exposure to light: translucent. **Streak** Scarlet-vermilion. **Lustre** Adamantine. **Distinguishing features** Colour, vermilion streak; lighter in colour than pyrargyrite. **Occurrence** Proustite occurs together with pyrargyrite in silver veins. It is less common than pyrargyrite and, like it, is an ore of silver. Proustite is named after J L Proust (1755–1826), a French chemist.

Proustite
Pyrargyrite
Smaltite
Skutterudite
Chloanthite
5 cm

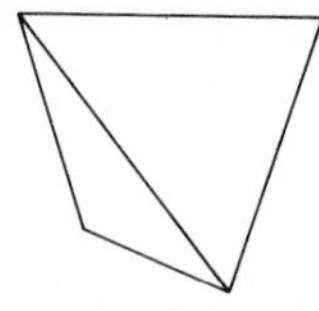

Tetrahedrite: tetrahedron

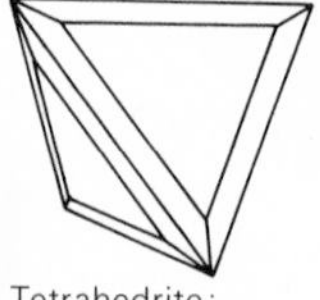

Tetrahedrite: modified tetrahedron

Tetrahedrite-tennantite series (Fahlerz) $(Cu,Fe)_{12}(Sb,As)_4S_{13}$ **Crystal system** Cubic. **Habit** Crystals commonly tetrahedral; also massive, granular, compact. **Twinning** Contact or penetration twins on tetrahedron. **SG** 4·6–5·1 **Hardness** 3–4½ **Cleavage** None. **Fracture** Subconchoidal to uneven. **Colour and transparency** Dark grey to black: opaque. **Streak** Dark grey or brown to black. **Lustre** Metallic. **Distinguishing features** Tetrahedral form, grey-black colour. There is a continuous series between tetrahedrite and tennantite as arsenic substitutes for antimony. **Alteration** Oxidizes to minerals such as malachite and azurite. **Occurrence** Commonly found in hydrothermal veins together with silver, copper, lead and zinc minerals; also in igneous contact metamorphic deposits. Tennantite is less widespread in occurrence than tetrahedrite and occurs in metasomatic deposits in limestone, whereas tetrahedrite is more commonly found in lead-silver veins.

Enargite Cu_3AsS_4 **Crystal system** Orthorhombic. **Habit** Crystals usually small, tabular or prismatic; often massive, granular, bladed or columnar. **Twinning** Sometimes produces star-shaped forms comprising three individuals. **SG** 4·4 **Hardness** 3 **Cleavage** Prismatic, perfect; pinacoidal, distinct. **Fracture** Uneven. **Colour and transparency** Dark grey to black: opaque. **Streak** Black. **Lustre** Metallic. **Distinguishing features** Colour, two cleavages. It fuses easily and will melt in a match flame. **Occurrence** Enargite is not a common mineral. It occurs in low temperature, near-surface deposits in association with chalcosine, bornite, covelline, pyrite, sphalerite, tetrahedrite, baryte and quartz.

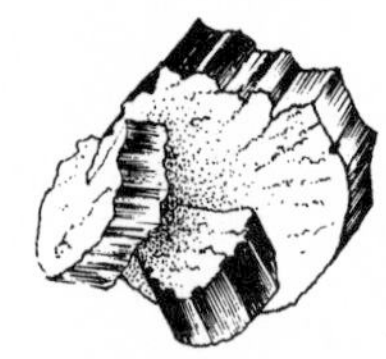

Bournonite cog-wheel shaped twins

Bournonite (Wheel-ore) $PbCuSbS_3$ **Crystal system** Orthorhombic. **Habit** Crystals tabular; also massive, granular, compact. **Twinning** Very common; repeated twinning produces crystals reminiscent of cog wheels. **SG** 5·7–5·9 **Hardness** 2½–3 **Cleavage** None, or poor. **Fracture** Uneven. **Colour and transparency** Grey to black; opaque. **Streak** Grey to black. **Lustre** Metallic. **Distinguishing features** Habit of twinned crystals, also high specific gravity. Fuses easily. **Alteration** Occasionally alters to cerussite, malachite or azurite. **Occurrence** In hydrothermal veins accompanying such minerals as galena, chalcopyrite, tetrahedrite, stibnite, sphalerite. It has been used as an ore of copper, lead and antimony. Bournonite is named in honour of Count J L de Bournon (1751–1825), a French mineralogist.

Boulangerite $Pb_5Sb_4S_{11}$ **Crystal system** Orthorhombic. **Habit** Crystals usually elongated prismatic; also as fibrous or plumose masses. **SG** 5·7–6·3 **Hardness** 2½–3 **Cleavage** One good cleavage. **Colour and transparency** Lead-grey; sometimes with yellow spots caused by oxidation: opaque. **Streak** Red-brown. **Lustre** Metallic. **Distinguishing features** Similar in appearance to stibnite and jamesonite and very difficult to distinguish from them. **Occurrence** Commonly occurs in veins together with stibnite, galena, sphalerite, pyrite, and with quartz, dolomite and calcite as gangue minerals.

Tetrahedrite

Tennantite

Enargite

Enargite

Boulangerite

Bournonite

5 cm

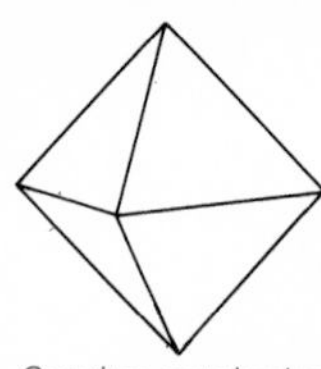
Cuprite: octahedron

Cuprite (Red copper ore) Cu_2O **Crystal system** Cubic. **Habit** Crystals usually octahedral, sometimes cubic or rhombdodecahedral, or combinations of these forms; also acicular. Also massive, granular. **SG** 5·8–6·1 **Hardness** 3½–4 **Cleavage** None. **Fracture** Uneven. **Colour and transparency** Red, though sometimes so dark as to be nearly black: subtranslucent; subtransparent when very thin. **Streak** Brownish red. **Lustre** Adamantine or sub-metallic. **Distinguishing features** Cuprite is similar in colour to hematite and cinnabar, but it is softer than hematite and harder than cinnabar; it differs also in colour of the streak. **Alteration** Malachite pseudomorphs after cuprite are not uncommon. **Occurrence** Cuprite is usually formed as a secondary mineral in the oxidized zone of copper deposits, and is commonly accompanied by malachite, azurite and chalcosine. Fine, hair-like crystals of cuprite are called chalcotrichite. It is an ore of copper, and the name derives from the Latin *cuprum* for copper.

Tungstite $WO_3.H_2O$ **Crystal system** Orthorhombic. **Habit** Powdery or earthy coatings. **Cleavage** Basal, perfect. **Colour and transparency** Yellow or yellowish green. **Lustre** Earthy. **Distinguishing features** Yellowish colour and association with other tungsten minerals. **Occurrence** Tungstite is a secondary mineral found in association with wolframite.

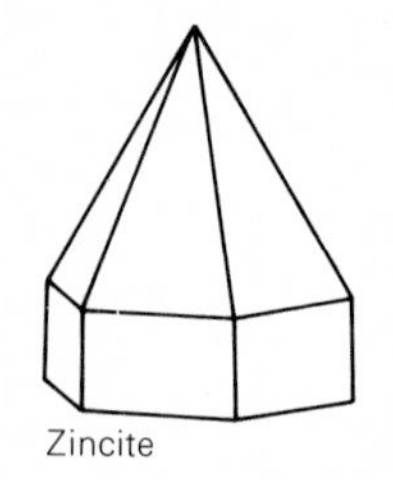
Zincite

Zincite ZnO **Crystal system** Hexagonal. **Habit** Crystals very rare; usually massive, foliated, granular. **SG** 5·4–5·7 **Hardness** 4–4½ **Cleavage** Prismatic, distinct; basal, parting. **Fracture** Subconchoidal. **Colour and transparency** Deep red to orange-yellow: translucent. **Streak** Orange-yellow. **Lustre** Subadamantine. **Distinguishing features** Red colour, orange-yellow streak. Dissolves in hydrochloric acid. **Occurrence** Zincite is a rare mineral. In the few places where it does occur, notably at Franklin, New Jersey, USA, it is associated with franklinite and willemite in a contact metamorphic deposit.

Franklinite $(Zn,Mn^{2+},Fe^{2+})(Fe^{3+},Mn^{3+})_2O_4$ **Crystal system** Cubic. **Habit** Crystals octahedral; also massive, granular. **SG** 5·0–5·2 **Hardness** 5½–6½ **Cleavage** None; octahedral parting. **Fracture** Uneven. **Colour and transparency** Black: opaque. **Streak** Reddish brown to dark brown. **Lustre** Metallic. **Distinguishing features** Franklinite, a member of the spinel group, most closely resembles magnetite, but is only slightly magnetic and has a dark brown streak. **Occurrence** Franklinite, zincite and willemite occur together in the zinc deposits of Franklin, New Jersey, USA. The deposits are associated with crystalline limestone and are probably of metasomatic origin. The deposit is worked as an ore of zinc and manganese, and the name of the mineral is taken from the locality.

Cuprite
Cuprite
Zincite
Zincite and franklinite
Tungstite
Franklinite

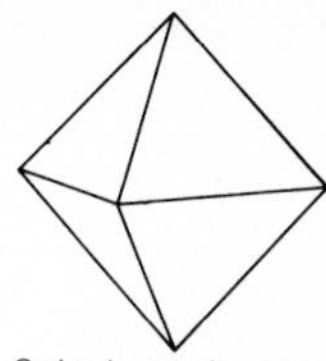
Spinel: octahedron

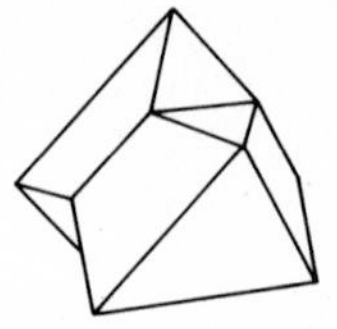
Spinel: twinned octahedron

Spinel $MgAl_2O_4$ **Crystal system** Cubic. **Habit** Crystals usually octahedral; also massive. **Twinning** Common on octahedron, giving 'spinel twins'. **SG** 3·5–4·1 **Hardness** 7½–8 **Cleavage** None; octahedral parting. **Fracture** Conchoidal. **Colour and transparency** Very variable; commonly red (ruby spinel), but also blue, green, brown, black or colourless: transparent to nearly opaque, usually translucent. **Streak** White, but can be grey or even brown. **Lustre** Vitreous. **Distinguishing features** Octahedral form, twinning, hardness. Spinel is the name of a series, rather than of a specific mineral. Iron, zinc and manganese atoms can substitute for magnesium in the structure and so give rise to variations in colour and physical properties. **Occurrence** Spinel occurs as an accessory mineral in igneous rocks such as gabbro. It also occurs in contact metamorphosed impure dolomitic limestones in association with phlogopite, graphite and chondrodite, and in aluminous metamorphic rocks. Gem quality spinels occur in contact altered limestones and in alluvial gravels derived from them in Burma, Ceylon and India. Its hardness and resistance to weathering result in spinel being found as rolled pebbles in river and beach sands. The origin of the name is unknown.

Magnetite Fe_3O_4 **Crystal system** Cubic. **Habit** Crystals most commonly octahedra, also rhombdodecahedra: also massive, granular. **Twinning** Common on octahedron. **SG** 5·2 **Hardness** 5½–6½ **Cleavage** None; octahedral parting. **Fracture** Subconchoidal to uneven. **Colour and transparency** Black: opaque. **Streak** Black. **Lustre** Metallic, shining; to submetallic, dull. **Distinguishing features** Colour and streak, strongly magnetic character. **Occurrence** Magnetite is very widely distributed in several environments. It is a common accessory mineral in igneous rocks, and its relatively high specific gravity sometimes results in it accumulating in them to form deposits of economic value. It is a common mineral in contact and regionally metamorphosed rocks and occurs in high-temperature mineral veins. It often occurs in association with corundum in emery deposits. Magnetite, by reason of its strongly magnetic properties, has attracted attention since early times. Specimens possessing polarity (lodestone) act as compass needles when free to swing. Magnetite is an important iron ore. There is some dispute as to the origin of the name. It probably comes from Magnesia, a locality in Asia Minor, but it is associated with the fable of Magnes, a shepherd who is alleged to have discovered the mineral on Mount Ida when the nails in his shoes and the iron ferrule of his staff adhered to the ground.

5 cm
Spinel
Spinel
Spinel
Spinel (Ceylonite)
Magnetite
Magnetite
Magnetite

Chromite $FeCr_2O_4$ (often with some Mg and Al) **Crystal system** Cubic. **Habit** Crystals rare, octahedral. Usually massive, granular. **SG** 4·1–5·1 **Hardness** $5\frac{1}{2}$ **Cleavage** None. **Fracture** Uneven. **Colour and transparency** Black to brownish black; opaque, translucent in thin fragments. **Streak** Dark brown. **Lustre** Metallic to submetallic. **Distinguishing features** Brown streak and weakly magnetic character distinguish chromite from magnetite. **Occurrence** Chromite is an accessory mineral in igneous rocks such as peridotite and serpentinite. It may be concentrated into layers or lenses in sufficient quantity to be worked as an ore and, in fact, chromite is the only ore of chromium. Its durability sometimes results in chromite being concentrated in alluvial sands and gravels. It is an extremely refractory mineral and chromite bricks are used to line blast furnaces.

Hematite

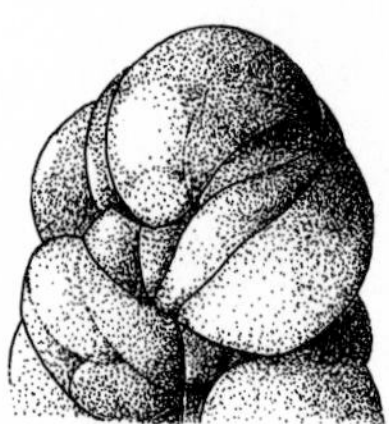
Hematite: mamillated form

Hematite Fe_2O_3 **Crystal system** Trigonal. **Habit** Crystals tabular or rhombohedral, sometimes with curved and striated rhombohedral faces. Also columnar, laminated or massive, often in striking mamillated or botryoidal forms. **Twinning** Penetration twins on basal pinacoid. **SG** 4·9–5·3 **Hardness** 5–6 **Cleavage** None. **Fracture** Uneven, brittle. **Colour and transparency** Steel-grey to black, sometimes iridescent. Massive compact varieties vary from dull to bright red. Opaque, except in very thin flakes. **Streak** Red to reddish brown. **Lustre** Metallic, sometimes dull. **Distinguishing features** Red streak, hardness. **Occurrence** Hematite is the most important ore of iron and is widely distributed. It occurs as an accessory mineral in igneous rocks and in hydrothermal veins. It is of widespread occurrence in sedimentary rocks in which it may be of primary origin, often occurring as ooliths or as a cementing material; or as a secondary mineral, precipitated from iron-bearing percolating waters and replacing other minerals. Bedded iron formations of Precambrian age have afforded huge quantities of hematite in North America and elsewhere. In addition to its use as an iron ore, hematite is used as a pigment and as a polishing powder. The name is derived from the Greek word for 'blood' and is descriptive of the colour of the powdered mineral.

Hematite
Hematite
Hematite
Hematite
Chromite
5 cm

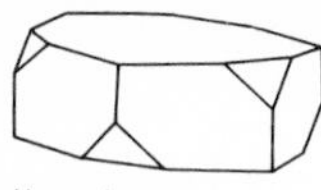

Ilmenite

Ilmenite $FeTiO_3$ **Crystal system** Trigonal. **Habit** Crystals thick tabular; often massive, compact. **Twinning** Common, on basal pinacoid. **SG** 4·5–5·0 **Hardness** 5–6 **Cleavage** None; basal parting. **Fracture** Conchoidal. **Colour and transparency** Black: opaque. **Streak** Black to brownish red. **Lustre** Submetallic to metallic. **Distinguishing features** Distinguished from magnetite by non-magnetic character, and from hematite by streak. **Occurrence** Ilmenite is an accessory mineral in igneous rocks such as gabbro and diorite. It occurs occasionally in quartz veins and pegmatites with hematite and chalcopyrite, and in some gneisses. Its resistance to weathering leads to its concentration in alluvial sands, together with magnetite, monazite and rutile. The name is taken from the Ilmen Mountains, USSR.

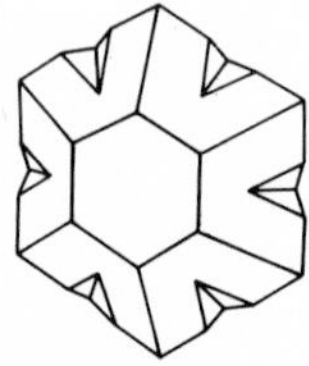

Chrysoberyl: twinned crystal

Chrysoberyl $BeAl_2O_4$ **Crystal system** Orthorhombic. **Habit** Crystals generally tabular. **Twinning** Common, often repeated, to give pseudo-hexagonal crystals. **SG** 3·5–3·8 **Hardness** 8½ **Cleavage** Prismatic, poor. **Fracture** Uneven to conchoidal. **Colour and transparency** Various shades of green and yellow: transparent to translucent. Chrysoberyl is used as a gemstone when transparent; the variety alexandrite is emerald green but red by artificial light. **Streak** White. **Lustre** Vitreous. **Distinguishing features** Colour and hardness; crystals tabular, in contrast to the prismatic habit of beryl. May be confused with olivine. **Occurrence** Chrysoberyl occurs in granitic rocks and pegmatites; also in mica schists. It is frequently found in alluvial sands and gravels. The name is taken from the Greek and alludes to the golden yellow colour. Alexandrite was named for Tsar Alexander II of Russia.

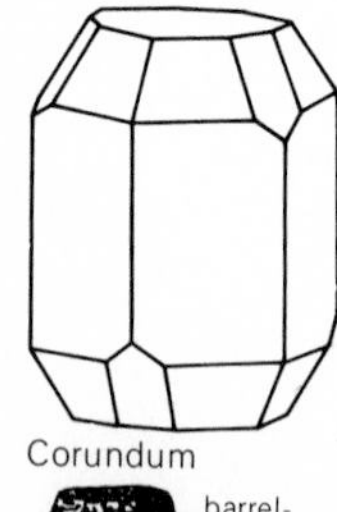

Corundum

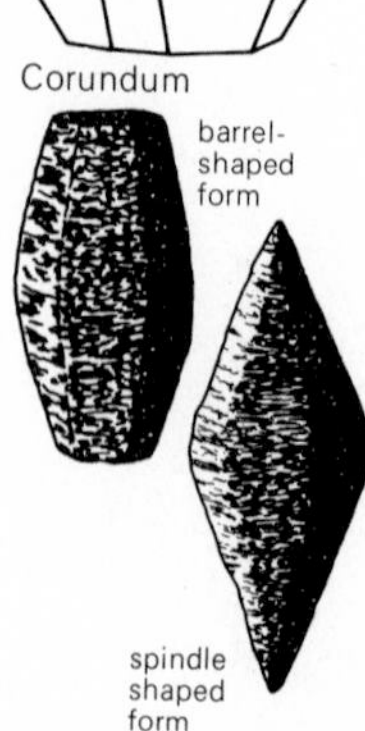

Corundum Al_2O_3 **Crystal system** Trigonal. **Habit** Crystals usually rough, barrel-shaped prismatic forms or tapering, spindle-shaped forms; also flat, tabular. Emery is a mixture of the massive black granular form with magnetite and spinel. **Twinning** Common, often repeated, giving rise to striations on basal pinacoid. **SG** 3·9–4·1 **Hardness** 9 **Cleavage** None; basal parting. **Fracture** Uneven to conchoidal. **Colour and transparency** Two main varieties: blue (sapphire), red (ruby). Also yellow, brown, green: crystals sometimes show variation in colour. Star sapphire and star ruby are opalescent and show a six-rayed star when polished and viewed in one particular direction. Transparent to translucent. **Streak** White. **Lustre** Adamantine to vitreous. **Distinguishing features** Great hardness, specific gravity, crystal form. **Occurrence** Corundum occurs in certain nepheline syenites and nepheline syenite pegmatites. It occurs also in metamorphic rocks such as marble, gneiss and schist. Large crystals occur in some pegmatites, and emery deposits occur in some regionally metamorphosed rocks. The hardness and durability of corundum lead to its occurrence in alluvial sands and gravels. Gem quality ruby occurs in Burma and Ceylon, and sapphire also occurs there and in India, Australia and elsewhere. Corundum is also used as an abrasive either as fragments produced by grinding massive corundum, or as the impure form, emery.

Corundum (star sapphire)

Corundum

Chrysoberyl

Chrysoberyl

Chrysoberyl

Corundum

Ilmenite

Pyrochlore - microlite series $(Ca,Na)_2(Nb,Ta)_2O_6(O,OH,F)$ **Crystal system** Cubic. **Habit** Crystals commonly octahedral; also as small, irregular grains. **SG** 4·2–6·4 Increasing with tantalum content. **Hardness** 5–5½ **Cleavage** Octahedral, distinct. **Fracture** Conchoidal. **Colour and transparency** Pyrochlore is brown to black; microlite yellow to brown, sometimes red: subtranslucent to opaque. **Streak** Light brown. **Lustre** Vitreous to resinous; sometimes greasy. **Distinguishing features** Crystal form. Pyrochlore is the name given to the niobium-rich members of the series and microlite to those in which tantalum is dominant. Many other elements can substitute for sodium and calcium in the structure, including uranium, thorium and the rare earth elements. Some specimens are therefore radioactive. **Occurrence** Usually in pegmatites in, or near to, alkaline rocks and associated with zircon and apatite. Pyrochlore occurs characteristically in pipe-like igneous intrusions composed essentially of calcium and magnesium carbonates and called carbonatite. Microlite is usually associated with granite pegmatites. The name pyrochlore is derived from the Greek in allusion to the green colour it acquires on heating, and microlite is so named because of the small size of the first described crystals.

Braunite $3Mn_2O_3.MnSiO_3$ **Crystal system** Tetragonal. **Habit** Pyramidal crystals; also massive, granular. Braunite departs only slightly from cubic symmetry and so the crystals appear octahedral. **SG** 4·7–4·8 **Hardness** 6–6½ **Cleavage** Pyramidal, perfect. **Fracture** Uneven. **Colour and transparency** Brownish black to steel-grey: opaque. **Streak** Brownish black to steel-grey. **Lustre** Submetallic. **Distinguishing features** Colour, crystal form. Soluble in hydrochloric acid, leaving a residue of silica. **Occurrence** Occurs in hydrothermal veins with other manganese oxides, and also as a secondary mineral. It sometimes forms as a result of metamorphism of manganese-bearing sediments. It is named after K Braun von Gotha.

Psilomelane: botryoidal stalactitic form

Psilomelane **No fixed composition: a mixture of manganese oxides. Crystal system** Monoclinic. **Habit** Massive, botryoidal, stalactitic. **SG** 3·3–4·7 **Hardness** 5–7 **Colour and transparency** Black to dark grey: opaque. **Streak** Brownish black to black. **Lustre** Submetallic. **Distinguishing features** Greater hardness distinguishes psilomelane from other manganese minerals; botryoidal habit. **Occurrence** A secondary manganese mineral, precipitated at atmospheric temperatures together with pyrolusite and limonite, in sediments or in quartz veins.

Wad **Crystal system** Amorphous. **Habit** Shapeless, stalactitic, or reniform masses, often earthy. **SG** 2·8–4·4 **Hardness** Usually soft. **Colour and transparency** Dull black to brownish black. **Streak** Black. **Lustre** Earthy. **Distinguishing features** Black colour; often loosely compacted and so feels light and soils the fingers. **Occurrence** Wad is not a single mineral but a mixture of several hydrous manganese oxides which occur in the oxidized zone of ore deposits, in lake or bog deposits, and shallow water marine sediments.

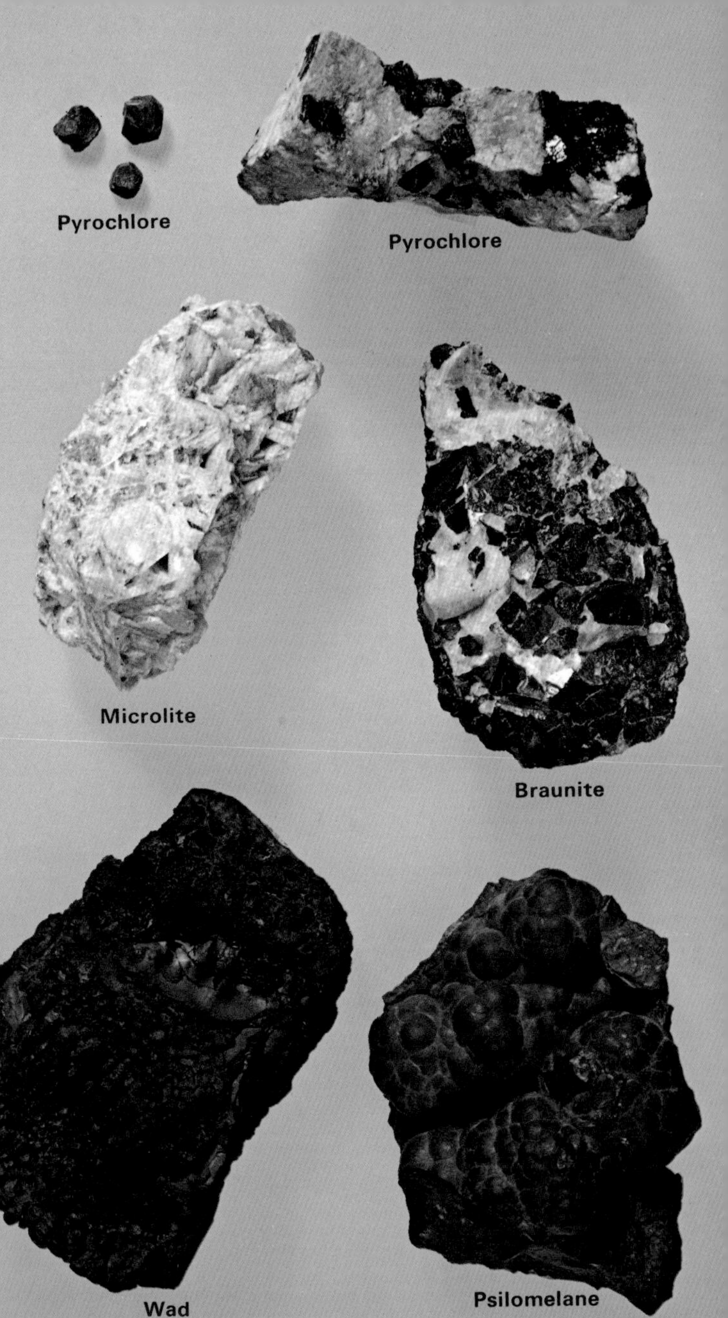
Pyrochlore
Pyrochlore
Microlite
Braunite
Wad
Psilomelane
5 cm

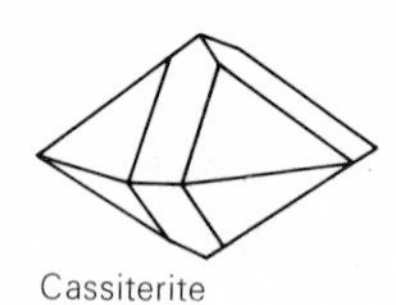
Cassiterite

Cassiterite (Tinstone) SnO_2 **Crystal system** Tetragonal. **Habit** Crystals pyramidal or short prismatic; also massive, granular. Reniform shapes with a fibrous structure are called wood tin. **Twinning** Common. **SG** 6·8–7·1 **Hardness** 6–7 **Cleavage** Prismatic, imperfect. **Fracture** Uneven. **Colour and transparency** Usually reddish brown to nearly black; sometimes yellowish. **Streak** White or greyish. **Lustre** Adamantine, splendent to submetallic. **Distinguishing features** High specific gravity, adamantine lustre, light streak, crystal form. **Occurrence** Cassiterite is one of the few tin minerals and is the principal ore of the metal. It occurs typically in high temperature hydrothermal veins and pegmatites located within, or close to, granite masses. Associated minerals are wolframite, arsenopyrite bismuthinite. topaz, quartz, tourmaline and mica. Rounded pebbles of cassiterite occur as stream tin in alluvial deposits.

Pyrolusite: dendritic form

Pyrolusite MnO_2 **Crystal system** Tetragonal. **Habit** Usually massive, often as reniform coatings or dendritic, plant-like shapes on joints or bedding planes in sedimentary rocks; also as divergent fibres or columns. Very rarely as crystals (polianite). **SG** 4·5–7·9 **Hardness** 1–2 (massive), 6–6½ (crystals) **Cleavage** Perfect, prismatic. **Fracture** Uneven, splintery. **Colour and transparency** Black to bluish steel-grey; opaque. **Streak** Black. **Lustre** Metallic. **Distinguishing features** Colour, softness when massive. **Occurrence** Pyrolusite is a common manganese mineral and forms under oxidizing conditions. It is often of secondary origin, being found in the oxidized zone of ore deposits containing manganese, and in bog or shallow marine sediments. It occurs also in quartz veins, and as nodules on the bottom of the sea. The name is derived from two Greek words meaning 'fire' and 'to wash', because it was at one time used to remove from glass the colour due to the presence of iron oxide.

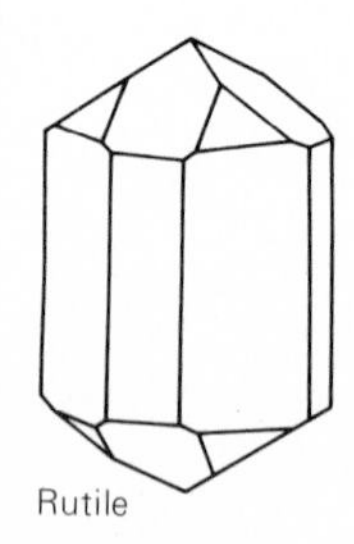
Rutile

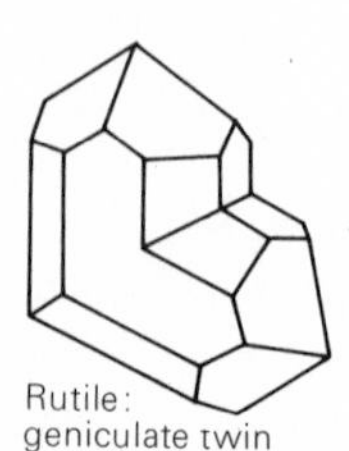
Rutile: geniculate twin

Rutile TiO_2 **Crystal system** Tetragonal. **Habit** Crystals prismatic and terminated by bipyramids. Prism faces often striated. Also massive. **Twinning** Common on bipyramid giving geniculate (knee-shaped) twins, or complex cyclic twins made up of six or eight individuals. **SG** 4·2–4·4 **Hardness** 6–6½ **Cleavage** Prismatic, distinct. **Fracture** Uneven. **Colour and transparency** Usually reddish brown: can be yellowish red or black. Transparent when thin, usually subtranslucent, occasionally nearly opaque. **Streak** Pale brown. **Lustre** Adamantine; submetallic when dark coloured. **Distinguishing features** Red colour, adamantine lustre, crystal form. **Occurrence** Rutile occurs as an accessory mineral in a variety of igneous rocks, and in schists, gneisses, metamorphosed limestones and quartzites. If often occurs as acicular crystals in quartz (rutilated quartz). It is also a secondary mineral produced by the breakdown of titanium-bearing minerals such as sphene and some micas. Rutile is also concentrated in alluvial deposits and beach sands.

Cassiterite

Cassiterite

Pyrolusite

Rutile

Rutile

Pyrolusite

Rutile

5 cm

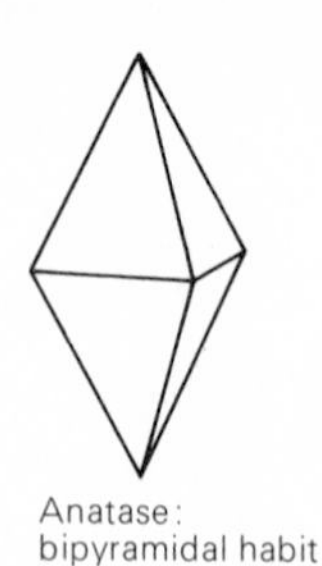

Anatase: bipyramidal habit

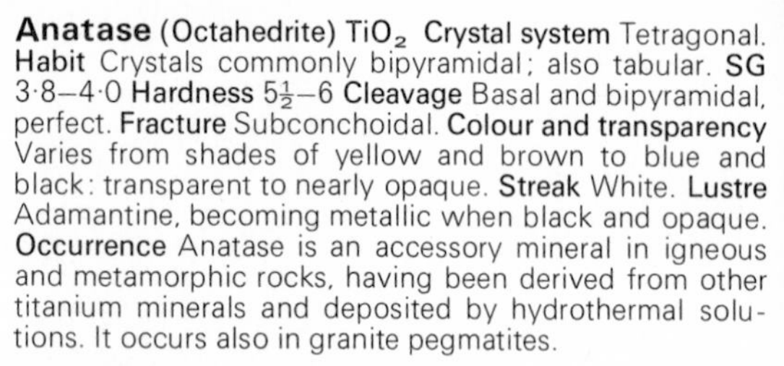

Anatase (Octahedrite) TiO_2 **Crystal system** Tetragonal. **Habit** Crystals commonly bipyramidal; also tabular. **SG** 3·8–4·0 **Hardness** 5½–6 **Cleavage** Basal and bipyramidal, perfect. **Fracture** Subconchoidal. **Colour and transparency** Varies from shades of yellow and brown to blue and black: transparent to nearly opaque. **Streak** White. **Lustre** Adamantine, becoming metallic when black and opaque. **Occurrence** Anatase is an accessory mineral in igneous and metamorphic rocks, having been derived from other titanium minerals and deposited by hydrothermal solutions. It occurs also in granite pegmatites.

Brookite TiO_2 **Crystal system** Orthorhombic. **Habit** Crystals vary in habit but often tabular or platy. **SG** 3·9–4·2 **Hardness** 5½–6 **Cleavage** Prismatic, poor. **Fracture** Subconchoidal to uneven. **Colour and transparency** Reddish brown to brownish black: translucent. **Streak** White. **Lustre** Metallic-adamantine. **Occurrence** As an accessory mineral in igneous and metamorphic rocks, and in hydrothermal veins. Rutile, anatase and brookite are polymorphs of TiO_2. Brookite is named after HJ Brooke (1771–1857), a British mineralogist.

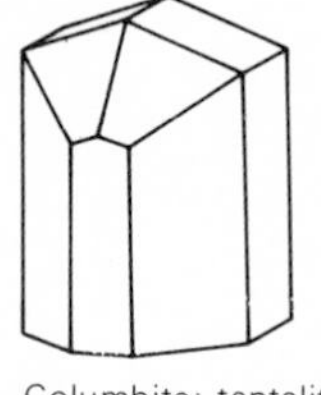

Columbite: tantalite

Columbite-tantalite series $(Fe,Mn)(Nb,Ta)_2O_6$ **Crystal system** Orthorhombic. **Habit** Crystals tabular or short prismatic. **Twinning** Common. **SG** 5·0–8·0 (increases with tantalum content). **Hardness** 6–6½ **Cleavage** Pinacoidal. **Fracture** Uneven. **Colour and transparency** Iron-black to brownish black: subtranslucent to opaque. **Streak** Dark red to black. **Lustre** Submetallic to subresinous. **Distinguishing features** High specific gravity, black colour, crystal form. As with the pyrochlore-microlite series there is continuous substitution of tantalum for niobium. When niobium dominates over tantalum, the name columbite is applied, and tantalite when tantalum is present in greater amount than niobium. **Occurrence** Usually in granite pegmatites in association with quartz, feldspar, tourmaline, beryl, spodumene, petalite, cassiterite and wolframite. Columbite and tantalite are sources of niobium and tantalum.

Uraninite (Pitchblende) UO_2 **Crystal system** Cubic. **Habit** Crystals rare: usually massive, botryoidal (pitchblende). **SG** 6·5–8·5 (massive), 8–10 (crystals). **Hardness** 5–6 **Fracture** Conchoidal to uneven. **Colour and transparency** Brownish black to black: opaque. **Streak** Brownish black or grey. **Lustre** Submetallic to greasy or pitch-like; dull. **Distinguishing features** High specific gravity, characteristic greasy or pitch-like lustre, black colour, radioactivity. The name uraninite is used for crystallized varieties of UO_2; massive varieties are called pitchblende. **Occurrence** Crystallized uraninite occurs in pegmatites associated with granitic or syenitic rocks and associated with monazite, zircon and tourmaline. Pitchblende is usually found as massive crusts in high- or moderate-temperature hydrothermal veins associated with cassiterite, pyrite, chalcopyrite, arsenopyrite and galena. It occurs also as a detrital mineral in alluvial deposits. It is an important ore of uranium, and was the source of radium, discovered by the Curies.

5 cm
Anatase
Anatase
Brookite
Brookite
Columbite-tantalite
Columbite-tantalite
Uraninite
Uraninite

Brucite $Mg(OH)_2$ **Crystal system** Trigonal. **Habit** Crystals broad tabular; also fibrous (nemalite) and massive, foliated. **SG** 2·4 **Hardness** 2½ **Cleavage** Basal, perfect. **Colour and transparency** White, shading to pale grey, blue or green: transparent to translucent. **Streak** White. **Lustre** Pearly parallel to cleavage, elsewhere waxy to vitreous. **Distinguishing features** Cleavage, softness, foliated habit. Distinguished from talc by greater hardness, and from gypsum by form. Readily soluble in hydrochloric acid, much more so than gypsum. **Occurrence** Brucite occurs in metamorphosed dolomitic limestones and also in hydrothermal veins together with calcite and talc, and in serpentinite. It is named in honour of A Bruce (1777–1818), an American mineralogist.

Gibbsite (Hydrargillite) $Al(OH)_3$ **Crystal system** Monoclinic. **Habit** Crystals tabular; also as stalactitic or encrusting forms, spheroidal concretions or foliated and earthy aggregates. **Twinning** Common. **SG** 2·3–2·4 **Hardness** 2½–3½ **Cleavage** Basal, perfect. **Colour and transparency** White or near white, sometimes pink or red: transparent to translucent. **Streak** White. **Lustre** Pearly parallel to cleavage; other faces vitreous. **Distinguishing features** Strong clay odour when breathed on. **Occurrence** Gibbsite occurs as crystals in low-temperature hydrothermal veins, and with boehmite and diaspore is a constituent mineral of bauxite. It is a secondary mineral resulting from the decomposition of aluminium silicates. It is named after Col G Gibbs (1776–1833), an American mineral collector.

Boehmite $AlO(OH)$ **Crystal system** Orthorhombic. **Habit** Crystals microscopic: usually as scattered grains or pisolitic aggregates. **SG** 3·0–3·1 **Hardness** 3½–4 **Cleavage** One very good cleavage. **Colour and transparency** White. **Occurrence** Boehmite, with gibbsite, diaspore and kaolinite, is an important constituent mineral of bauxite, but because of its occurrence as tiny crystals or grains, it cannot be recognized by the naked eye. It is named after J G Böhm, a nineteenth-century German chemist.

Diaspore $AlO(OH)$ **Crystal system** Orthorhombic. **Habit** Crystals platy or tabular; also massive, foliated, sometimes acicular. **SG** 3·2–3·5 **Hardness** 6½–7 **Cleavage** One perfect cleavage. **Colour and transparency** Variable, from colourless, through white and grey, to brown or pink: translucent. **Lustre** Pearly on cleavage faces, vitreous elsewhere. **Distinguishing features** Cleavage, hardness, platy habit. **Occurrence** Diaspore is an important constituent of bauxite together with boehmite and gibbsite. It occurs in close association with corundum in emery deposits, and in chlorite schist.

Lepidocrocite $FeO(OH)$ **Crystal system** Orthorhombic. **Habit** Scaly, fibrous or massive aggregates. **SG** 4·1 **Hardness** 5 **Cleavage** One perfect cleavage. **Colour** Red to reddish brown. **Streak** Orange. **Distinguishing features** Colour. **Occurrence** Lepidocrocite and goethite are similar in chemistry and occurrence and it is difficult to distinguish them.

Gibbsite
Gibbsite
Lepidocrocite
Diaspore
Brucite
Boehmite
5 cm

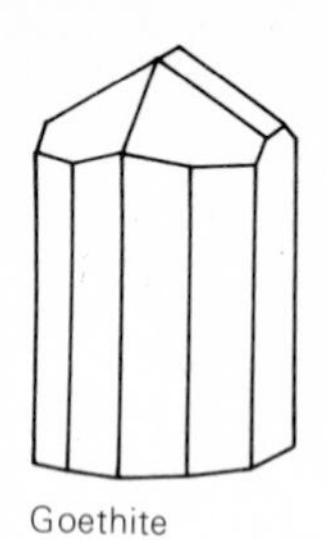
Goethite

Goethite $FeO(OH)$ **Crystal system** Orthorhombic. **Habit** Crystals rare but platy, bladed or prismatic; usually massive as mamillated, botryoidal or stalactitic masses with a fibrous radiating structure. **SG** 3·3–4·3 **Hardness** 5–5½ **Cleavage** One perfect cleavage. **Fracture** Uneven. **Colour and transparency** Usually very dark brown; earthy forms ochrous yellow-brown: subtranslucent; transparent in thin fragments. **Streak** Brownish yellow. **Lustre** Adamantine (crystals); massive goethite is often silky by reason of its fibrous structure; sometimes dull. **Distinguishing features** Colour, streak. The presence of cleavage, radial growth, and other indications of crystallinity distinguish goethite from limonite. **Occurrence** Goethite is produced by the oxidation of iron-bearing minerals such as pyrite and magnetite. It is of widespread occurrence, though at one time it was thought to be a rare mineral. In fact much of the ochrous brown ferric oxide described as 'limonite' is composed in large part of crystalline goethite. Limonite has a yellowish brown streak and a vitreous lustre which, together with the lack of cleavage, serves to distinguish it from crystalline goethite. Goethite occurs as a secondary mineral in the oxidized zone (iron hat) of veins containing iron-bearing minerals. It replaces other minerals, and goethite pseudomorphs after pyrite are common. Oolitic 'limonitic' iron ores of sedimentary origin occur in eastern France (the minette ores), and goethite is precipitated from marine or fresh water in bogs or lagoons to form bog iron ore. The mineral is named after the German poet JW von Goethe (1749–1832), who was also a mineral collector.

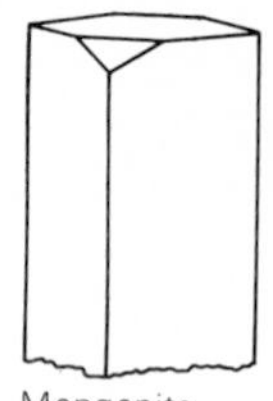
Manganite

Manganite $MnO(OH)$ **Crystal system** Monoclinic, pseudo-orthorhombic. **Habit** Crystals prismatic, often striated and frequently grouped in bundles or as radiating aggregates. **Twinning** Penetration twins on prism. **SG** 4·2–4·4 **Hardness** 4 **Cleavage** Pinacoidal, perfect; prismatic, less so. **Fracture** Uneven. **Colour and transparency** Dark steel-grey to black: opaque. **Streak** Reddish brown to black. **Lustre** Submetallic. **Distinguishing features** Colour, prismatic habit, brown streak, soluble in concentrated hydrochloric acid. **Alteration** To pyrolusite and other manganese oxides. **Occurrence** Manganite occurs in association with such minerals as pyrolusite, baryte and goethite, in deposits precipitated from waters under oxidizing conditions. It occurs also in low-temperature hydrothermal veins, associated with granitic rocks. Manganite has been used as an ore of manganese.

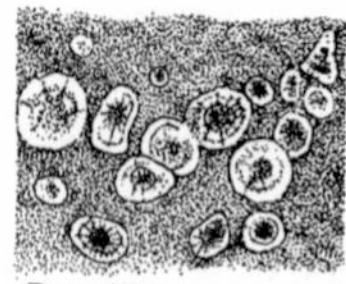
Bauxite:
pisolitic structure

Bauxite **Habit** Massive, oolitic, pisolitic, earthy; or as concretionary masses. **Colour** Ochre-yellow, brown, red; also grey. **Occurrence** Bauxite is not a single mineral but a mixture of several minerals, mainly diaspore, gibbsite, boehmite and iron oxides (see previous page). It is of secondary origin, forming under tropical conditions by the prolonged weathering and leaching of rocks containing aluminium silicates. The leaching by tropical rains removes the silica, leaving aluminous hydroxides.

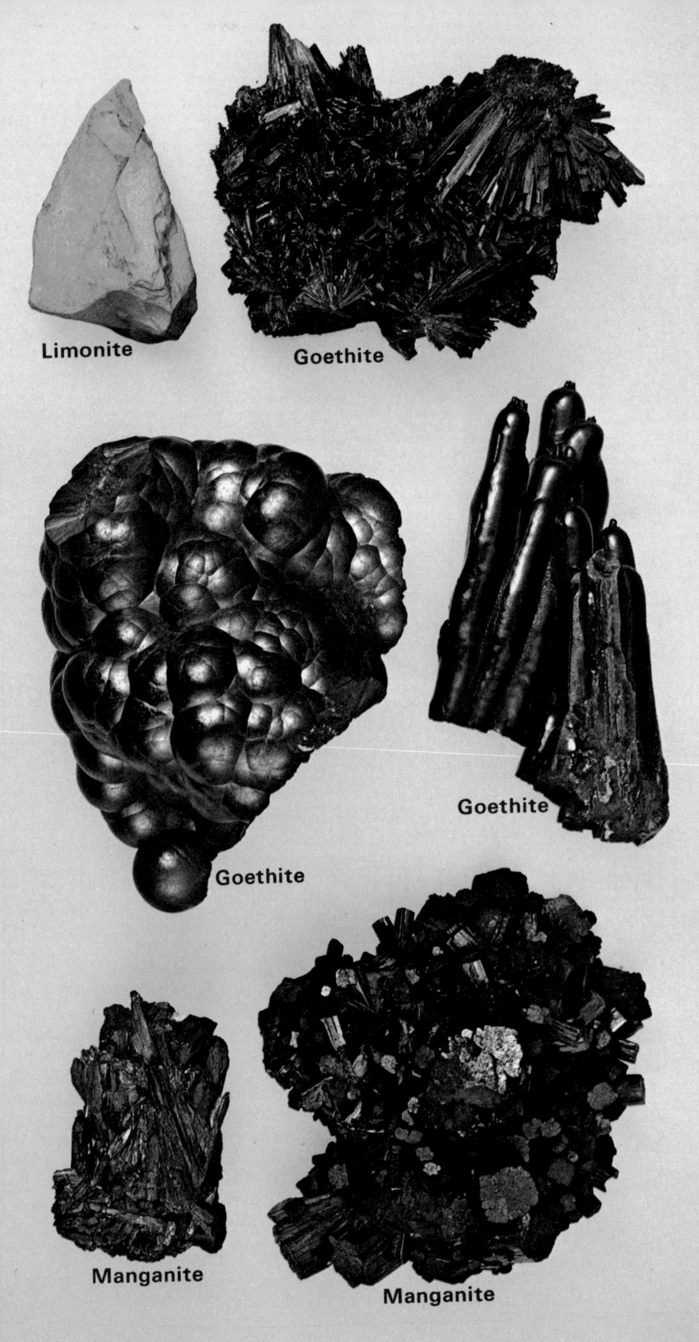
Limonite
Goethite
Goethite
Goethite
Manganite
Manganite
5 cm

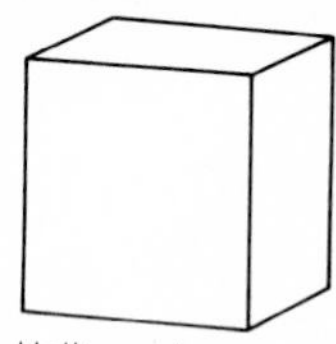
Halite: cube

Halite: hopper crystal

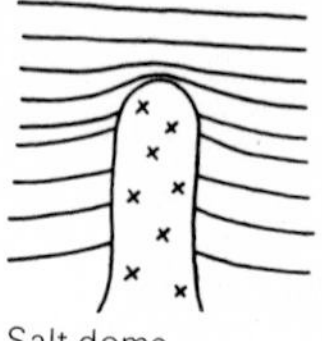
Salt dome

Halite (Rock salt) NaCl **Crystal system** Cubic. **Habit** Usually as cubes, often with concave faces (hopper crystals): also massive, granular, compact (rock salt). **SG** 2·1–2·2 (2·16 when pure) **Hardness** 2½ **Cleavage** Cubic, perfect. **Fracture** Conchoidal. **Colour and transparency** Colourless or white but also shades of yellow, red and sometimes blue: transparent to translucent. **Streak** White. **Lustre** Vitreous. **Distinguishing features** Ready solubility in water, perfect cubic cleavage, salty taste. Taste, for obvious reasons, is rarely used as a test in mineral identification. It is, however, a useful confirmatory test for halite. **Occurrence** Halite is widely distributed in stratified evaporite deposits formed when enclosed saline waters evaporate, for example around present-day playa lakes. Beds of halite, associated with other water-soluble minerals, such as sylvine, gypsum and anhydrite, occur in sedimentary basins of various ages, having formed by the evaporation of land-locked seas in the geological past. Not infrequently plug-like masses (salt domes) rise from the salt layer and intrude and arch up the overlying sediments, sometimes forming oil traps.

Sylvine KCl **Crystal system** Cubic. **Habit** Crystals usually cubic but often as combination of cube and octahedron; also massive, compact. **SG** 2·0 **Hardness** 2 **Cleavage** Cubic, perfect. **Fracture** Uneven. **Colour and transparency** Colourless or white, sometimes also shades of blue, yellow or red: transparent to translucent. **Streak** White. **Lustre** Vitreous. **Distinguishing features** Sylvine is similar to halite, with which it is commonly associated. It is best distinguished from halite by its bitter taste. **Occurrence** Sylvine, like halite, occurs in bedded evaporite deposits, but is present in lesser amounts because of its greater solubility in water.

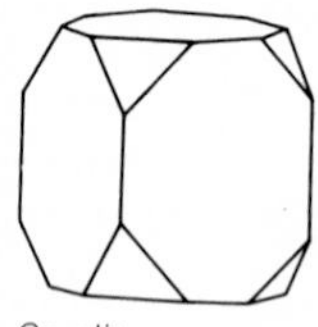
Cryolite

Cryolite Na_3AlF_6 **Crystal system** Monoclinic. **Habit** Crystals rare and pseudo-cubic in appearance, combinations of prisms and pinacoids resembling cube and octahedron; also massive, granular. **Twinning** Common; complex. **SG** 3·0 **Hardness** 2½ **Cleavage** None; basal and prismatic parting. **Fracture** Uneven. **Colour and transparency** Colourless to white, sometimes brownish to reddish: transparent to translucent. **Streak** White. **Lustre** Vitreous to greasy. **Distinguishing features** Pseudo-cubic cleavage, greasy lustre, fuses readily when heated. Cryolite has a low refractive index (about 1·34) so that it becomes almost invisible when its powder is placed in water. **Occurrence** Cryolite is a rare mineral which is used as a flux in the production of aluminium by the electrolytic process. The only notable locality is in Greenland, where it occurs in pegmatites associated with granite, and in company with siderite, quartz, galena, sphalerite, chalcopyrite, fluorite, cassiterite and many other minerals. The name is derived from Greek words meaning 'ice-stone' and refers to its ice-like appearance.

Halite
Halite
Halite
Cryolite
Sylvine
5 cm

Carnallite $KMgCl_3.6H_2O$ **Crystal system** Orthorhombic. **Habit** Crystals rare and pseudo-hexagonal; usually massive, granular. **SG** 1·6 **Hardness** 1–2 **Cleavage** None. **Fracture** Conchoidal. **Colour and transparency** White, sometimes reddish or yellowish: transparent to translucent. **Lustre** Greasy. **Distinguishing features** Lack of cleavage, conchoidal fracture, deliquescence (absorbs moisture and becomes wet), bitter taste, fuses readily when heated. **Occurrence** Carnallite occurs in evaporite deposits together with minerals such as halite and sylvine. Specimens of carnallite should be kept in sealed bottles because of their deliquescent nature. The mineral is named after R von Carnall, a nineteenth-century German mining engineer.

Chlorargyrite (Horn silver, Cerargyrite) $AgCl$ **Crystal system** Cubic. **Habit** Crystals rare but usually cubes; commonly massive, resembling wax or horn. **Twinning** On octahedron. **SG** 5·5–5·6 **Hardness** $1\frac{1}{2}$–$2\frac{1}{2}$ **Cleavage** None. **Fracture** Subconchoidal. **Colour and transparency** Colourless when pure; usually pearl-grey becoming violet-brown on exposure to light: translucent. **Lustre** Resinous to adamantine. **Distinguishing features** Horn-like appearance, can be cut with a knife, melts readily when heated, soluble in ammonium hydroxide. **Occurrence** Chlorargyrite occurs as a secondary mineral in the oxidized zone of silver deposits, together with native silver and cerussite.

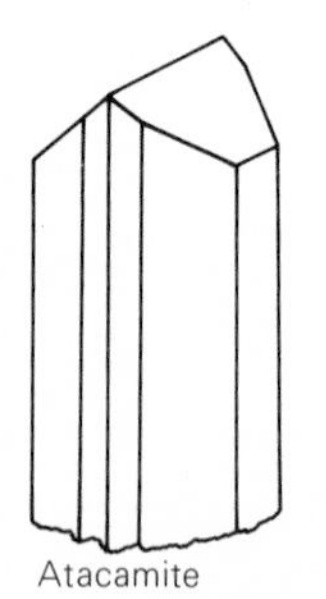

Atacamite

Atacamite $Cu_2Cl(OH)_3$ **Crystal system** Orthorhombic. **Habit** Crystals slender, prismatic and striated; or tabular; also massive, fibrous, granular. **Twinning** Complex. **SG** 3·8 **Hardness** 3–$3\frac{1}{2}$ **Cleavage** Pinacoidal, perfect. **Fracture** Conchoidal. **Colour and transparency** Bright green to dark green: transparent to translucent. **Streak** Apple-green. **Lustre** Adamantine to vitreous. **Distinguishing features** Colour; distinguished from malachite by lack of effervescence in hydrochloric acid, though readily soluble. **Occurrence** Atacamite is always of secondary origin, forming in the oxidized zone of copper deposits, and commonly associated with cuprite and malachite. Named after Atacama, Chile.

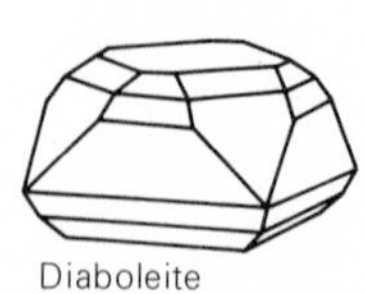

Diaboleite

Diaboleite $Pb_2CuCl_2(OH)_4$ **Crystal system** Tetragonal. **Habit** Crystals tabular; also massive, granular. **SG** 5·5 **Hardness** $2\frac{1}{2}$ **Cleavage** Basal. **Colour and transparency** Bright blue: translucent to nearly opaque. **Distinguishing features** Colour. **Occurrence** Diaboleite is a rare but colourful mineral that occurs in the oxidized parts of some copper-lead ore deposits.

Chlorargyrite
Carnallite
Atacamite
Diaboleite
Atacamite
5 cm

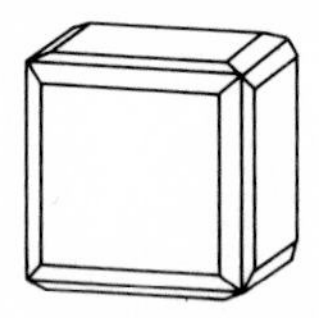
Fluorite: modified cube

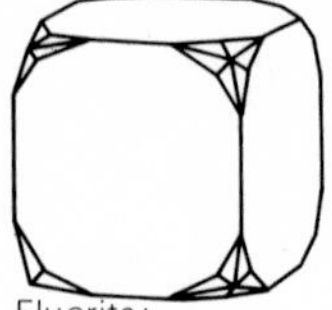
Fluorite: modified cube

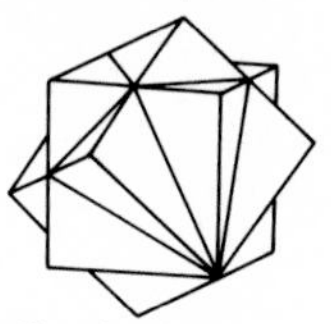
Fluorite: interpenetration twin

Fluorite (Fluorspar) CaF_2 **Crystal system** Cubic. **Habit** Crystals commonly cubic, less frequently octahedral or rhombdodecahedral. Combinations of cube with octahedron or rhombdodecahedron often have cube faces smooth and others dull, or rough, being formed of tiny cube faces in parallel arrangement. **Twinning** Interpenetration twins common. **SG** 3·2 **Hardness** 4 **Cleavage** Octahedral, perfect. **Fracture** Subconchoidal. **Colour and transparency** Colour varies greatly; it is often yellow, green, blue, purple; more rarely colourless, pink, red and black: transparent to translucent. Single crystals may vary in colour and, like some fluorite masses, are often colour banded. **Streak** White. **Lustre** Vitreous. **Distinguishing features** Cubic crystal form, octahedral cleavage, harder than calcite, lack of effervescence with hydrochloric acid. Dissolves in sulphuric acid giving off fumes of hydrogen fluoride which etch glass. Fluorescent. **Occurrence** Fluorite is a widely distributed mineral It occurs in mineral veins, either alone or as a gangue mineral with metallic ores, and in association with quartz, baryte, calcite, celestine, dolomite, galena, cassiterite, sphalerite, topaz and many other minerals. Although fluorite is too soft and too readily cleavable to be used as a precious stone, the colour variation, particularly in the colour-banded variety known as Blue John, has made it prized as a semi-precious ornamental stone from which vases and ornaments have been fashioned since ancient times. Fluorite has given its name to the phenomenon of fluorescence but it shows this effect only weakly. Many other minerals are more spectacular in this respect. Fluorite is worked mainly for use as a flux in the smelting of iron and in the chemical industry; smaller amounts are used as decorative stones and in the manufacture of specialized optical equipment. The name comes from the Latin word *fluere* meaning 'to flow' in reference to its low melting point and use as a flux in the smelting of metals.

Fluorite

5 cm

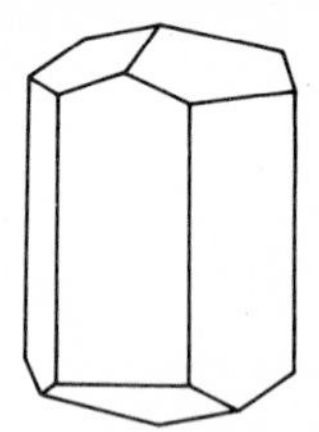
Calcite: prism and flat rhombohedron

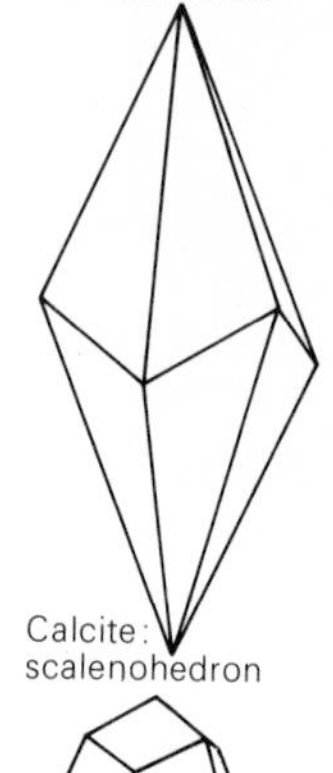
Calcite: scalenohedron

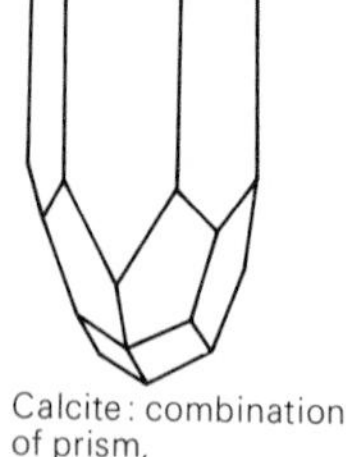
Calcite: combination of prism, scalenohedron and rhombohedron

Calcite $CaCO_3$ **Crystal system** Trigonal. **Habit** Crystals common and very varied in habit, more so than any other mineral. The commonest habits are tabular; prismatic; acute or obtuse rhombohedral; and scalenohedral (dog tooth spar). Calcite also occurs as parallel or fibrous aggregates, or as granular, stalactitic or massive aggregates. **Twinning** Common: there are two main laws, in the first the twin plane is the basal pinacoid, and in the second the twin plane is a rhombohedral face. Lamellar twinning may also be produced by pressure. **SG** 2·7 (when pure) **Hardness** 3 **Cleavage** Rhombohedral, perfect. Although the habit of calcite is so variable, it always cleaves into rhombohedral cleavage fragments. The phenomenon known as double refraction is well shown by calcite, in that if a clear cleavage rhomb of calcite is placed over a dark spot on paper, two images are seen. If the calcite rhomb is rotated, one image remains stationary, and the other rotates with the rhomb. **Fracture** Conchoidal but rarely seen owing to perfection of cleavage. **Colour and transparency** Usually colourless (Iceland spar) or white; also shades of grey, yellow, green, red, purple, blue and even brown or black: transparent to translucent; some deeply coloured forms nearly opaque. **Streak** White. **Lustre** Vitreous, sometimes pearly parallel to cleavage. **Distinguishing features** Perfect rhombohedral cleavage, hardness, dissolves readily with effervescence in cold dilute hydrochloric acid. **Occurrence** Calcite is a common and widely distributed mineral. It is a rock-forming mineral that is a major constituent in calcareous sedimentary rocks (limestone) and metamorphic rocks (marble). It may be precipitated directly from sea water, and it forms the shells of many living organisms which, on death accumulate to form limestone. Metamorphosed limestone, when pure, forms white granular marble; the presence of other minerals results in coloured, figured marble. It occurs also as a primary mineral in carbonatites, which are calcareous igneous rocks that form intrusive plugs. It is of common occurrence in veins, either as the main constituent or as a gangue mineral accompanying metallic ores. Secondary calcite sometimes replaces primary minerals such as pyroxenes or feldspars in igneous rocks. In areas of hot springs it is deposited as travertine or tufa, and stalactites and stalagmites of calcite are common in caves in limestone areas. Calcite, in the form of limestone, is quarried on a large scale for use in the making of cement, as a flux in the smelting of metallic ores, as a fertilizer and as a building stone.

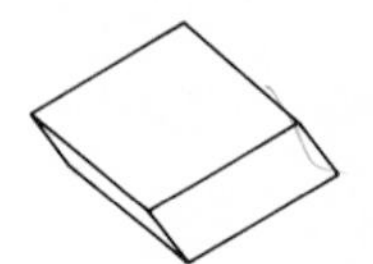
Calcite: rhombohedron

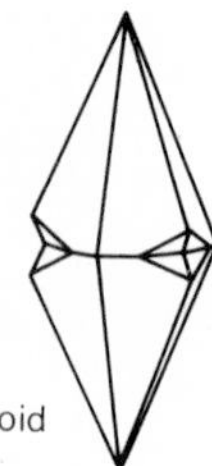
Calcite: scalenohedron twinned on basal pinacoid (right)

Calcite: scalenohedron twinned on rhombohedron

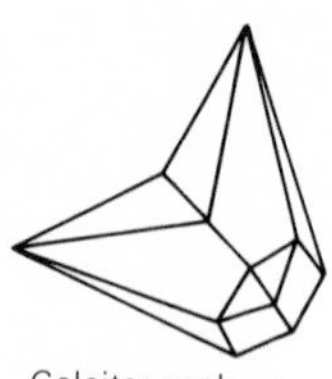
Calcite: scalenohedron twinned on rhombohedron

Calcite

Magnesite $MgCO_3$ **Crystal system** Trigonal. **Habit** Crystals uncommon but rhombohedral or prismatic; usually massive, granular, compact or fibrous. **SG** 3·0–3·2 (increasing with iron content) **Hardness** 3½–4½ **Cleavage** Rhombohedral, perfect. **Fracture** Conchoidal. **Colour and transparency** White or colourless when pure; often in greyish, yellowish or brownish shades when iron is present: transparent to translucent. **Streak** White. **Lustre** Vitreous. **Distinguishing features** Similar to calcite; hardly affected by cold dilute hydrochloric acid but dissolves with effervescence when warmed. **Occurrence** Magnesite is much less common than calcite and does not usually form sedimentary rocks. It occurs as replacement deposits formed by the action of carbonate-bearing waters on rocks containing magnesium minerals, or by the action of magnesium-rich solutions on calcite-bearing rocks. Magnesite occurs also as veins in magnesium-rich metamorphic rocks such as talc schists and serpentinites. Extensive deposits of replacement origin are worked for use in making cement and refractory bricks.

Siderite (Chalybite) $FeCO_3$ **Crystal system** Trigonal. **Habit** Crystals rhombohedral, usually with curved faces which are composite, being an aggregate of small individuals. Also massive, granular, fibrous, compact, botryoidal or earthy. **Twinning** On rhombohedron; often lamellar. **SG** 3·8–4·0 (decreasing with magnesium content) **Hardness** 3½–4½ **Cleavage** Rhombohedral, perfect. **Fracture** Uneven. **Colour and transparency** Grey to grey-brown and yellowish brown: transparent to translucent. **Streak** White. **Lustre** Vitreous. **Distinguishing features** Rhombohedral form and cleavage; its brown colour and higher specific gravity distinguish it from calcite and dolomite. Dissolves slowly in cold dilute hydrochloric acid but dissolves with effervescence when warmed. **Occurrence** Massive siderite is widespread in sedimentary rocks, particularly in clays and shales where it forms clay ironstones which are usually of concretionary origin (see page 204). It also occurs as a gangue mineral in hydrothermal veins accompanying metallic ore minerals such as pyrite, chalcopyrite and galena; and it also forms where limestone has been replaced by the action of iron-bearing solutions.

Rhodochrosite $MnCO_3$ **Crystal system** Trigonal. **Habit** Crystals rare but rhombohedral and with curved faces; usually massive, compact, cleavable or granular. **SG** 3·4–3·7 **Hardness** 3½–4½ **Cleavage** Rhombohedral, perfect. **Fracture** Uneven. **Colour and transparency** Rose-pink, sometimes light grey to brown: translucent. **Streak** White. **Lustre** Vitreous. **Distinguishing features** Colour, rhombohedral cleavage; dissolves with effervescence in hot dilute hydrochloric acid. It is distinguished from rhodonite by its inferior hardness, and often develops a brown or black crust on exposure to air. **Occurrence** Rhodochrosite occurs in hydrothermal mineral veins containing ores of silver, lead and copper. It has been also noted in metamorphic and metasomatic rocks of sedimentary origin, and in sedimentary deposits of manganese oxide, where it is of secondary origin.

5 cm
Magnesite
Siderite
Siderite
Rhodochrosite
Rhodochrosite

Smithsonite (Calamine) $ZnCO_3$ **Crystal system** Trigonal. **Habit** Crystals rare but rhombohedral with rough, curved faces; usually as botryoidal, reniform, stalactitic or encrusting masses; also massive. **SG** 4·3–4·5 **Hardness** 4–4½ **Cleavage** Rhombohedral, perfect. **Fracture** Uneven. **Colour and transparency** Shades of grey, brown or greyish white; but green, brown and yellow varieties also occur: translucent. **Streak** White. **Lustre** Vitreous. **Distinguishing features** Rhombohedral cleavage, high density for a carbonate, soluble in dilute hydrochloric acid with effervescence. **Occurrence** The main occurrence of smithsonite is in the oxidized zone of ore deposits carrying zinc minerals. It is commonly associated with sphalerite, hemimorphite, galena and calcite. It is recorded also in some hydrothermal veins accompanying sphalerite, and as a replacement of limestone. The translucent green variety of smithsonite is used as an ornamental stone. It is named after J L M Smithson (1754–1829), a British mineralogist, and founder of the Smithsonian Institution in Washington.

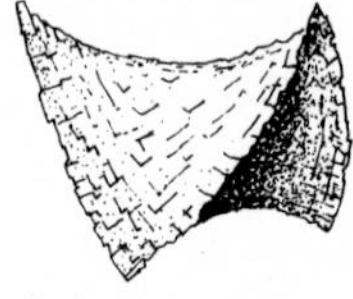

Dolomite: showing curved composite faces

Dolomite $CaMg(CO_3)_2$ **Crystal system** Trigonal. **Habit** Crystals usually rhombohedral with curved composite faces; also occurs in massive, granular aggregates and as a rock-forming mineral in dolomitic limestones. **Twinning** Common. **SG** 2·8–2·9 **Hardness** 3½–4 **Cleavage** Rhombohedral, perfect. **Fracture** Subconchoidal. **Colour and transparency** Usually white, though sometimes colourless; also yellowish to brown, occasionally pink: transparent to translucent. **Streak** White. **Lustre** Vitreous to pearly. **Distinguishing features** Dolomite is similar to calcite but dissolves only slowly in cold dilute hydrochloric acid, but effervesces readily when warmed. **Occurrence** Dolomite occurs widely as a rock-forming mineral. It is usually of secondary occurrence, having formed by the action of magnesium-bearing solutions on limestone. It also occurs as a gangue mineral in hydrothermal veins, particularly those containing galena and sphalerite. Dolomitic limestones are used as building stones, and the mineral is used in the manufacture of refractory bricks for furnace linings. Dolomite is named after D Dolomieu (1750–1801), a French mineralogist.

Ankerite $Ca(Mg,Fe)(CO_3)_2$ **Crystal system** Trigonal. **Habit** Crystals rhombohedral; also massive, granular. **SG** 2·9–3·2 **Hardness** 3½–4 **Cleavage** Rhombohedral. **Colour and transparency** White, yellow, yellowish brown, sometimes grey; becomes dark brown on weathering: translucent. **Streak** White **Lustre** Vitreous. **Distinguishing features** Brown colour. Ferrous iron substitutes for magnesium in dolomite, and there is a series from dolomite to ankerite, with the brown colour usually becoming more pronounced with increasing iron content. **Occurrence** Ankerite occurs in similar ways to dolomite: it is often a gangue mineral accompanying iron ores, and it frequently fills joints in coal seams. Ankerite is named after M J Anker (1772–1843), an Austrian mineralogist.

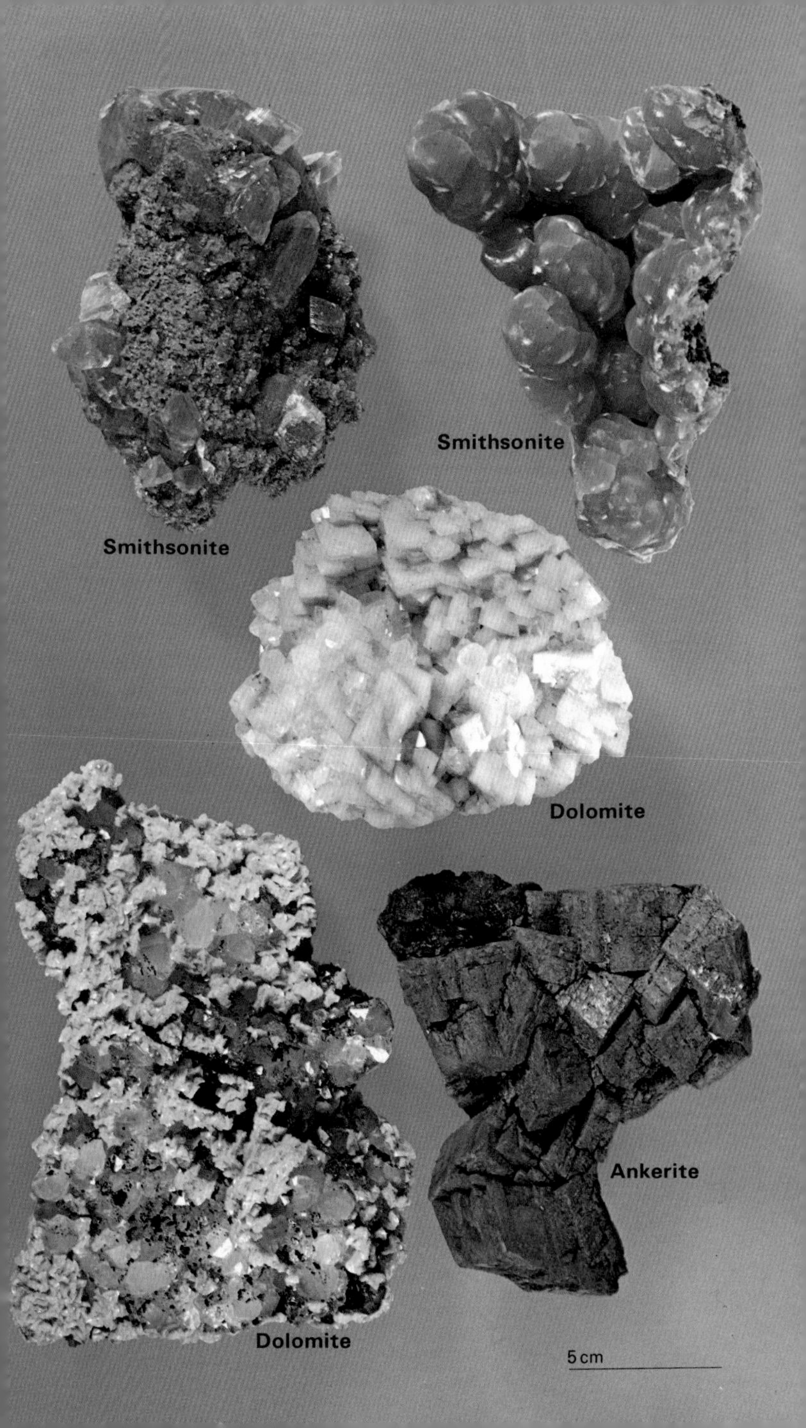
Smithsonite
Smithsonite
Dolomite
Dolomite
Ankerite
5 cm

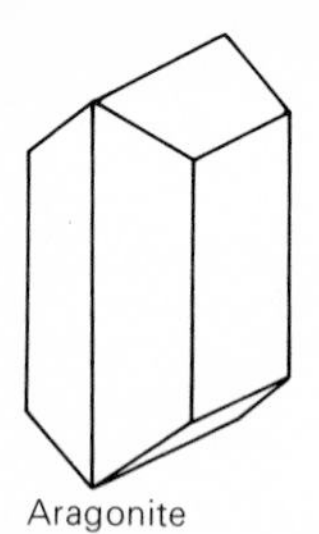

Aragonite

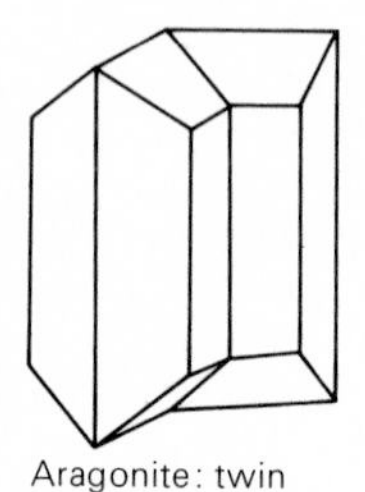

Aragonite: twin

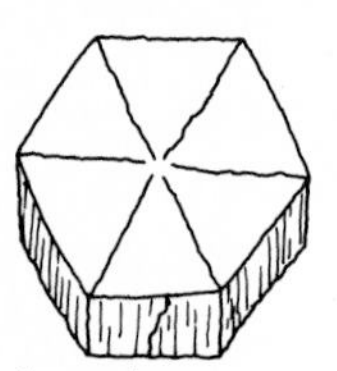

Aragonite:
repeated twin

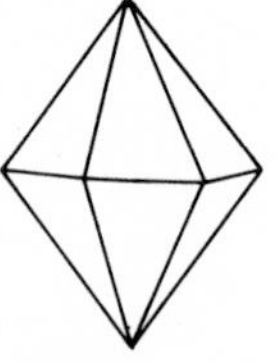

Witherite:
pseudo-hexagonal
twin

Aragonite $CaCO_3$ **Crystal system** Orthorhombic. **Habit** Untwinned crystals, which are rare, are often acicular, though sometimes tabular. Twins stout, prismatic, with marked pseudo-hexagonal symmetry. Also as fibrous, stalactitic and encrusting masses. **Twinning** Very common. Repeated twinning produces pseudo-hexagonal forms. **SG** 2·9 **Hardness** $3\frac{1}{2}$–4 **Cleavage** Pinacoidal, imperfect. **Fracture** Subconchoidal. **Colour and transparency** Colourless, grey, white; also yellowish: transparent to translucent. **Streak** White. **Lustre** Vitreous. **Distinguishing features** Soluble with effervescence in cold dilute hydrochloric acid. Aragonite is a polymorph of $CaCO_3$ and is distinguished from calcite by its form, lack of rhombohedral cleavage and greater specific gravity. **Occurrence** Aragonite is not as widespread as its polymorph, calcite. It occurs as a deposit from hot springs and in association with beds of gypsum. It has been noted in veins and cavities with calcite and dolomite, and in the oxidized zone of ore deposits together with secondary minerals such as malachite and smithsonite. The shells of certain molluscs are made of aragonite and many fossil shells now composed of calcite were probably formed originally of aragonite. It also occurs in some glaucophane schists in association with jadeite and glaucophane. The name comes from the province of Aragon, in Spain, where it was first noted.

Witherite $BaCO_3$ **Crystal system** Orthorhombic. **Habit** Crystals invariably twinned, with pseudo-hexagonal form; also massive, granular, columnar, botryoidal. **Twinning** Ubiquitous. **SG** 4·3 **Hardness** 3–$3\frac{1}{2}$ **Cleavage** Pinacoidal, distinct. **Fracture** Uneven. **Colour and transparency** White; sometimes grey or pale yellow to brown: transparent to translucent. **Streak** White. **Lustre** Vitreous. **Distinguishing features** High specific gravity, soluble in dilute hydrochloric acid with effervescence. Distinguished from strontianite by the flame test; witherite colours the flame green. **Occurrence** Witherite is not of wide occurrence. It sometimes accompanies galena in hydrothermal veins, together with anglesite and baryte. It is named after W Withering (1741–1799), the British mineralogist who first recognized and analyzed the mineral.

Strontianite $SrCO_3$ **Crystal system** Orthorhombic. **Habit** Crystals prismatic or acicular; also massive, fibrous, columnar, granular. **Twinning** Common. **SG** 3·7 **Hardness** $3\frac{1}{2}$–4 **Cleavage** Prismatic, good. **Fracture** Uneven. **Colour and transparency** White, pale green, grey, pale yellow: transparent to translucent. **Streak** White **Lustre** Vitreous. **Distinguishing features** High specific gravity, soluble with effervescence in dilute hydrochloric acid, colours the flame crimson. **Occurrence** Strontianite occurs in low-temperature hydrothermal veins, often in limestone, together with celestine, baryte and calcite. It is a source of strontium and is used in fireworks and red flares. It is named after the locality of Strontian in Argyllshire, Scotland, where it was first found.

Aragonite
Aragonite
Witherite
Witherite
Strontianite
Strontianite
5 cm

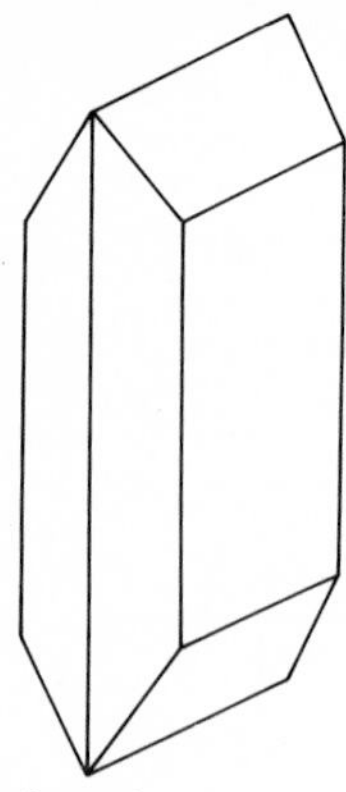

Cerussite

Cerussite $PbCO_3$ **Crystal system** Orthorhombic. **Habit** Crystals often prismatic, or tabular parallel to side pinacoid; sometimes bipyramidal or pseudo-hexagonal star-like twins. Also acicular; or granular, massive, compact. **Twinning** Very common: either contact or penetration twins often of arrow-head shape. **SG** 6·4–6·6 **Hardness** 3–$3\frac{1}{2}$ **Cleavage** Prismatic in two directions, distinct. **Fracture** Conchoidal. **Colour and transparency** Usually white or grey; sometimes darker colours: transparent to translucent. **Streak** White. **Lustre** Adamantine. **Distinguishing features** High specific gravity, adamantine lustre, dissolves with effervescence in warm dilute nitric acid, which distinguishes it from anglesite. **Occurrence** Cerussite is usually of secondary origin, occurring in the oxidized zone of lead veins. It is frequently found in association with anglesite, galena, smithsonite, pyromorphite and sphalerite. Cerussite is a lead ore, and its name comes from the Latin for 'white lead'.

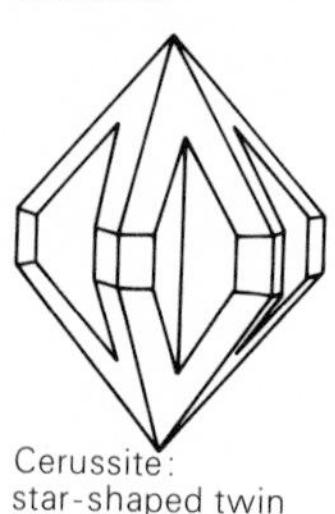

Cerussite:
star-shaped twin

Malachite $Cu_2CO_3(OH)_2$ **Crystal system** Monoclinic. **Habit** Crystals rare. Generally as botryoidal encrusting masses, in bands of varying colour, and frequently of fibrous radiating habit. Also granular or earthy. **Twinning** Common. **SG** 3·9–4·0 (massive varieties as low as 3·5) **Hardness** $3\frac{1}{2}$–4 **Cleavage** Pinacoidal, perfect. **Fracture** Subconchoidal, uneven. **Colour and transparency** Bright green: translucent. **Streak** Pale green. **Lustre** Fibrous varieties silky; rather dull when massive; crystals adamantine. **Distinguishing features** Colour, botryoidal form, soluble with effervescence in dilute hydrochloric acid. **Occurrence** Malachite is a common secondary copper mineral, occurring typically in the oxidized zone of copper deposits. It is frequently associated with azurite, native copper and cuprite, which it sometimes replaces. Other accompanying minerals are calcite, chrysocolla and limonite.

Azurite

Azurite (Chessylite) $Cu_3(CO_3)_2(OH)_2$ **Crystal system** Monoclinic. **Habit** Crystals often either tabular or short prismatic; as radiating aggregates; also massive or earthy. **SG** 3·8–3·9 **Hardness** $3\frac{1}{2}$–4 **Cleavage** Prismatic, perfect; pinacoidal, less so. **Fracture** Conchoidal. **Colour and transparency** Various shades of deep azure-blue: transparent to translucent. **Streak** Light blue. **Lustre** Vitreous. **Distinguishing features** Colour, soluble with effervescence in nitric or hydrochloric acid. **Occurrence** Azurite, like malachite, is a secondary copper mineral and occurs with it in the oxidized zone of copper deposits. Azurite is not as widely distributed as malachite, though it is sometimes interbanded with it when massive. Pseudomorphs of malachite after azurite are common. Azurite often forms good, sharp crystals in contrast to malachite.

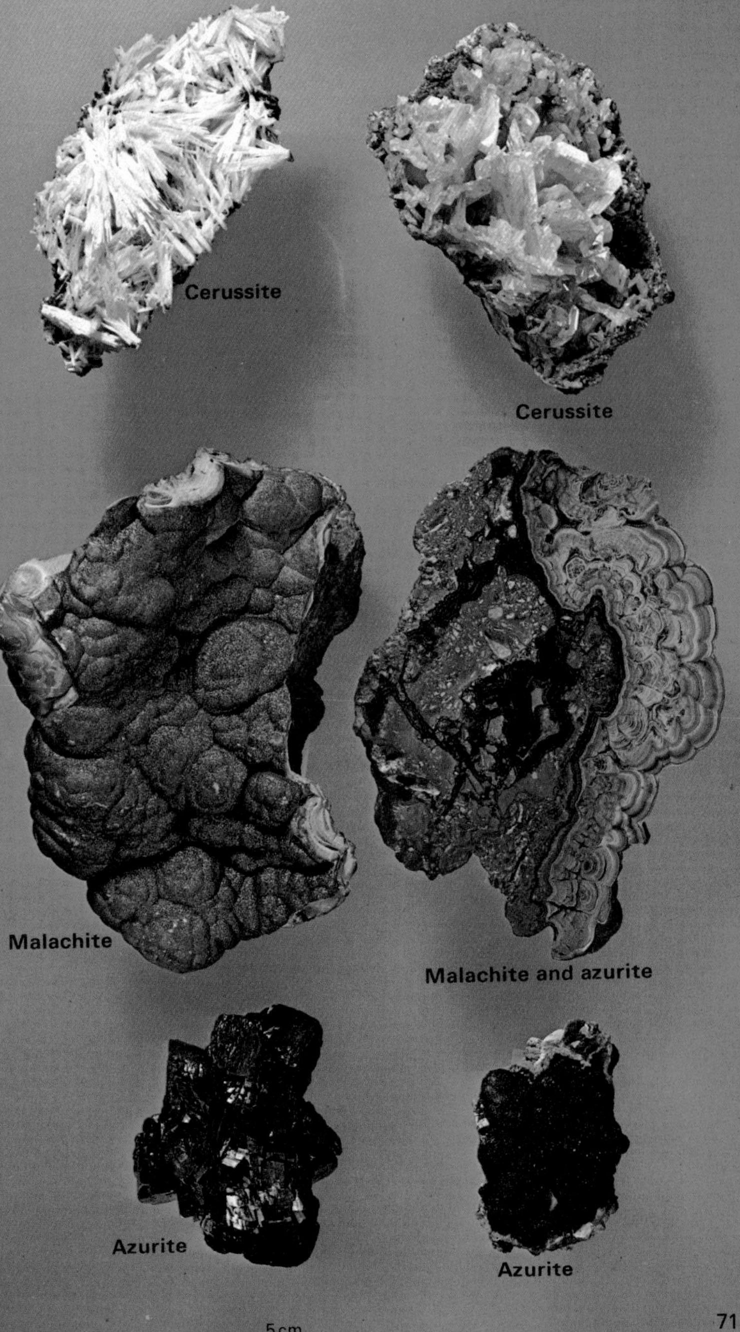

Cerussite
Cerussite
Malachite
Malachite and azurite
Azurite
Azurite
5 cm

Nitratine (Chile saltpetre, Soda nitre) $NaNO_3$ **Crystal system** Trigonal. **Habit** Crystals rare, rhombohedral; usually massive. **Twinning** Common. **SG** 2·2–2·3 **Hardness** 1–2 **Cleavage** Rhombohedral, perfect. **Fracture** Conchoidal; rarely seen owing to perfect cleavage. **Colour and transparency** Colourless or white; may be darker owing to impurities: transparent. **Streak** White. **Lustre** Vitreous. **Distinguishing features** Low specific gravity and hardness, fuses easily, soluble in water, deliquescent. Nitratine resembles calcite but is lighter in weight and less hard. **Occurrence** Because of its ready solubility nitratine occurs in arid regions as surface deposits associated with gypsum, halite and other soluble nitrates and sulphates. Nitre, KNO_3, occurs together with nitratine under similar conditions but is less common. Nitratine is worked as a source of nitrate.

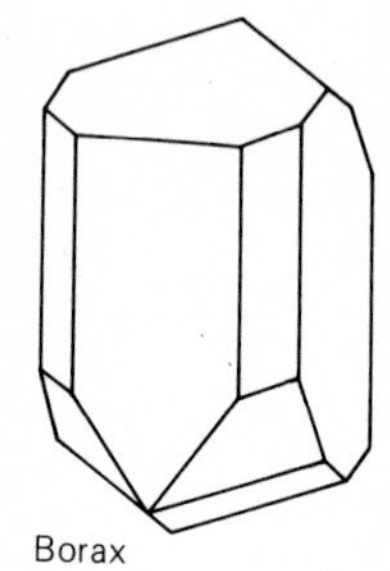
Borax

Borax $Na_2B_4O_7.10H_2O$ **Crystal system** Monoclinic. **Habit** Crystals prismatic; also massive. **SG** 1·7 **Hardness** 2–2½ **Cleavage** Pinacoidal, perfect. **Fracture** Conchoidal. **Colour and transparency** Colourless or white, sometimes greyish or tinged with blue: translucent. **Streak** White. **Lustre** Vitreous to resinous, sometimes dull. **Distinguishing features** Crystal form, low specific gravity, soluble in water, fuses easily. **Occurrence** Borax is an evaporite mineral, precipitated by the evaporation of the water of saline lakes. It occurs in association with other evaporite minerals such as halite, sulphates, carbonates and other borates, in dried lakes in arid regions.

Colemanite $Ca_2B_6O_{11}.5H_2O$ **Crystal system** Monoclinic. **Habit** Crystals variable in habit, but usually short prismatic; also massive, compact, granular. **SG** 2·4 **Hardness** 4–4½ **Cleavage** One perfect cleavage. **Fracture** Uneven. **Colour and transparency** Colourless to white, also yellowish or grey: transparent to translucent. **Lustre** Vitreous. **Distinguishing features** Crystal form, perfect cleavage, fuses easily, relatively hard for a borate. **Occurrence** Colemanite occurs in association with borax, but principally as a lining to cavities in sedimentary rocks where it was probably deposited from waters passing through primary borates. It is named after WT Coleman, a Californian industrialist.

Ulexite $NaCaB_5O_9.8H_2O$ **Crystal system** Triclinic. **Habit** Usually as rounded masses of fine fibrous crystals (cotton balls) and as parallel fibrous aggregates. **SG** 1·9–2·0 **Hardness** 2½ (aggregates have an apparent hardness of 1) **Colour and transparency** White: transparent. **Lustre** Silky. **Distinguishing features** Soft 'cotton ball' habit, low specific gravity, insoluble in cold water, slightly soluble in hot, fuses easily. **Occurrence** Ulexite is an evaporite mineral that sometimes accompanies colemanite in geodes in sedimentary rocks in areas of borax deposits. It occurs also with borax in the surface deposits of arid areas. It is named after GL Ulex, a nineteenth-century German chemist who discovered the mineral.

Ulexite
Borax
Colemanite
Nitratine
5 cm

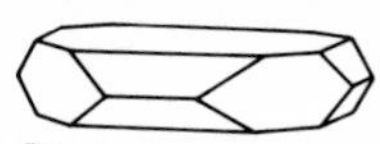

Baryte: tabular habit

Baryte: cockscomb mass

Baryte: 'desert rose'

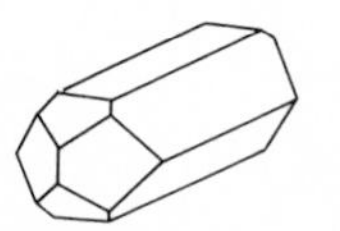

Celestine: prismatic habit

Baryte (Barytes, Barite) $BaSO_4$ **Crystal system** Orthorhombic. **Habit** Crystals commonly tabular, sometimes prismatic giving a diamond-shaped outline; also fibrous or lamellar, and in cockscomb masses; also granular and stalagmitic. **SG** 4·3–4·6 **Hardness** $2\frac{1}{2}$–$3\frac{1}{2}$ **Cleavage** Basal, perfect; prismatic, very good. **Fracture** Uneven. **Colour and transparency** Colourless to white, often tinged with yellow, brown, blue, green or red: transparent to translucent. **Streak** White. **Lustre** Vitreous. **Distinguishing features** High specific gravity, cleavage, crystal form, insoluble in acids, colours the flame green. **Occurrence** Baryte is the most common mineral of barium. It occurs as a vein filling and as a gangue mineral accompanying ores of lead, copper, zinc, silver, iron and nickel, together with calcite, quartz, fluorite, dolomite and siderite. Baryte also occurs as a replacement deposit of limestone, and as the cement in certain sandstones. Baryte concretions in some sandstones have a characteristic rosette-like form and are called 'desert roses'. The name comes from a Greek word meaning 'heavy'.

Celestine (Celestite) $SrSO_4$ **Crystal system** Orthorhombic. **Habit** Crystals tabular or prismatic, resembling baryte; also fibrous or granular. **SG** 3·9–4·0 **Hardness** 3–$3\frac{1}{2}$ **Cleavage** Basal, perfect; prismatic, good. **Fracture** Uneven. **Colour and transparency** Colourless to faint bluish white; sometimes reddish: transparent to translucent. **Streak** White. **Lustre** Vitreous. **Distinguishing features** High specific gravity, cleavage. Distinguished from baryte, though often with difficulty, by lower specific gravity. As barium substitutes for strontium in the structure celestine grades into baryte, but intermediate members are rare. Colours the flame crimson. **Occurrence** Celestine occurs in sedimentary rocks, particularly dolomite, as cavity linings associated with baryte, gypsum, halite, anhydrite, calcite, dolomite and fluorite. It occurs along with anhydrite in evaporite deposits. It is often associated with sulphur, both in the sedimentary environment and also in volcanic areas. It also occurs as a gangue mineral in hydrothermal veins with galena and sphalerite, and it forms concretionary masses in clay and marl. The name is taken from the Latin *celestis*, meaning celestial, and alludes to the pale blue colour of many crystals.

Baryte
Baryte
Baryte
Celestine
Celestine
5 cm

Anglesite $PbSO_4$ **Crystal system** Orthorhombic. **Habit** Crystals sometimes tabular, often prismatic or pyramidal; also massive, compact, granular. **SG** 6·2–6·4 **Hardness** 2½–3 **Cleavage** Basal, good; prismatic, distinct. **Fracture** Conchoidal. **Colour and transparency** Colourless to white, sometimes with a yellow, grey or bluish tinge: transparent to translucent. **Streak** White. **Lustre** Adamantine. **Distinguishing features** High specific gravity (higher than baryte), lustre, association with galena. Distinguished from cerussite by lack of reaction with warm dilute nitric acid. **Occurrence** Anglesite is a secondary lead mineral, most commonly occurring in the oxidized zone of lead deposits; masses of anglesite often surround a core of galena.

Anhydrite $CaSO_4$ **Crystal system** Orthorhombic. **Habit** Crystals rare; usually massive, granular, fibrous. **SG** 2·9–3·0 **Hardness** 3–3½ **Cleavage** Three good cleavages, at right-angles. **Fracture** Uneven. **Colour and transparency** Colourless to white, frequently with a bluish tinge, sometimes grey or reddish: transparent to translucent. **Streak** White. **Lustre** Vitreous to pearly. **Distinguishing features** Three cleavages at right-angles, harder than gypsum, higher specific gravity than calcite. **Occurrence** Anhydrite is an evaporite mineral which occurs with gypsum and halite. It is deposited directly from sea water at temperatures in excess of 42°C, or it may form by the dehydration of gypsum. It occurs also in the 'cap rock' above salt domes, and as a minor gangue mineral in hydrothermal metallic ore veins.

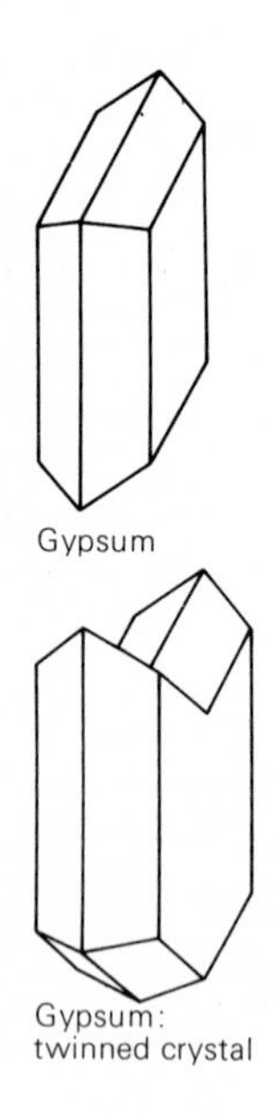
Gypsum

Gypsum: twinned crystal

Gypsum $CaSO_4.2H_2O$ **Crystal system** Monoclinic. **Habit** Crystals tabular, often with curved faces. The colourless, transparent variety is called selenite. Also fibrous (satin spar), massive, granular. The fine-grained granular variety is called alabaster. **Twinning** Very common, giving swallow-tail contact twins. **SG** 2·3 **Hardness** 2 **Cleavage** One perfect; two others, good. **Colour and transparency** Colourless to white but sometimes in shades of yellow, grey, red and brown: transparent to translucent. **Streak** White. **Lustre** Vitreous, pearly parallel to cleavage. **Distinguishing features** Low hardness (can be scratched with the finger nail), cleavage. **Occurrence** Gypsum is an evaporite mineral, and so occurs in bedded deposits together with halite and anhydrite. Having a low solubility, it is the first mineral to be precipitated from evaporating sea water, followed by anhydrite and then halite. It occurs in much smaller quantities in volcanic areas where sulphuric acid fumes have reacted with limestone, and in mineral veins where sulphuric acid produced by the oxidation of pyrite has reacted with calcareous wall rocks. Much gypsum is produced by the secondary hydration of anhydrite.

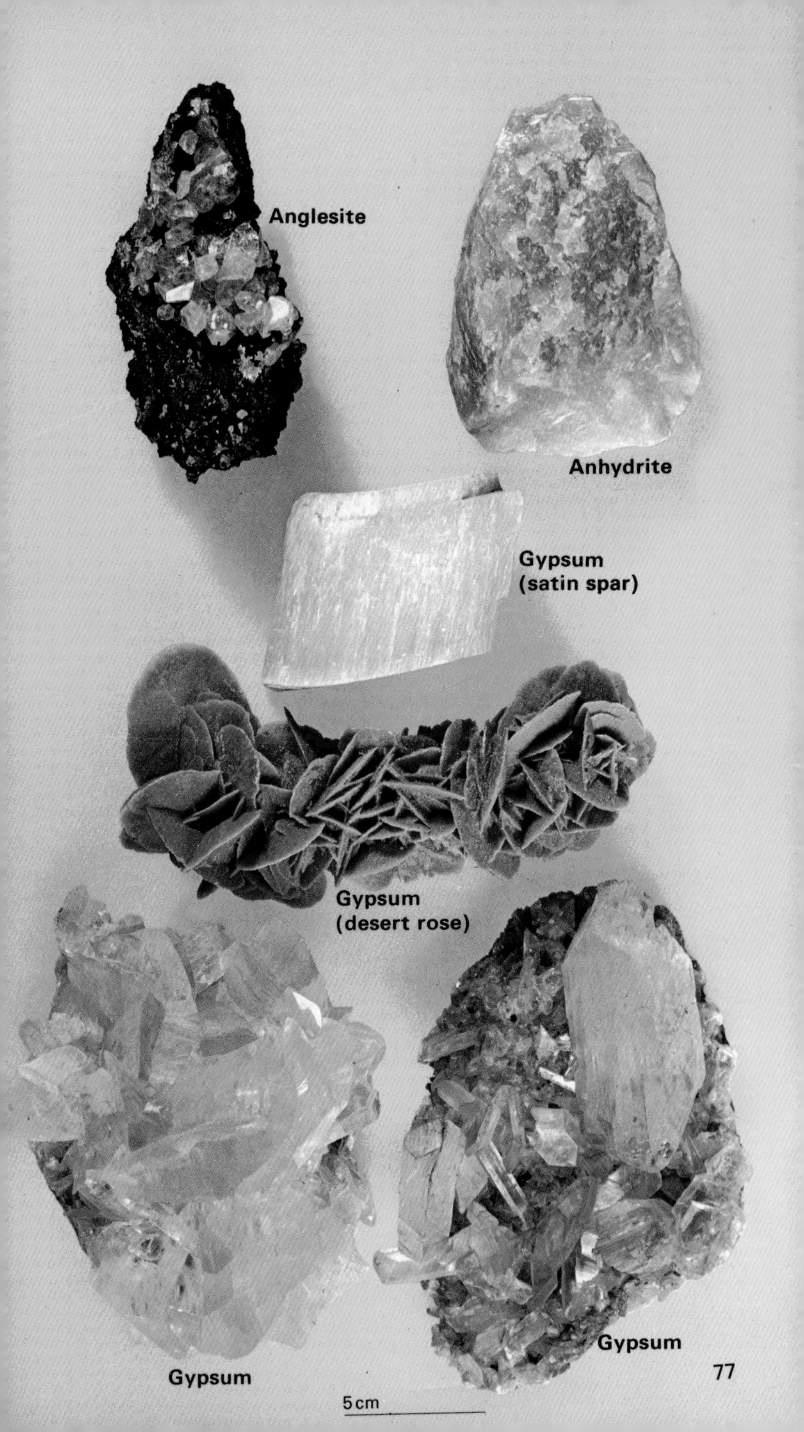
Anglesite
Anhydrite
Gypsum
(satin spar)
Gypsum
(desert rose)
Gypsum
Gypsum
5 cm

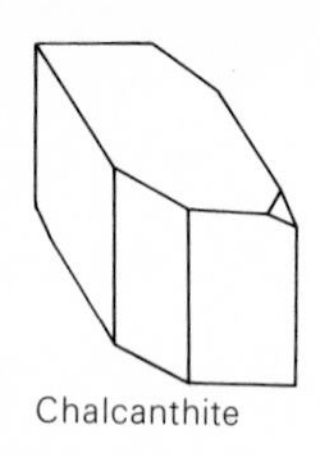
Chalcanthite

Chalcanthite $CuSO_4.5H_2O$ **Crystal system** Triclinic. **Habit** Crystals usually stout, prismatic; also massive, stalactitic, or fibrous. **SG** 2·1–2·3 **Hardness** 2½ **Cleavage** Pinacoidal, imperfect. **Fracture** Conchoidal. **Colour and transparency** Deep sky-blue: transparent to translucent. **Streak** White. **Lustre** Vitreous. **Distinguishing features** Colour, soluble in water. **Occurrence** Chalcanthite is rare, and is a secondary mineral of copper found usually in the oxidized zone of copper sulphide ore deposits. Because of its ready solubility it is most commonly preserved in arid regions. It is also deposited from mine waters having been recorded coating the walls of abandoned mines.

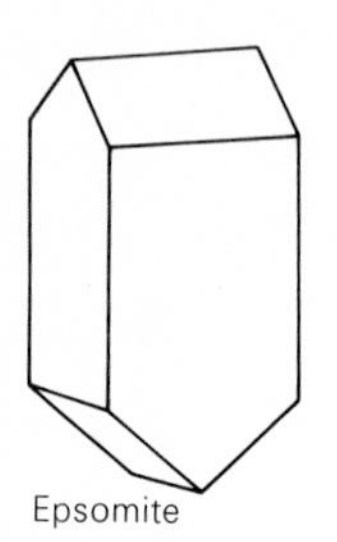
Epsomite

Epsomite (Epsom salt) $MgSO_4.7H_2O$ **Crystal system** Orthorhombic. **Habit** Natural crystals rare, crystals can be grown artificially; usually forms botryoidal encrusting masses with a fibrous structure. **SG** 1·7 **Hardness** 2–2½ **Cleavage** One perfect cleavage. **Fracture** Conchoidal. **Colour and transparency** Colourless to white; transparent to translucent. **Streak** White. **Lustre** Vitreous; fibrous varieties silky to earthy. **Distinguishing features** Fibrous habit, readily soluble in water, bitter taste. **Occurrence** Epsomite usually occurs as encrusting masses on the walls of caves or mine workings where rocks rich in magnesium are exposed. It also occurs in the oxidized zone of pyrite deposits in arid regions.

Alunite (Alumstone) $KAl_3(SO_4)_2(OH)_6$ **Crystal system** Trigonal. **Habit** Crystals rare; rhombohedral and pseudo-cubic; usually massive. **SG** 2·6–2·8 **Hardness** 3½–4 **Cleavage** Basal, distinct. **Fracture** Uneven, conchoidal. **Colour and transparency** White, sometimes grey or reddish: transparent to translucent. **Streak** White. **Lustre** Vitreous; pearly parallel to cleavage. **Distinguishing features** Difficult to distinguish from massive dolomite, anhydrite or magnesite without chemical tests. Somewhat astringent taste. **Occurrence** Alunite is usually found as a secondary mineral in areas where volcanic rocks containing potassic feldspars have been altered by solutions containing sulphuric acid.

Jarosite $KFe_3(SO_4)_2(OH)_6$ **Crystal system** Trigonal. **Habit** Crystals minute pseudo-cubic rhombohedra; also fibrous, massive, granular, encrusting or nodular; often earthy. **SG** 3·2 **Hardness** 2½–3½ **Cleavage** Basal, distinct. **Fracture** Uneven. **Colour** Yellow ochre to dark brown. **Streak** Yellow. **Lustre** Vitreous. **Distinguishing features** Colour. **Occurrence** Jarosite forms under similar conditions to alunite, particularly where rocks contain ferric iron, and commonly in association with decomposing pyrite. It is most commonly found in volcanic areas around volcanic gas vents. It is named after Jaroso, a locality in Spain.

Chalcanthite
Chalcanthite
Jarosite
Epsomite
Alunite
Alunite
5 cm

Thenardite Na_2SO_4 **Crystal system** Orthorhombic. **Habit** Crystals prismatic, tabular or pyramidal. **SG** 2·7 **Hardness** 2–3 **Cleavage** One perfect cleavage. **Colour and transparency** White to brownish white. **Streak** White. **Occurrence** Thenardite is a rare evaporite mineral found with borates in evaporated salt lakes. It is named after LJ Thénard (1777–1857), a French chemist.

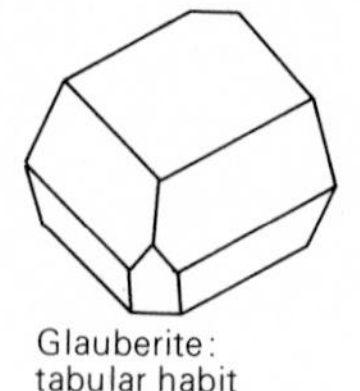
Glauberite: tabular habit

Glauberite $Na_2SO_4.CaSO_4$ **Crystal system** Monoclinic. **Habit** Crystals prismatic or tabular. **SG** 2·7–2·8 **Hardness** 2½–3 **Cleavage** Basal, perfect. **Fracture** Conchoidal. **Colour and transparency** Pale yellow to grey: transparent to translucent. **Streak** White. **Lustre** Vitreous. **Distinguishing features** Thin tabular crystals, soluble in hydrochloric acid, partially soluble in water with loss of transparency. **Occurrence** Glauberite is an evaporite mineral that occurs in bedded salt deposits, in association with halite, thenardite and polyhalite.

Polyhalite $K_2Ca_2Mg(SO_4)_4.2H_2O$ **Crystal system** Triclinic. **Habit** Crystals rare; usually as fibrous or lamellar masses. **Twinning** Common. **SG** 2·8 **Hardness** 2½–3 **Cleavage** Pinacoidal, distinct. **Colour and transparency** Flesh-pink to brick-red: translucent. **Lustre** Silky in fibrous masses, otherwise resinous. **Distinguishing features** Pink colour, bitter taste. **Occurrence** Polyhalite occurs with glauberite in bedded evaporite deposits. It is one of the last minerals to be precipitated from saline waters owing to its high solubility.

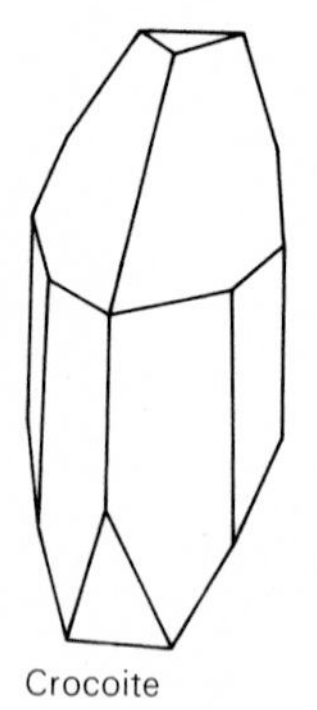
Crocoite

Crocoite $PbCrO_4$ **Crystal system** Monoclinic. **Habit** Crystals usually prismatic or acicular, sometimes short prismatic; also massive, granular. **SG** 5·9–6·1 **Hardness** 2½–3 **Cleavage** Prismatic, distinct. **Fracture** Uneven. **Colour and transparency** Orange-red to various shades of brown: translucent. **Streak** Orange-yellow. **Lustre** Adamantine to vitreous. **Distinguishing features** Orange-red colour, lustre, high specific gravity, fuses fairly easily. **Occurrence** Crocoite is a rare secondary mineral that occurs in the oxidized zone of lead mineral veins together with other secondary lead minerals such as cerussite and pyromorphite. Chromium was first discovered in crocoite, and the name comes from the Greek word for 'saffron'.

Linarite $(Pb,Cu)_2SO_4(OH)_2$ **Crystal system** Monoclinic. **Habit** Crystals prismatic. **SG** 5·3–5·4 **Hardness** 2½–3 **Cleavage** Pinacoidal, perfect; basal, distinct. **Colour and transparency** Deep blue: translucent. **Lustre** Vitreous. **Distinguishing features** Colour, cleavage, association. Distinguished from azurite by lack of effervescence in dilute hydrochloric acid; instead, a white coating is developed. **Occurrence** Linarite is a rare but colourful secondary mineral that occurs in association with some lead-copper ores. The name comes from Linares, a locality in Spain.

Thenardite
Glauberite
Polyhalite
Crocoite
Crocoite
Linarite
5 cm

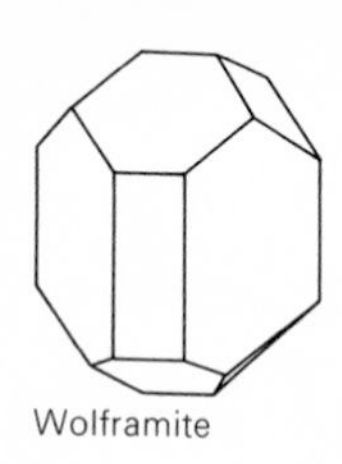

Wolframite

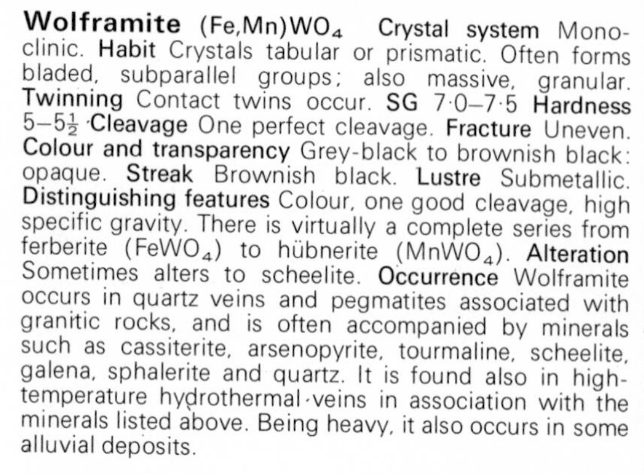

Wolframite (Fe,Mn)WO_4 **Crystal system** Monoclinic. **Habit** Crystals tabular or prismatic. Often forms bladed, subparallel groups; also massive, granular. **Twinning** Contact twins occur. **SG** 7·0–7·5 **Hardness** 5–5½ **Cleavage** One perfect cleavage. **Fracture** Uneven. **Colour and transparency** Grey-black to brownish black: opaque. **Streak** Brownish black. **Lustre** Submetallic. **Distinguishing features** Colour, one good cleavage, high specific gravity. There is virtually a complete series from ferberite ($FeWO_4$) to hübnerite ($MnWO_4$). **Alteration** Sometimes alters to scheelite. **Occurrence** Wolframite occurs in quartz veins and pegmatites associated with granitic rocks, and is often accompanied by minerals such as cassiterite, arsenopyrite, tourmaline, scheelite, galena, sphalerite and quartz. It is found also in high-temperature hydrothermal veins in association with the minerals listed above. Being heavy, it also occurs in some alluvial deposits.

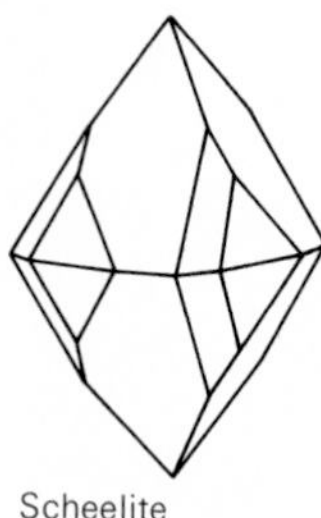

Scheelite

Scheelite $CaWO_4$ **Crystal system** Tetragonal. **Habit** Crystals usually bipyramidal; also massive, granular. **Twinning** Penetration twins are common. **SG** 5·9–6·1 **Hardness** 4½–5 **Cleavage** Pyramidal, distinct. **Colour and transparency** White, sometimes in shades of yellow, green, brown, red: transparent to translucent. **Streak** White. **Lustre** Vitreous. **Distinguishing features** Pyramidal habit, white colour together with high specific gravity. Scheelite is commonly fluorescent. **Occurrence** Scheelite often accompanies wolframite in pegmatites and high-temperature hydrothermal veins. Associated minerals are cassiterite, molybdenite, fluorite and topaz. It also occurs in contact metamorphic deposits together with idocrase, axinite, garnet and wollastonite. Scheelite, with wolframite, is an ore of tungsten. It is named after KW Scheele, the eighteenth-century Swedish chemist who discovered tungsten.

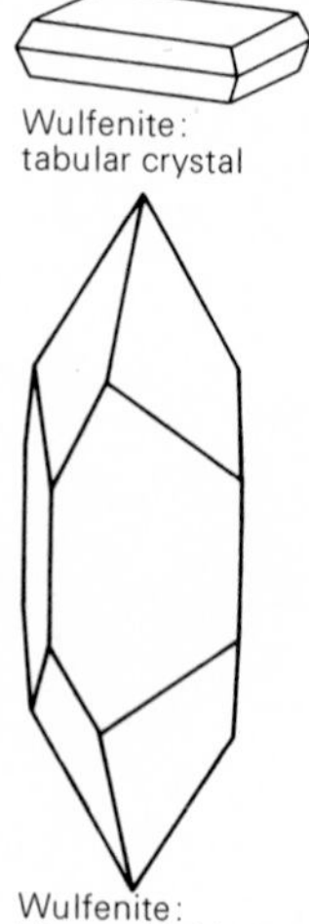

Wulfenite: tabular crystal

Wulfenite: bipyramidal habit

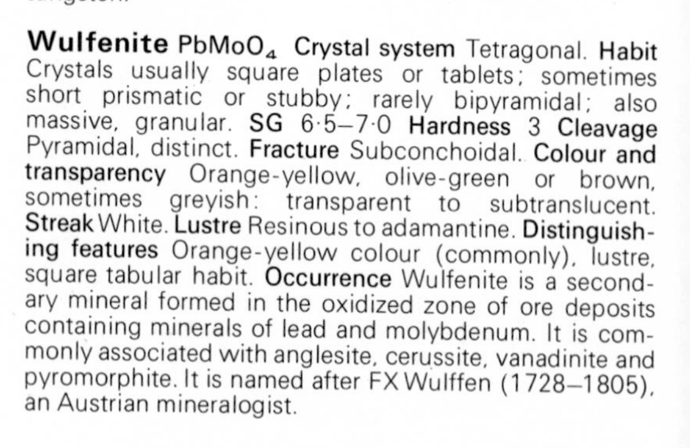

Wulfenite $PbMoO_4$ **Crystal system** Tetragonal. **Habit** Crystals usually square plates or tablets; sometimes short prismatic or stubby; rarely bipyramidal; also massive, granular. **SG** 6·5–7·0 **Hardness** 3 **Cleavage** Pyramidal, distinct. **Fracture** Subconchoidal. **Colour and transparency** Orange-yellow, olive-green or brown, sometimes greyish: transparent to subtranslucent. **Streak** White. **Lustre** Resinous to adamantine. **Distinguishing features** Orange-yellow colour (commonly), lustre, square tabular habit. **Occurrence** Wulfenite is a secondary mineral formed in the oxidized zone of ore deposits containing minerals of lead and molybdenum. It is commonly associated with anglesite, cerussite, vanadinite and pyromorphite. It is named after FX Wulffen (1728–1805), an Austrian mineralogist.

Wolframite
Wolframite
5 cm
Scheelite
Wulfenite
Wulfenite

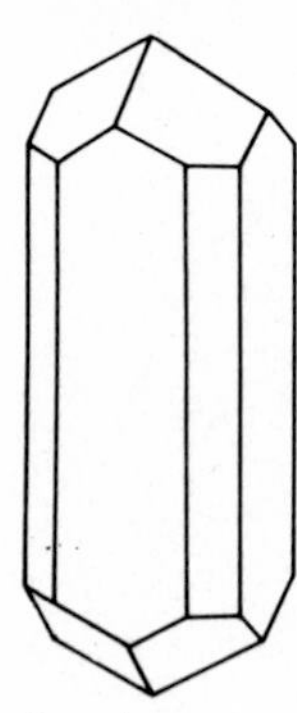
Xenotime: prismatic habit

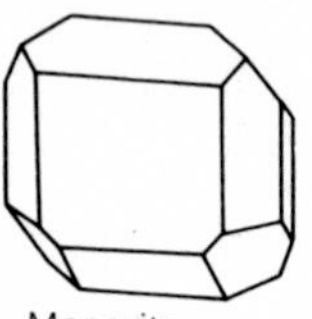
Monazite

Xenotime YPO_4 **Crystal system** Tetragonal. **Habit** Crystals prismatic, resembling zircon with which it is sometimes associated in parallel growth. **SG** 4·4–5·1 **Hardness** 4–5 **Cleavage** Prismatic, perfect. **Fracture** Uneven. **Colour and transparency** Yellowish brown, also greyish white, pale yellow: translucent to opaque. **Streak** Pale brown. **Lustre** Resinous to vitreous. **Distinguishing features** Very similar to zircon but less hard and has a good prismatic cleavage. **Occurrence** Xenotime is an accessory mineral in granitic and alkaline igneous rocks. It also occurs in some pegmatites and gneisses.

Monazite $(Ce,La,Th)PO_4$ **Crystal system** Monoclinic. **Habit** Crystals small, short prismatic or tabular. Large crystals often have striated faces. **Twinning** Common. **SG** 4·9–5·4 **Hardness** 5–5½ **Cleavage** Pinacoidal, distinct. **Fracture** Uneven. **Colour and transparency** Clove-brown to reddish brown, sometimes green: translucent. **Streak** Off-white. **Lustre** Resinous to waxy. **Distinguishing features** Similar to zircon but softer. **Occurrence** Monazite is an accessory mineral of granitic rocks and associated pegmatites, and it also occurs in gneisses and carbonatites. It is concentrated in some detrital sands in sufficient quantity to merit commercial exploitation for cerium and thorium.

Vivianite $Fe_3(PO_4)_2.8H_2O$ **Crystal system** Monoclinic. **Habit** Crystals prismatic; also as reniform or encrusting masses, often with a fibrous structure; sometimes blue, earthy. **SG** 2·6–2·7 **Hardness** 1½–2 **Cleavage** One perfect cleavage. **Colour and transparency** Colourless when fresh and unaltered, becoming blue or green on oxidation, the colour deepening with exposure. **Streak** White, becoming dark blue or brown. **Lustre** Vitreous, pearly parallel to cleavage. **Distinguishing features** Blue colour. **Occurrence** Vivianite is a secondary phosphate which occurs in the oxidized zone of metallic ore deposits containing pyrrhotine and pyrite, in the weathered zone of certain phosphate-bearing pegmatites, and in sedimentary rocks, particularly those containing bone or other organic fragments.

Amblygonite $(Li,Na)Al(PO_4)(F,OH)$ **Crystal system** Triclinic. **Habit** Crystals usually rough and ill-formed, sometimes large; also massive, compact, or as cleavable masses. **Twinning** Lamellar twinning common. **SG** 3·0–3·1 **Hardness** 5½–6 **Cleavage** Two good cleavages. **Fracture** Uneven. **Colour and transparency** White to pale green or bluish white; sometimes pinkish or yellowish: subtransparent to translucent. **Streak** White. **Lustre** Vitreous to greasy; pearly parallel to best cleavage. **Distinguishing features** Two cleavages, specific gravity. The name amblygonite is used for the fluorine-rich end member of the series, and montebrasite for the more common hydroxyl-rich member. **Occurrence** Amblygonite is a rare mineral which occurs in granite pegmatites together with other lithium minerals such as spodumene, tourmaline and lepidolite, and with albite, for which it may be mistaken.

Xenotime
Monazite
Vivianite
Vivianite
Amblygonite
5 cm

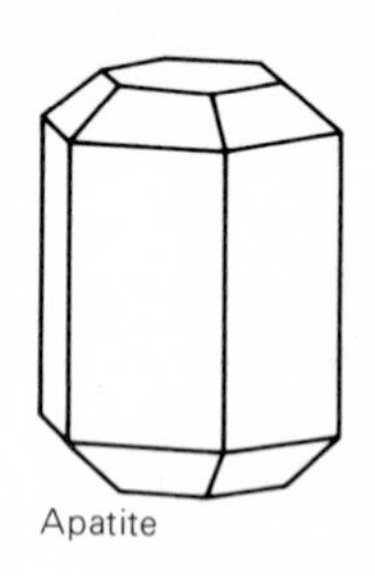

Apatite

Apatite $Ca_5(PO_4)_3(F,Cl,OH)$ **Crystal system** Hexagonal. **Habit** Crystals common, and are usually prismatic or tabular; also massive, granular. **SG** 3·1–3·3 **Hardness** 5 **Cleavage** Basal, imperfect. **Fracture** Conchoidal, uneven. **Colour and transparency** Usually in shades of green to grey-green; also white, brown, yellow, bluish or reddish: transparent to translucent. **Streak** White. **Lustre** Vitreous to subresinous. **Distinguishing features** Hexagonal crystal form, hardness. Distinguished from beryl, for which it may be mistaken, by inferior hardness; apatite can be scratched with a steel knife blade. **Occurrence** Apatite is a widely distributed phosphate mineral. It occurs as small crystals as an accessory mineral in a wide range of igneous rocks. Large crystals occur in pegmatites and in some high-temperature hydrothermal veins. It occurs also in both regional and contact metamorphic rocks, especially in metamorphosed limestones and skarns. In sedimentary rocks, apatite is a principal constituent of fossil bones and other organic matter. The name collophane is sometimes used for such phosphatic material. The name comes from a Greek word meaning 'to deceive' because apatite, particularly the gem variety, is readily mistaken for other minerals.

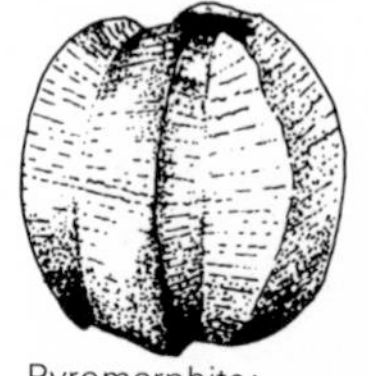

Pyromorphite: mimetite, variety campylite, showing barrel-shaped form

Pyromorphite $Pb_5(PO_4)_3Cl$ **Crystal system** Hexagonal. **Habit** Crystals are usually of simple prismatic form, often barrel-shaped (campylite), or as hollow prismatic forms. Also fibrous, granular, globular. **SG** 6·5–7·1 **Hardness** 3½–4 **Fracture** Subconchoidal. **Colour and transparency** Shades of green, yellow and brown: subtransparent to translucent. **Streak** White. **Lustre** Resinous. **Distinguishing features** Colour, hexagonal form, high specific gravity, resinous lustre. **Occurrence** Pyromorphite is a secondary lead phosphate that occurs, often with mimetite, in the oxidized zone of mineral veins containing lead minerals, such as galena and anglesite.

Mimetite $Pb_5(AsO_4)_3Cl$ **Crystal system.** Hexagonal. **Habit** Crystals, like those of pyromorphite, are commonly simple hexagonal forms of prismatic habit; also as rounded, globular forms (campylite). **SG** 7·0–7·2 **Hardness** 3½–4 **Fracture** Subconchoidal. **Colour and transparency** Pale yellow to yellow-brown: subtransparent to translucent. **Streak** White. **Lustre** Resinous. **Distinguishing features** Colour, hexagonal form, high specific gravity, resinous lustre. Mimetite and pyromorphite are often difficult to distinguish without chemical tests. **Occurrence** Mimetite is a rather rare secondary mineral that occurs in the oxidized parts of lead ores, especially those containing arsenic. Like pyromorphite, it occurs in association with galena, anglesite and hemimorphite. The name comes from a Greek word meaning 'imitator', because of the close resemblance between pyromorphite and mimetite.

Pyromorphite
Apatite
Pyromorphite
Pyromorphite
Mimetite
Mimetite
5 cm

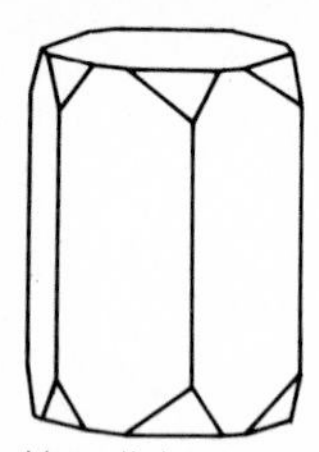
Vanadinite

Vanadinite $Pb_5(VO_4)_3Cl$ **Crystal system** Hexagonal. **Habit** Crystals frequently sharp and prismatic; sometimes as hollow prisms; also as rounded forms similar to pyromorphite. **SG** 6·7–7·1 **Hardness** 3 **Fracture** Subconchoidal. **Colour and transparency** Orange-red, brownish red to yellow: transparent to subtranslucent. **Streak** White to yellowish. **Lustre** Resinous. **Distinguishing features** Like pyromorphite and mimetite it has hexagonal form, resinous lustre and high specific gravity, but it is distinguished from them by its orange-red colour. **Occurrence** Vanadinite is a rare mineral and, like pyromorphite, it occurs in the oxidized zone of sulphide ore deposits carrying galena and other lead minerals.

Vanadinite: hollow prismatic crystals

Erythrite (Cobalt bloom) $Co_3(AsO_4)_2.8H_2O$: **Annabergite** (Nickel bloom) $Ni_3(AsO_4)_2.8H_2O$ **Crystal system** Monoclinic. **Habit** Crystals usually prismatic, often acicular; also as radiating groups and reniform masses with a columnar structure; also as powdery coatings. **SG** 3·0–3·1 **Hardness** $1\frac{1}{2}$–$2\frac{1}{2}$ **Cleavage** One perfect cleavage. **Colour and transparency** Erythrite, crimson-red to pink, becoming paler with increasing nickel content; annabergite, apple-green: transparent to subtranslucent. **Streak** Erythrite, red, but paler than colour. **Lustre** Adamantine to vitreous; pearly parallel to cleavage. **Distinguishing features:** Pink colour (erythrite); green colour (annabergite), association with cobalt and nickel minerals. **Occurrence** Erythrite and annabergite are secondary minerals produced by the surface oxidation of primary cobalt and nickel minerals. Their occurrence as pink or green powdery coatings give rise to the names 'cobalt bloom' and 'nickel bloom'.

Turquoise $CuAl_6(PO_4)_4(OH)_8.5H_2O$ **Crystal system** Triclinic. **Habit** Crystals rare and minute; usually massive, granular to cryptocrystalline as reniform or encrusting masses, or in veins. **SG** 2·6–2·8 **Hardness** 5–6 **Fracture** Conchoidal. **Colour and transparency** Sky-blue, blue-green to greenish grey: nearly opaque. **Streak** White or greenish. **Lustre** Waxy when massive; crystals vitreous. **Distinguishing features** Blue colour, distinguished from chrysocolla by its greater hardness. **Occurrence** Turquoise is a secondary mineral occurring in veins in association with aluminous igneous or sedimentary rocks that have undergone considerable alteration, usually in arid regions. Prized as a semi-precious stone.

Scorodite

Scorodite $FeAsO_4.2H_2O$ **Crystal system** Orthorhombic. **Habit** Crystals pyramidal and pseudo-octahedral or prismatic; also nodular or earthy. **SG** 3·1–3·3 **Hardness** $3\frac{1}{2}$–4 **Cleavage** Prismatic, imperfect. **Fracture** Uneven. **Colour and transparency** Pale green, blue-green to blue, brown: subtransparent to translucent. **Streak** White. **Lustre** Vitreous to adamantine. **Distinguishing features** Crystal habit and association with arsenic minerals. **Occurrence** Scorodite is an alteration product of arsenic minerals, and of arsenopyrite in particular. It is also deposited from the waters of some hot springs. The name comes from the Greek word for 'garlic' and refers to the odour it emits when heated.

5 cm
Vanadinite
Vanadinite
Annabergite
Erythrite
Turquoise
Turquoise
Scorodite

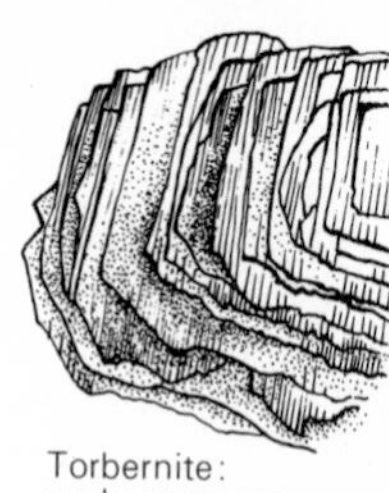
Torbernite: scaly aggregate

Torbernite (and Metatorbernite) $Cu(UO_2)_2(PO_4)_2.8\text{–}12H_2O$ **Crystal system** Tetragonal. **Habit** Crystals tabular, often with square outline; also as foliated or scaly aggregates. **SG** 3·2 (increasing to 3·7 with alteration to metatorbernite) **Hardness** 2–2½ **Cleavage** Basal, perfect. **Colour and transparency** Bright emerald-green, sometimes dark green: transparent to translucent. **Streak** Pale green. **Lustre** Vitreous; pearly parallel to cleavage. **Distinguishing features** Colour, cleavage. **Occurrence** Torbernite and autunite occur together as secondary minerals in the oxidized parts of veins containing uraninite and copper minerals. At atmospheric temperatures torbernite loses some of its water and forms metatorbernite $Cu(UO_2)_2(PO_4)_2.8H_2O$.

Autunite (and Meta-autunite) $Ca(UO_2)_2(PO_4)_2.10\text{–}12H_2O$ **Crystal system** Tetragonal. **Habit** Tabular crystals with square outline, very similar in shape to torbernite; also as foliated and scaly masses. **SG** 3·1–3·2 **Hardness** 2–2½ **Cleavage** Basal, perfect. **Colour and transparency** Lemon-yellow to greenish yellow. **Streak** Yellow. **Lustre** Vitreous; pearly parallel to cleavage. **Distinguishing features** Yellow-green colour, cleavage, crystal form. Easily distinguished from torbernite by colour, but autunite can be distinguished from other secondary uranium minerals only by chemical or X-ray methods. Fluorescent. **Occurrence** Autunite, like torbernite, is a secondary mineral that occurs in the oxidized parts of veins and pegmatites carrying uranium minerals. Autunite may lose some of its water forming meta-autunite.

Carnotite $K_2(UO_2)_2(VO_4)_2.3H_2O$ **Crystal system** Monoclinic. **Habit** Usually powdery; rarely as minute, thin tabular crystals. **SG** 4–5 **Hardness** about 2 **Cleavage** Basal, perfect. **Colour and transparency** Bright yellow to greenish yellow. **Lustre** Dull, earthy. **Distinguishing features** Yellow colour, though it is difficult to distinguish from tyuyamunite other than by chemical or X-ray means. Its powdery habit and lack of fluorescence distinguish it from autunite. **Occurrence** Carnotite and tyuyamunite are secondary minerals frequently found together in sedimentary rocks, having been deposited from waters which have been in contact with primary uranium and vanadium minerals.

Tyuyamunite (and Metatyuyamunite) $Ca(UO_2)_2(VO_4)_2.5\text{–}10H_2O$ **Crystal system** Orthorhombic. **Habit** Scales, laths or radial aggregates; also massive or powdery. **SG** 3·6 (increasing to 4·4 with alteration to metatyuyamunite) **Hardness** 2–2½ **Cleavage** Basal, perfect. **Colour** Greenish yellow. **Lustre** Earthy; massive material waxy. **Distinguishing features** Closely resembles carnotite but has more greenish colour and is fluorescent. **Occurrence** Tyuyamunite, like carnotite, with which it is commonly associated, is a secondary mineral that occurs in sedimentary rocks, notably certain sandstones. Together with carnotite, it is an ore of uranium. The strange-sounding name is taken from Tyuya Muyun, a locality in Turkestan, USSR.

Torbernite
Torbernite
Autunite
Autunite
Carnotite
Tyuyamunite
5 cm

Descloizite $Pb(Zn,Cu)VO_4(OH)$ **Crystal system** Orthorhombic. **Habit** Crystals platy, prismatic or wedge-shaped; also mamillated with fibrous radiating structure. **SG** 5·9–6·2 **Hardness** $3\frac{1}{2}$ **Colour and transparency** Usually clove-brown but varies from cherry-red to black: translucent. **Streak** Orange to brownish red. **Distinguishing features** Colour, orange streak, crystal form. **Occurrence** Descloizite is a secondary mineral found occasionally in lead-zinc deposits.

Olivenite $Cu_2AsO_4(OH)$ **Crystal system** Orthorhombic. **Habit** Crystals prismatic or acicular; also reniform, fibrous, radiating or granular. **SG** 4·1–4·4 **Hardness** 3 **Cleavage** Poor. **Fracture** Conchoidal to uneven. **Colour and transparency** Olive-green in various shades (hence the name), but can range from white to nearly black: subtransparent to opaque. **Streak** Olive-green to brown. **Lustre** Vitreous; some fibrous varieties, pearly. **Distinguishing features** Olive-green colour. **Occurrence** Olivenite is a rare secondary mineral which occurs in the oxidized parts of copper sulphide deposits, sometimes in association with adamite.

Adamite $Zn_2AsO_4(OH)$ **Crystal system** Orthorhombic. **Habit** Crystals usually small; more often as radiating and encrusting aggregates. **SG** 4·3–4·4 **Hardness** $3\frac{1}{2}$ **Colour and transparency** Yellowish green to green, sometimes reddish brown: translucent. **Distinguishing features** Yellowish green colour. **Occurrence** Adamite is a rare secondary zinc mineral found as a weathering product in the oxidized zone of zinc deposits.

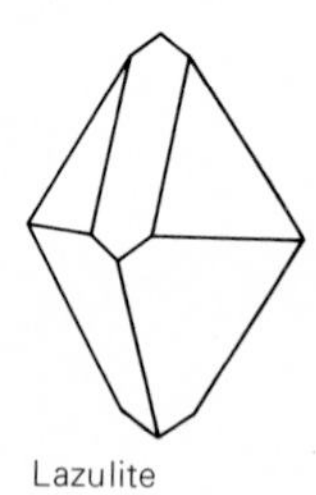

Lazulite

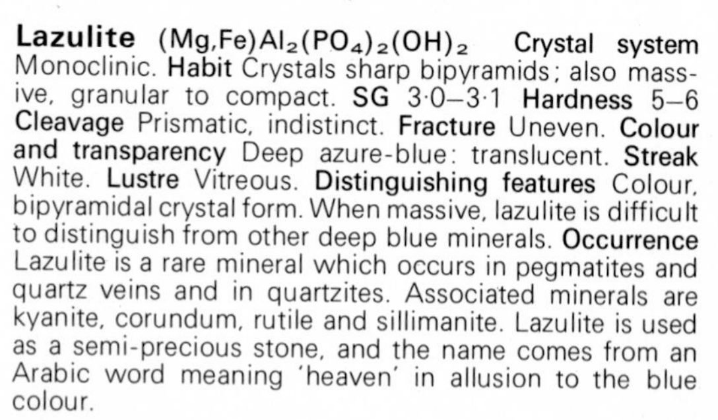

Lazulite $(Mg,Fe)Al_2(PO_4)_2(OH)_2$ **Crystal system** Monoclinic. **Habit** Crystals sharp bipyramids; also massive, granular to compact. **SG** 3·0–3·1 **Hardness** 5–6 **Cleavage** Prismatic, indistinct. **Fracture** Uneven. **Colour and transparency** Deep azure-blue: translucent. **Streak** White. **Lustre** Vitreous. **Distinguishing features** Colour, bipyramidal crystal form. When massive, lazulite is difficult to distinguish from other deep blue minerals. **Occurrence** Lazulite is a rare mineral which occurs in pegmatites and quartz veins and in quartzites. Associated minerals are kyanite, corundum, rutile and sillimanite. Lazulite is used as a semi-precious stone, and the name comes from an Arabic word meaning 'heaven' in allusion to the blue colour.

Wavellite $Al_3(PO_4)_2(OH)_3.5H_2O$ **Crystal system** Orthorhombic. **Habit** Crystals rare; characteristically as hemispherical or globular aggregates with a fibrous, radiating structure. **SG** 2·3–2·4 **Hardness** $3\frac{1}{2}$–4 **Cleavage** Prismatic, good. **Fracture** Uneven. **Colour and transparency** White; often greenish, yellow, grey, brown: translucent. **Streak** White. **Lustre** Vitreous. **Distinguishing features** The radiating habit is the characteristic feature. **Occurrence** Wavellite is a secondary mineral found on joint surfaces and in cavities in rocks, particularly slates. It occurs in limonitic ore bodies and in association with phosphorite deposits. Wavellite is named after W Wavell, the discoverer of the mineral.

5cm
Descloizite
Descloizite
Olivenite
Adamite
Wavellite
Lazulite

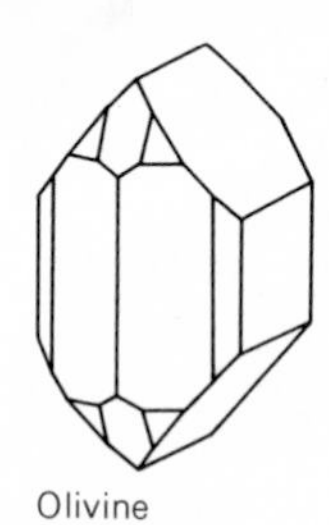

Olivine

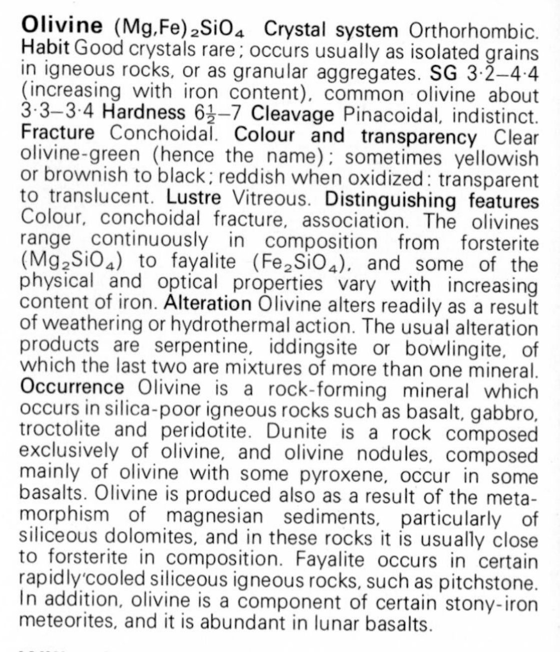

Olivine $(Mg,Fe)_2SiO_4$ **Crystal system** Orthorhombic. **Habit** Good crystals rare; occurs usually as isolated grains in igneous rocks, or as granular aggregates. **SG** 3·2–4·4 (increasing with iron content), common olivine about 3·3–3·4 **Hardness** $6\frac{1}{2}$–7 **Cleavage** Pinacoidal, indistinct. **Fracture** Conchoidal. **Colour and transparency** Clear olivine-green (hence the name); sometimes yellowish or brownish to black; reddish when oxidized: transparent to translucent. **Lustre** Vitreous. **Distinguishing features** Colour, conchoidal fracture, association. The olivines range continuously in composition from forsterite (Mg_2SiO_4) to fayalite (Fe_2SiO_4), and some of the physical and optical properties vary with increasing content of iron. **Alteration** Olivine alters readily as a result of weathering or hydrothermal action. The usual alteration products are serpentine, iddingsite or bowlingite, of which the last two are mixtures of more than one mineral. **Occurrence** Olivine is a rock-forming mineral which occurs in silica-poor igneous rocks such as basalt, gabbro, troctolite and peridotite. Dunite is a rock composed exclusively of olivine, and olivine nodules, composed mainly of olivine with some pyroxene, occur in some basalts. Olivine is produced also as a result of the metamorphism of magnesian sediments, particularly of siliceous dolomites, and in these rocks it is usually close to forsterite in composition. Fayalite occurs in certain rapidly cooled siliceous igneous rocks, such as pitchstone. In addition, olivine is a component of certain stony-iron meteorites, and it is abundant in lunar basalts.

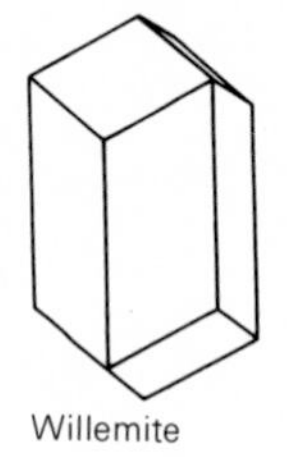

Willemite

Willemite Zn_2SiO_4 **Crystal system** Trigonal. **Habit** Crystals prismatic; usually massive, granular. **SG** 3·9–4·2 **Hardness** $5\frac{1}{2}$ **Cleavage** Basal, good. **Fracture** Uneven. **Colour and transparency** Greenish yellow is typical; but varies from near white to brown: transparent to nearly opaque. **Lustre** Vitreous to resinous. **Distinguishing features** Greenish colour, association. Willemite is usually strongly fluorescent. **Occurrence** Willemite occurs in the oxidized zone of zinc ore deposits but never in large amounts. It is named in honour of King William I (Willem Frederik) of the Netherlands.

Monticellite $CaMgSiO_4$ **Crystal system** Orthorhombic. **Habit** Crystals small and prismatic; also as grains. **SG** 3·1–3·3 **Hardness** $5\frac{1}{2}$ **Colour** Colourless to grey. **Occurrence** Monticellite occurs in metamorphosed calcareous rocks, usually impure dolomites. It is named after T Monticelli (1759–1846), an Italian mineralogist.

5 cm

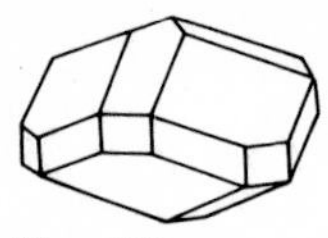

Phenakite: rhombohedral habit

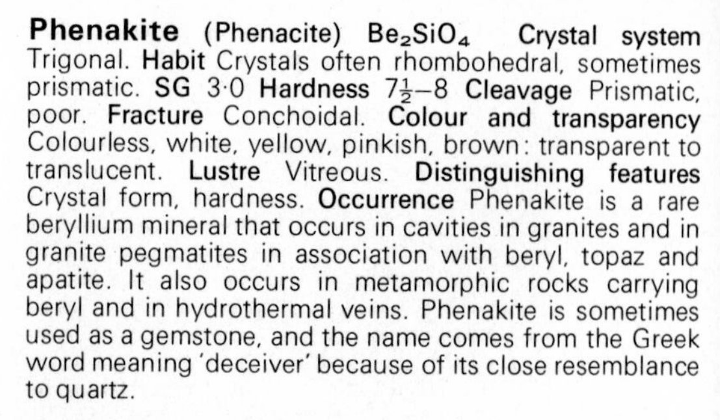

Phenakite (Phenacite) Be_2SiO_4 **Crystal system** Trigonal. **Habit** Crystals often rhombohedral, sometimes prismatic. **SG** 3·0 **Hardness** 7½–8 **Cleavage** Prismatic, poor. **Fracture** Conchoidal. **Colour and transparency** Colourless, white, yellow, pinkish, brown: transparent to translucent. **Lustre** Vitreous. **Distinguishing features** Crystal form, hardness. **Occurrence** Phenakite is a rare beryllium mineral that occurs in cavities in granites and in granite pegmatites in association with beryl, topaz and apatite. It also occurs in metamorphic rocks carrying beryl and in hydrothermal veins. Phenakite is sometimes used as a gemstone, and the name comes from the Greek word meaning 'deceiver' because of its close resemblance to quartz.

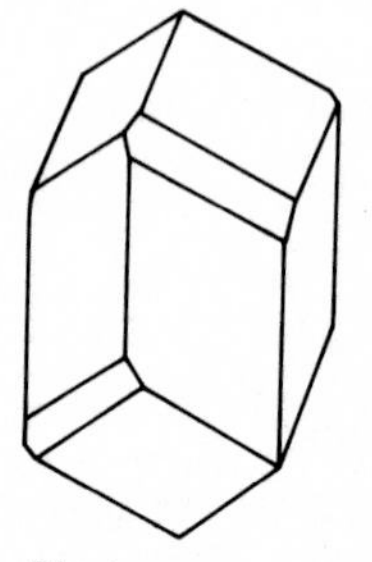

Dioptase

Dioptase $CuSiO_2(OH)_2$ **Crystal system** Trigonal. **Habit** Crystals usually short prismatic, often terminated by rhombohedra; also massive. **SG** 3·3 **Hardness** 5 **Cleavage** Rhombohedral, perfect. **Fracture** Conchoidal to uneven. **Colour and transparency** Emerald-green: transparent to translucent. **Lustre** Vitreous. **Distinguishing features** Colour, crystal form, association with copper minerals. **Occurrence** Dioptase is not a common mineral but is found in the oxidized parts of copper sulphide deposits.

Humite series $Mg(OH,F)_2.1–4Mg_2SiO_4$ **Crystal system** Orthorhombic and monoclinic. **Habit** Crystals usually of stubby but varied habit; also massive. **Twinning** Common. **SG** 3·1–3·3 **Hardness** 6–6½ **Cleavage** One poor cleavage. **Fracture** Uneven. **Colour and transparency** White, pale yellow, brown: translucent. **Lustre** Vitreous to resinous. **Distinguishing features** Light yellow or brownish colour, association with metamorphosed limestone. Colourless varieties difficult to distinguish from olivine. The humite group comprises four minerals, norbergite, chondrodite, humite and clinohumite, which differ only in the amount of magnesia and silica they contain. Humite and norbergite are orthorhombic, and chondrodite and clinohumite monoclinic, but the monoclinic minerals depart only slightly from orthorhombic symmetry. Individual members of the series are difficult to distinguish in hand specimen. **Occurrence** Members of the humite group occur typically in metamorphosed dolomitic limestones and spinel, phlogopite, garnet, idocrase, diopside, graphite and calcite are associated minerals. The name chondrodite is taken from a Greek word meaning 'a grain'. Humite is named after Sir Abraham Hume (1748–1838); and norbergite is named after Norberg, a locality in Sweden.

Phenakite
Dioptase
Dioptase
Chondrodite
Humite
5 cm

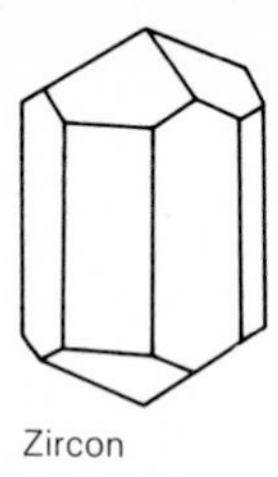
Zircon

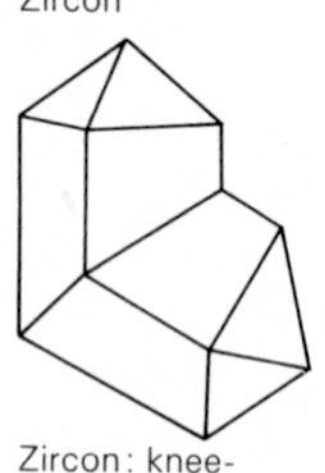
Zircon: knee-shaped twin

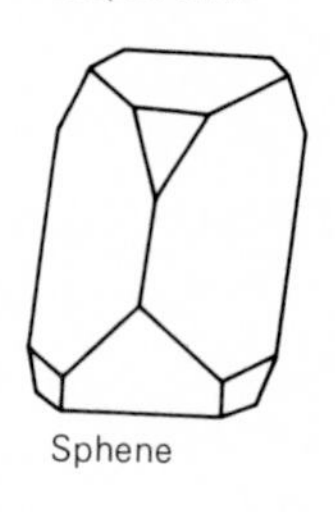
Sphene

Zircon $ZrSiO_4$ **Crystal system** Tetragonal. **Habit** Crystals usually prismatic, with bipyramidal terminations. **Twinning** Common, giving knee-shaped twins. **SG** 4·6–4·7 **Hardness** 7½ **Cleavage** Prismatic, indistinct. **Fracture** Conchoidal, very brittle. **Colour and transparency** Variable, though most commonly light brown to reddish brown; also colourless, grey, yellow, green: transparent to translucent, occasionally nearly opaque. **Lustre** Vitreous to adamantine. **Distinguishing features** Square prismatic habit, brownish colour, hardness, high specific gravity. **Occurrence** Zircon is one of the most widely distributed accessory minerals in igneous rocks such as granite, syenite and nepheline syenite. In pegmatites, the crystals sometimes reach a considerable size. It occurs also in metamorphic rocks such as schists and gneisses and, owing to its specific gravity and durability, it becomes concentrated as a detrital mineral in beach and river sands. Transparent zircon is used as a gemstone, and brownish varieties are called hyacinth or jacinth. It is a source of the metal zirconium, which took its name from the mineral. The name zircon is very old and may come from Persian words meaning 'golden colour'.

Sphene (Titanite) $CaTiSiO_5$ **Crystal system** Monoclinic. **Habit** Crystals commonly flattened and wedge-shaped; sometimes massive. **Twinning** Common. **SG** 3·4–3·6 **Hardness** 5–5½ **Cleavage** Prismatic, distinct. **Fracture** Conchoidal. **Colour and transparency** Brown and greenish yellow are the commonest colours, sometimes grey or nearly black: transparent to translucent, occasionally nearly opaque. **Lustre** Resinous to adamantine. **Distinguishing features** Sharp, wedge-shaped habit, adamantine lustre, greenish yellow colour. **Occurrence** Sphene is widely distributed as an accessory mineral, particularly in coarse-grained igneous rocks such as syenite, nepheline syenite, diorite and granodiorite. It occurs similarly in schists and gneisses and in some metamorphosed limestones. The name comes from a Greek word meaning 'wedge' and refers to the crystal habit.

Dumortierite $(Al,Fe)_7BSi_3O_{18}$ **Crystal system** Orthorhombic. **Habit** Crystals rare; usually in fibrous radiating aggregates. **SG** 3·3–3·4 **Hardness** 7 **Cleavage** One poor cleavage. **Colour and transparency** Bright greenish blue, violet, pink: transparent to translucent. **Lustre** Vitreous. **Distinguishing features** Colour, fibrous habit. **Occurrence** Dumortierite is a rare mineral that occurs in some schists, gneisses and pegmatites. It is named after E Dumortier, a French palaeontologist.

Eudialyte $Na_4(Ca,Fe)_2ZrSi_6O_{17}(OH,Cl)_2$ **Crystal system** Trigonal. **Habit** Crystals rhombohedral or tabular; also massive, granular. **SG** 2·8–3·0 **Hardness** 5–5½ **Cleavage** Basal, indistinct. **Colour and transparency** Red to brown: transparent to translucent. **Distinguishing features** Colour, association with nepheline syenite. **Occurrence** Eudialyte occurs typically in nepheline syenites and nepheline syenite pegmatites.

Zircon
Zircon
Sphene
Sphene
Dumortierite
Dumortierite
Eudialyte
5 cm

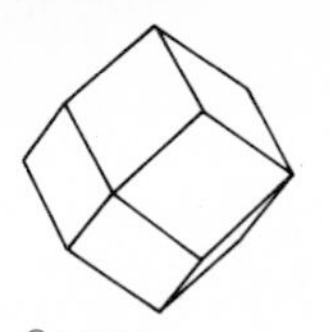

Garnet: rhombdodecahedron

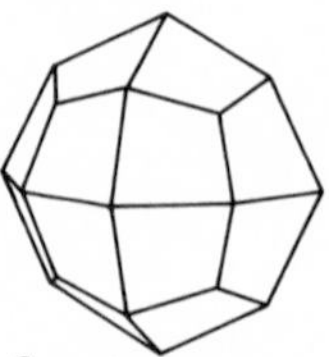

Garnet: icositetrahedron

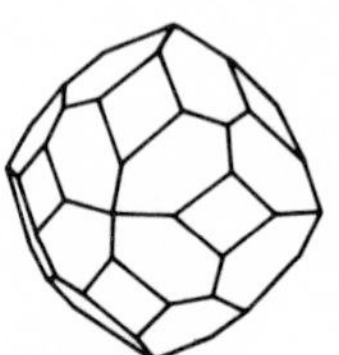

Garnet: combination of rhombdodecahedron and icositetrahedron

Garnet group General formula $X_3Y_2Si_3O_{12}$, where X is commonly Ca, Mn, Mg, or Fe^{2+}; and Y is Al, Cr, or Fe^{3+} Specific names are given to garnets of simple composition, though natural garnets rarely conform to such simple end members owing to substitution of one atom for another. The following names are in common use:

Pyrope	$Mg_3Al_2Si_3O_{12}$
Almandine	$Fe_3Al_2Si_3O_{12}$
Spessartine	$Mn_3Al_2Si_3O_{12}$
Grossular	$Ca_3Al_2Si_3O_{12}$
Uvarovite	$Ca_3Cr_2Si_3O_{12}$
Andradite	$Ca_3Fe_2Si_3O_{12}$

There are, in effect, two main groups of garnets; the pyrope-almandine-spessartine group, and the grossular-uvarovite-andradite group. Continuous atomic substitutions take place within these groups, but there is no continuous substitution between them. **Crystal system** Cubic. **Habit** Crystals common; usually rhombdodecahedra or icositetrahedra, or combinations of the two. Other forms occur but more rarely. Sometimes massive, granular. **SG** 3·6–4·3 (varying with composition) **Hardness** 6–7½ **Cleavage** None. **Fracture** Subconchoidal. **Colour and transparency** Colour varies with composition: transparent to translucent. Pyrope, almandine and spessartine are usually shades of deep red and brown to nearly black; uvarovite is a clear green; grossular is brown, pale green or white; and andradite is yellow, brown or black. **Lustre** Vitreous to resinous. **Distinguishing features** Hardness, rhombdodecahedral or icositetrahedral crystal form. Individual members may be distinguished by colour and specific gravity. However, chemical analysis is needed for precise determination. **Occurrence** Garnets are widely distributed in metamorphic and some igneous rocks. There is a link between composition and occurrence of garnets. Pyrope occurs in igneous rocks such as peridotite and in associated serpentinites, and also in kimberlite. Almandine is the common garnet of schists and gneisses; spessartine occurs in low grade metamorphic rocks, particularly if they contain manganese, and in some granites and pegmatites. Uvarovite is the rarest of the six garnet varieties listed here, occurring mainly in chromium-bearing serpentinites; grossular is characteristic of metamorphosed impure limestones; andradite occurs in metamorphosed limestones and in metasomatic calcareous rocks; and the black variety, called melanite, occurs in some feldspathoidal igneous rocks, such as phonolite and leucitophyre. Garnet is often a constituent of beach and river sands. Some varieties of garnet are used as gemstones. Hesonite (cinnamon stone) is yellow to brownish red and is a variety of grossular; demantoid is green andradite and is the best of the gem garnets; and rhodolite is a rose-coloured or purplish garnet of the pyrope-almandine series.

Grossular
Hessonite
Spessartine
Almandine
Álmandine
5 cm
Melanite
Pyrope
Uvarovite

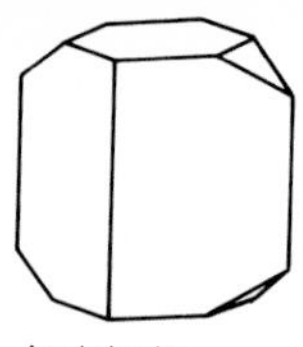
Andalusite

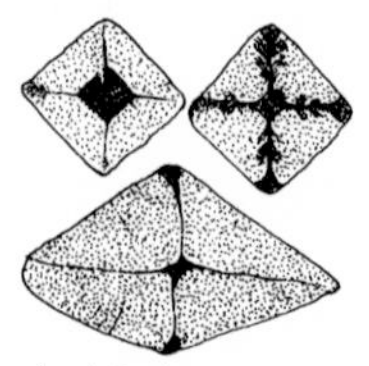
Andalusite: variety chiastolite

Andalusite Al_2SiO_5 **Crystal system** Orthorhombic. **Habit** Crystals prismatic and pseudo-tetragonal with a square cross-section; also massive. Some crystals have carbonaceous inclusions arranged so that in cross-section they form a dark cross. This variety is called chiastolite. **SG** 3·1–3·2 **Hardness** 6½–7½ **Cleavage** Prismatic, distinct. **Fracture** Uneven. **Colour and transparency** Commonly pink or red; also grey, brown and green: transparent to nearly opaque. **Lustre** Vitreous. **Distinguishing features** Square prismatic form, hardness, occurrence in metamorphic rocks. **Alteration** Andalusite alters to an aggregate of white mica flakes which often coat the crystals. **Occurrence** Andalusite occurs typically in thermally metamorphosed pelitic rocks, and in pelites that have been regionally metamorphosed under low pressure conditions. It occurs also in some pegmatites, together with corundum, tourmaline, topaz and other minerals. Transparent green andalusite is used as a gemstone. The name is derived from the Spanish province of Andalusia.

Sillimanite (Fibrolite) Al_2SiO_5 **Crystal system** Orthorhombic. **Habit** Commonly as elongated prismatic crystals, often fibrous and as felted (interwoven) masses. **SG** 3·2–3·3 **Hardness** 6½–7½ **Cleavage** Pinacoidal, good. **Fracture** Uneven. **Colour and transparency** Colourless, white, yellowish or brownish: transparent to translucent. **Lustre** Vitreous. **Distinguishing features** Fibrous habit, though in this it resembles other fibrous silicates, from which it may be distinguished with the microscope, or by its association. **Occurrence** Sillimanite occurs typically in schists and gneisses produced by high grade regional metamorphism. It is named after B Silliman, an American chemist.

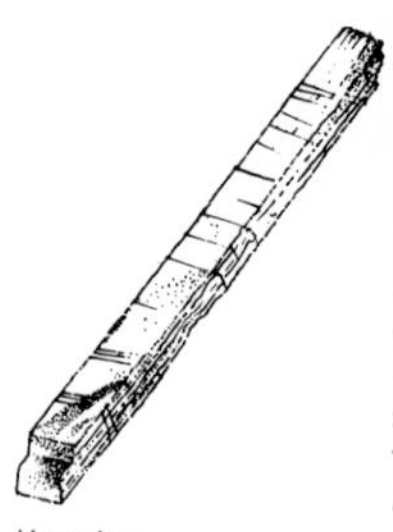
Kyanite: bladed crystal

Kyanite (Disthene) Al_2SiO_5 **Crystal system** Triclinic. **Habit** Crystals usually of flat, bladed habit, also as radiating bladed aggregates. **SG** 3·5–3·7 **Hardness** 5½–7 (hardness is variable, being 5½ along the length of the crystals and 6–7 across them) **Cleavage** Two good cleavages. **Colour and transparency** Blue to white but may be grey or green. Crystals are often unevenly coloured, the darkest tints being at the centres of crystals, or as streaks and patches: transparent to translucent. **Lustre** Vitreous, sometimes pearly on cleavage surfaces. **Distinguishing features** Blue colour, bladed habit, good cleavage, variable hardness. **Occurrence** Kyanite occurs typically in regionally metamorphosed schists and gneisses, together with garnet, staurolite, mica and quartz. It occurs also in pegmatites and quartz veins associated with schists and gneisses. The name kyanite comes from a Greek word meaning 'blue'.

Andalusite, sillimanite and kyanite provide an example of polymorphism. There is a relationship between their structure (and hence specific gravity) and their mode of formation, the least dense andalusite forming under low pressure metamorphism and kyanite, the densest, with a closely packed structure, forming under high pressure conditions.

5 cm
Chiastolite
Andalusite
Sillimanite
Kyanite
Kyanite

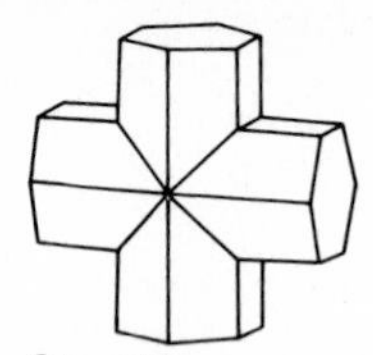
Staurolite: cruciform twin

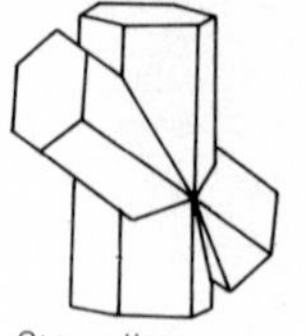
Staurolite: cruciform twin

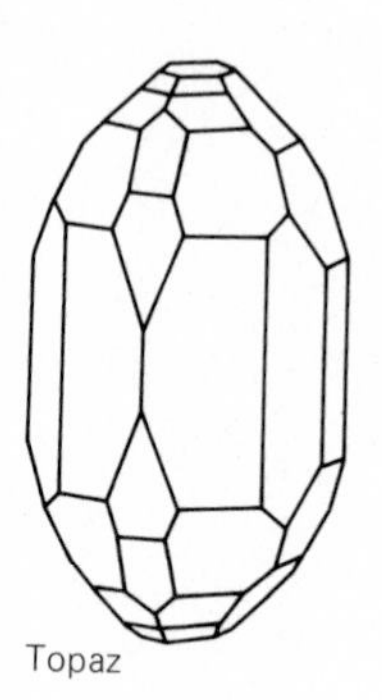
Topaz

Staurolite $(Fe,Mg)_2(Al,Fe)_9Si_4O_{20}(O,OH)_2$ **Crystal system** Monoclinic, pseudo-orthorhombic. **Habit** Crystals usually prismatic; rarely massive. **Twinning** Common, giving rise to cruciform twins with the two individuals crossing either at right-angles or obliquely. **SG** 3·7–3·8 **Hardness** 7–7½ **Cleavage** One, distinct cleavage. **Fracture** Subconchoidal. **Colour and transparency** Reddish brown to brown-black: translucent to nearly opaque. **Lustre** Vitreous to resinous. **Distinguishing features** Brown colour, crystal form (particularly if twinned). **Occurrence** Staurolite occurs typically as porphyroblasts in medium grade schists and gneisses often in association with garnet, kyanite and mica. The name comes from a Greek word meaning 'cross' in allusion to the form of the twins.

Topaz $Al_2SiO_4(OH,F)_2$ **Crystal system** Orthorhombic. **Habit** Crystals usually prismatic, often with two or more vertical prism forms, or with striated prism faces. Also massive, granular. **SG** 3·5–3·6 **Hardness** 8 **Cleavage** Basal, perfect. **Fracture** Subconchoidal to uneven. **Colour and transparency** Colourless; also pale yellow, pale blue, greenish and, rarely, pink: transparent to translucent. **Lustre** Vitreous. **Distinguishing features** Crystal form, hardness, perfect basal cleavage, high specific gravity. **Occurrence** Topaz occurs typically in granite pegmatites, rhyolites and quartz veins. It also occurs as grains in granites which have been subjected to alteration by fluorine-bearing solutions, and is accompanied by fluorite, tourmaline, apatite, beryl and cassiterite. Topaz also occurs as worn grains and pebbles in alluvial deposits. It is used as a gemstone.

Euclase $BeAlSiO_4(OH)$ **Crystal system** Monoclinic. **Habit** Crystals prismatic. **SG** 3·0–3·1 **Hardness** 7½ **Cleavage** One perfect cleavage, hence the name. **Colour and transparency** Colourless to pale blue-green: transparent to translucent. **Lustre** Vitreous. **Occurrence** Euclase is a rare mineral found in pegmatites in association with other beryllium minerals, notably beryl. It is sometimes used as a gemstone.

Topaz
Topaz
Staurolite
5 cm
Euclase
Topaz

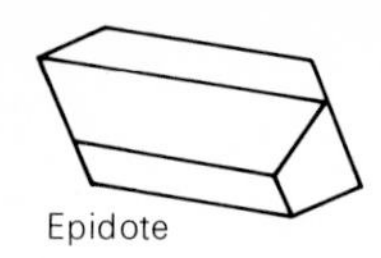
Epidote

Epidote group Epidotes have the general formula $X_2Y_3Si_3O_{12}(OH)$, in which X is commonly Ca, and Y is usually Al and Fe^{3+}, partly replaced by Mg and Fe^{2+} in some species.

Zoisite $Ca_2Al_3Si_3O_{12}(OH)$ **Crystal system** Orthorhombic. **Habit** Crystals prismatic; also massive. **SG** 3·2–3·4 (increasing with iron content) **Hardness** 6 **Cleavage** One perfect cleavage. **Fracture** Uneven. **Colour and transparency** Grey; sometimes pale green or brown. A pink, manganese-bearing variety is called thulite: transparent to subtranslucent. **Lustre** Vitreous, pearly on cleavage surfaces. **Distinguishing features** Colour, single perfect cleavage. **Occurrence** Zoisite occurs in schists and gneisses and in metasomatic rocks, together with garnet, idocrase and actinolite. It occurs occasionally in hydrothermal veins. The recently discovered blue variety called tanzanite is a valuable gemstone. Zoisite is named after Baron von Zois, an Austrian.

Clinozoisite $Ca_2Al_3Si_3O_{12}(OH)$ **Epidote** (Pistacite) $Ca_2(Al,Fe)_3Si_3O_{12}(OH)$ **Crystal system** Monoclinic. **Habit** Crystals prismatic and often striated parallel to their length; also massive, fibrous or granular. **Twinning** Uncommon. **SG** 3·2–3·5 (increasing with iron content) **Hardness** 6–7 **Cleavage** One perfect cleavage, parallel to the length of the crystals. **Fracture** Uneven. **Colour and transparency** Clinozoisite is usually greenish grey; epidote is yellowish green to black: transparent to nearly opaque. **Lustre** Vitreous. **Distinguishing features** Distinctive yellow-green colour, prismatic habit. Epidote can be mistaken for tourmaline, but the latter lacks cleavage and has a hexagonal or triangular cross-section. **Occurrence** Clinozoisite and epidote are widespread in medium to low grade metamorphic rocks, especially those derived from igneous rocks such as basalt and diabase, or from calcareous sediments. They also occur in contact metamorphosed limestones and in veins in igneous rocks.

Allanite (Orthite) $(Ca,Ce,Y,La,Th)_2(Al,Fe)_3Si_3O_{12}(OH)$ **Crystal system** Monoclinic. **Habit** Crystals prismatic, sometimes tabular; also massive. **SG** 3·4–4·2 **Hardness** 5–6½ **Cleavage** Two poor cleavages. **Fracture** Conchoidal to uneven. **Colour and transparency** Light brown to black: subtranslucent to opaque. **Lustre** Vitreous or pitchy to submetallic. **Distinguishing features** Dark colour, pitchy lustre, weak radioactivity. **Occurrence** Allanite occurs as an accessory mineral in many granites, syenites, pegmatites, gneisses and skarns. It is named after T Allan (1777–1833), a British mineralogist.

Piemontite (Piedmontite) $Ca_2(Al,Fe,Mn)_3Si_3O_{12}(OH)$ **Crystal system** Monoclinic. **Habit** As for epidote. **SG** 3·4–3·5 **Hardness** 6 **Cleavage** One perfect cleavage. **Fracture** Uneven. **Colour and transparency** Reddish or purplish brown to black. **Lustre** Vitreous. **Distinguishing features** Reddish brown colour. **Occurrence** Piemontite is a rare mineral which occurs in some low grade schists and manganese ore deposits.

5 cm
Zoisite
Thulite
Epidote
Epidote
Allanite
Piemontite

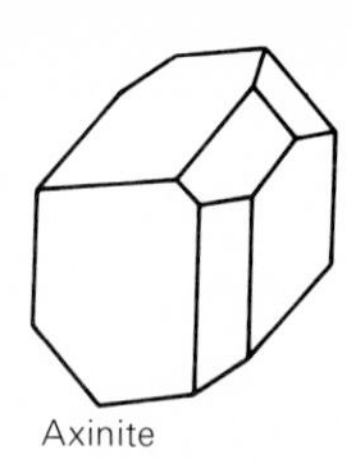
Axinite

Axinite $Ca_2(Fe,Mn)Al_2BSi_4O_{15}(OH)$ **Crystal system** Triclinic. **Habit** Crystals commonly broad and with sharp edges; also massive, lamellar or granular. **SG** 3·3–3·4 **Hardness** 6½–7 **Cleavage** One good cleavage. **Fracture** Conchoidal. **Colour and transparency** Most crystals are a distinctive clove-brown colour; also yellowish or grey: transparent to translucent. **Lustre** Vitreous. **Distinguishing features** Colour, sharp-edged crystal form. The name axinite, derived from the Greek word for 'axe', is descriptive of the crystal form. **Occurrence** Axinite occurs in calcareous rocks that have undergone contact metamorphism and metasomatism, and in cavities in granites, especially close to their contacts.

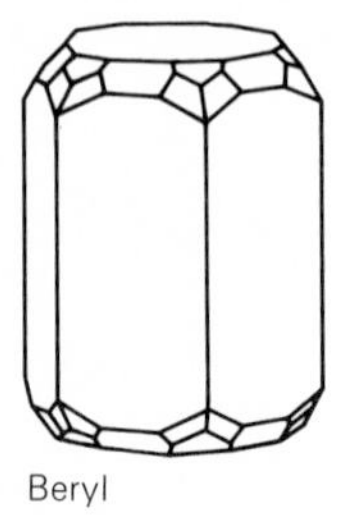
Beryl

Beryl $Be_3Al_2Si_6O_{18}$ **Crystal system** Hexagonal. **Habit** Crystals usually prismatic often with striations parallel to their length; also massive. **SG** 2·6–2·8 **Hardness** 7½–8 **Cleavage** Basal, poor. **Fracture** Conchoidal to uneven. **Colour and transparency** Green, blue, yellow, pink but rather variable: transparent to translucent. Transparent, gem quality beryl may be dark or light green (emerald), bluish green (aquamarine), yellow (heliodor) or pink (morganite). **Lustre** Vitreous. **Distinguishing features** Hexagonal crystal form, green colour (usually). It resembles apatite but the greater hardness of beryl is distinctive. Massive beryl can be mistaken for quartz. **Occurrence** Beryl most commonly occurs as an accessory mineral in granites, and is usually found in cavities and in granite pegmatites. Beryl crystals in some pegmatites grow to very large sizes. It occurs also in mica schists and gneisses in association with phenakite, rutile and chrysoberyl. Some varieties are prized as gemstones and beryl is a source of beryllium.

Cordierite $(Mg,Fe)_2Al_4Si_5O_{18}$ **Crystal system** Orthorhombic. **Habit** Crystals prismatic and pseudo-hexagonal but rather rare; usually as grains, or massive. **Twinning** Common, giving pseudo-hexagonal forms. **SG** 2·5–2·8 (increasing with iron content) **Hardness** 7 **Cleavage** One poor cleavage, basal parting. **Fracture** Subconchoidal to uneven. **Colour and transparency** Dark blue, greyish blue: transparent to translucent. **Lustre** Vitreous. **Distinguishing features** Dark blue colour, but granular cordierite closely resembles quartz. The gem variety, called iolite or dichroite, is deep blue or yellow depending on the direction in which it is viewed. **Alteration** To an aggregate of chlorite and muscovite called pinite. **Occurrence** Cordierite occurs in aluminous rocks that have undergone medium to high grade contact or regional metamorphism. It is found in hornfelses, schists and gneisses in association with andalusite, spinel, quartz and biotite. It also occurs in igneous rocks which have assimilated aluminous sediments. Cordierite is named after PLA Cordier, a French geologist.

5 cm
Heliodor
Aquamarine
Axinite
Aquamarine
Emerald
Emerald
Beryl
Cordierite

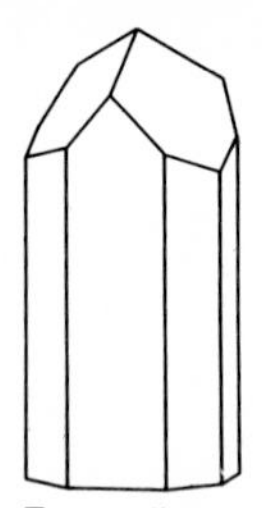
Tourmaline

Tourmaline $Na(Mg,Fe,Li,Al,Mn)_3Al_6(BO_3)_3Si_6O_{18}(OH,F)_4$ **Crystal system** Trigonal. **Habit** Crystals usually prismatic, often of curved triangular cross-section. The prism faces are often strongly striated parallel to their length, and the two ends of a crystal are often differently terminated. Parallel or radiating crystal groups are common; also massive. **SG** 3·0–3·2 (increasing with iron content) **Hardness** 7 **Cleavage** Very poor. **Fracture** Conchoidal to uneven. **Colour and transparency** Usually black or bluish black; also colourless, blue, pink and green. Some crystals are pink at one end and green at the other. Pink varieties are sometimes called rubellite; schorl is black and dravite brown. Transparent to nearly opaque. **Lustre** Vitreous. **Distinguishing features** Prismatic habit, striations, colour, triangular cross-section. Care is needed in distinguishing tourmaline from epidote. **Occurrence** Tourmaline commonly occurs in granite pegmatites, or in granites which have undergone metasomatism by boron-bearing fluids. It also occurs in sediments adjacent to such granites, and as an accessory mineral in schists and gneisses.

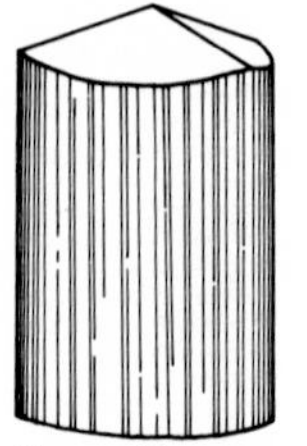
Tourmaline: showing striations

Hemimorphite (Calamine) $Zn_4Si_2O_7(OH)_2.H_2O$ **Crystal system** Orthorhombic. **Habit** Crystals tabular; also massive, fibrous, mamillated. **SG** 3·4–3·5 **Hardness** 4½–5 **Cleavage** Prismatic, perfect. **Fracture** Conchoidal to uneven. **Colour and transparency** White, sometimes bluish, greenish or brownish: transparent to translucent. **Lustre** Vitreous. **Distinguishing features** Crystal form. **Occurrence** Hemimorphite is a secondary mineral found in the oxidized zone of zinc-bearing ore bodies, and in limestones adjacent to such bodies, in association with sphalerite, smithsonite, cerussite and anglesite. The alternative name, calamine, is also applied to smithsonite.

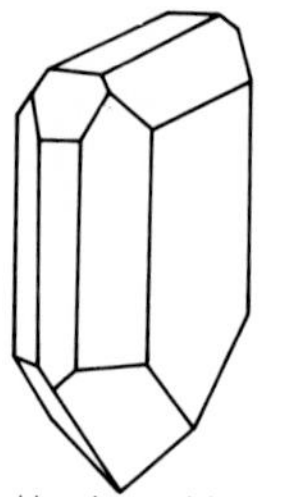
Hemimorphite

Idocrase (Vesuvianite) $Ca_{10}(Mg,Fe)_2Al_4Si_9O_{34}(OH,F)_4$ **Crystal system** Tetragonal. **Habit** Prismatic crystals often with striations parallel to their length; also massive, granular or columnar. **SG** 3·3–3·4 **Hardness** 6–7 **Cleavage** Poor. **Fracture** Subconchoidal to uneven. **Colour and transparency** Usually dark green or brown; also yellow. Blue varieties are called cyprine: subtransparent to translucent. **Lustre** Vitreous to resinous. **Distinguishing features** Prismatic, striated crystal form. Massive varieties may be mistaken for garnet, epidote or diopside. **Occurrence** Idocrase occurs in impure limestones that have undergone contact metamorphism. It occurs in blocks of dolomitic limestone erupted from Vesuvius, hence its alternative name. It is frequently accompanied by grossular, wollastonite, diopside and calcite.

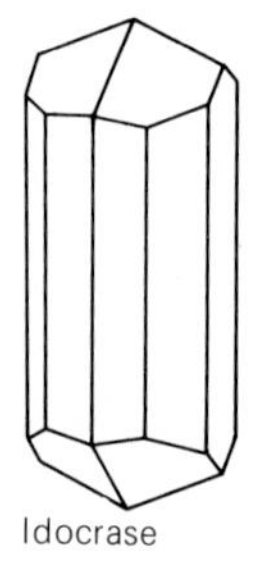
Idocrase

Ilvaite $CaFe^{2+}{}_2Fe^{3+}Si_2O_8(OH)$ **Crystal system** Orthorhombic. **Habit** Crystals prismatic; also columnar or massive. **SG** 4·1 **Hardness** 5½–6 **Cleavage** One good cleavage. **Fracture** Uneven. **Colour and transparency** Black: opaque. **Streak** Black. **Lustre** Submetallic. **Occurrence** Ilvaite occurs with magnetite in magmatic ore bodies, and in contact metasomatic deposits.

Tourmaline
Tourmaline (rubellite)
Tourmaline
Hemimorphite
Hemimorphite
Idocrase
Ilvaite
5 cm

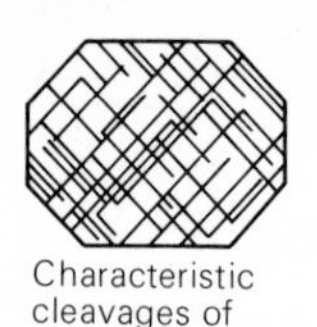
Characteristic cleavages of pyroxenes

Pyroxene group The pyroxenes are an important and widely distributed group of rock-forming silicates. They have a general formula $X_2Si_2O_6$, in which X is usually Mg,Fe,Mn,Li,Ti,Al,Ca or Na. The commonest pyroxenes are Ca,Mg,Fe silicates. The pyroxenes are characterized by two cleavages which intersect almost at right-angles. There are two main groups of pyroxenes: the orthopyroxenes crystallize in the orthorhombic system and contain very little calcium; the clinopyroxenes are monoclinic and contain either Ca, or Na, Al, Fe^{3+} or Li.

Orthopyroxenes: Enstatite $MgSiO_3$; **Hypersthene** (Mg,Fe)SiO_3 **Crystal system** Orthorhombic. **Habit** Crystals prismatic; usually as grains or massive. **SG** 3·2–4·0 (increasing with iron content) **Hardness** 5–6 **Cleavage** Prismatic, good. **Fracture** Uneven. **Colour and transparency** Pale green to dark brownish green, the colour usually deepening with iron content. Bronzite is intermediate in composition between enstatite and hypersthene and, as its name implies, it has a bronze lustre. Translucent to opaque. **Lustre** Vitreous. **Distinguishing features** Two cleavages intersecting almost at right-angles. Pale green colour and bronze lustre characterizes enstatite and bronzite respectively; hypersthene may resemble clinopyroxene very closely. **Occurrence** Orthopyroxenes are common constituents of igneous rocks such as gabbro and pyroxenite. Orthopyroxenes also occur in some andesitic volcanic rocks and in stony meteorites.

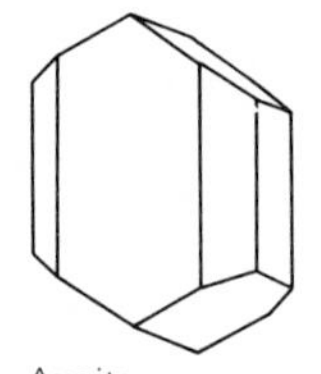
Augite

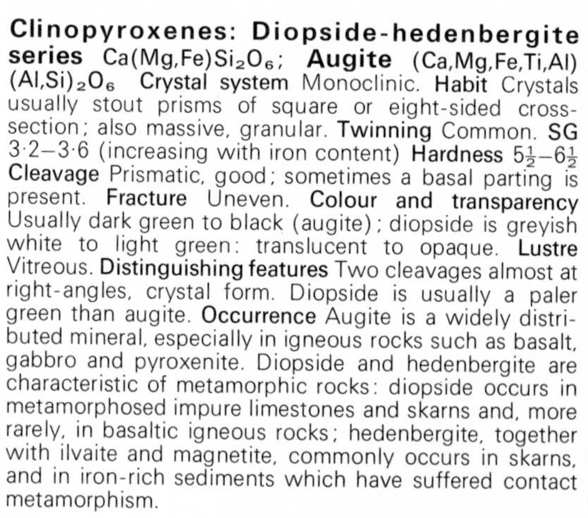
Clinopyroxenes: Diopside-hedenbergite series Ca(Mg,Fe)Si_2O_6; **Augite** (Ca,Mg,Fe,Ti,Al)(Al,Si)$_2O_6$ **Crystal system** Monoclinic. **Habit** Crystals usually stout prisms of square or eight-sided cross-section; also massive, granular. **Twinning** Common. **SG** 3·2–3·6 (increasing with iron content) **Hardness** $5\frac{1}{2}$–$6\frac{1}{2}$ **Cleavage** Prismatic, good; sometimes a basal parting is present. **Fracture** Uneven. **Colour and transparency** Usually dark green to black (augite); diopside is greyish white to light green: translucent to opaque. **Lustre** Vitreous. **Distinguishing features** Two cleavages almost at right-angles, crystal form. Diopside is usually a paler green than augite. **Occurrence** Augite is a widely distributed mineral, especially in igneous rocks such as basalt, gabbro and pyroxenite. Diopside and hedenbergite are characteristic of metamorphic rocks: diopside occurs in metamorphosed impure limestones and skarns and, more rarely, in basaltic igneous rocks; hedenbergite, together with ilvaite and magnetite, commonly occurs in skarns, and in iron-rich sediments which have suffered contact metamorphism.

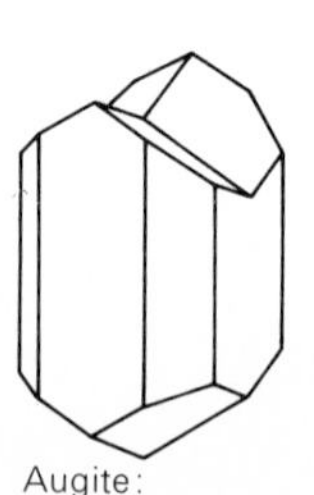
Augite: twinned crystal

Hypersthene
Bronzite
Diopside
Diopside
Augite
Augite
5 cm

Clinopyroxenes: Aegirine $NaFeSi_2O_6$ **Crystal system** Monoclinic. **Habit** Crystals usually slender prisms sometimes terminated by steeply inclined faces giving a pointed appearance; also as discrete grains, or radiating aggregates. **Twinning** Common. **SG** 3·5–3·6 **Hardness** 6 **Cleavage** Prismatic, good. **Fracture** Uneven. **Colour and transparency** Dark green or brown often nearly black: subtransparent to opaque. **Lustre** Vitreous. **Distinguishing features** Association. **Occurrence** Aegirine occurs typically in sodium-rich igneous rocks such as syenites, nepheline syenites and associated pegmatites. The name is derived from Aegir, the Scandinavian god of the sea, because the mineral was first described from Norway.

Clinopyroxenes: Jadeite $NaAlSi_2O_6$ **Crystal system** Monoclinic. **Habit** Crystals rare, usually massive, granular, compact or columnar. **SG** 3·2–3·4 **Hardness** 6 **Cleavage** Prismatic, good. **Fracture** Splintery. **Colour and transparency** Usually various shades of light or dark green; sometimes white: translucent. **Lustre** Vitreous; inclined to pearly on cleavage surfaces. **Distinguishing features** Green colour, massive habit, tough nature of massive material. The name 'jade' is applied to two distinct minerals: jadeite is one, and nephrite, an amphibole, is the other. They are best distinguished by their specific gravity. Softer material, such as serpentine, is often sold as jade. **Occurrence** Jadeite has long been used as a semi-precious and ornamental stone. Jadeite is formed at high pressures: it occurs as grains in metamorphosed sodic sediments and volcanic rocks, and is associated with glaucophane and aragonite.

Clinopyroxenes: Spodumene $LiAlSi_2O_6$ **Crystal system** Monoclinic. **Habit** Crystals usually prismatic, often striated along their length, and commonly etched or corroded; also massive, columnar. **Twinning** Common. **SG** 3·0–3·2 **Hardness** $6\frac{1}{2}$–7 **Cleavage** Prismatic, perfect. **Fracture** Uneven, splintery. **Colour and transparency** Usually white or greyish white. Hiddenite is a transparent green variety and kunzite is a transparent lilac form; both are used as gemstones. Transparent to translucent. **Lustre** Vitreous. **Distinguishing features** Two cleavages. **Alteration** Spodumene is prone to alteration to clay minerals. **Occurrence** Spodumene occurs typically in lithium-bearing granite pegmatites, together with minerals such as lepidolite, tourmaline and beryl. Very large crystals have been recorded, some reaching 15 metres (nearly 50 feet) in length and weighing up to 90 tons.

Aegirine
Aegirine
Jadeite
Jadeite
5 cm
Jadeite

Spodumene (kunzite)

Spodumene

Wollastonite $CaSiO_3$ **Crystal system** Triclinic. **Habit** Crystals tabular or short prismatic; also massive, compact, fibrous or as cleavable masses. **Twinning** Common. **SG** 2·8–3·1 **Hardness** 4½–5 **Cleavage** In three directions, one perfect, two others good. **Fracture** Uneven. **Colour and transparency** White to grey: subtransparent to translucent. **Lustre** Vitreous; pearly on cleavage surfaces, inclined to silky when fibrous. **Distinguishing features** Colour, cleavages, association, dissolves in hydrochloric acid with separation of silica. **Occurrence** Wollastonite occurs in metamorphosed siliceous limestones, either in contact aureoles, in high grade regionally metamorphosed rocks, or in xenoliths in igneous rocks. It also occurs in certain alkaline igneous rocks. Associated minerals are calcite, epidote, idocrase, grossular and tremolite. It is named after WH Wollaston (1766–1828), a British mineralogist.

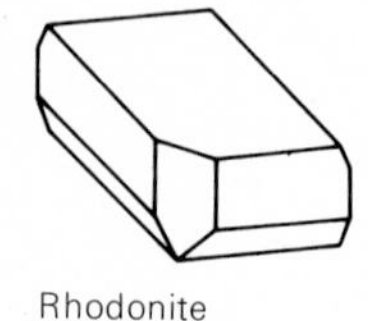
Rhodonite

Rhodonite $(Mn,Fe,Ca)SiO_3$ **Crystal system** Triclinic. **Habit** Crystals prismatic or tabular but uncommon; usually massive, compact, cleavable or granular. **SG** 3·5–3·7 **Hardness** 5½–6½ **Cleavage** In three directions, two perfect, one good. **Fracture** Conchoidal to uneven. **Colour and transparency** Pink to brown, weathers to black: transparent to translucent. **Lustre** Vitreous. **Distinguishing features** Pink colour, good cleavages. It resembles rhodochrosite but is harder and unaffected by warm dilute hydrochloric acid. **Occurrence** Rhodonite commonly occurs in association with manganese ore deposits in hydrothermal or metasomatic veins, or in regionally metamorphosed manganese-bearing sediments. It is used as a decorative stone.

Pectolite $Ca_2NaHSi_3O_9$ **Crystal system** Triclinic. **Habit** Aggregates of fibrous or acicular crystals, often radiating or stellate. **SG** 2·8–2·9 **Hardness** 4½–5 **Cleavage** Two perfect cleavages. **Fracture** Uneven. **Colour and transparency** White: subtranslucent to opaque. **Lustre** Silky when fibrous, otherwise vitreous. **Distinguishing features** Acicular form, two cleavages. **Occurrence** Pectolite occurs typically, along with zeolites, in cavities in basalts and similar rocks.

Petalite $LiAlSi_4O_{10}$ **Crystal system** Monoclinic. **Habit** Crystals rare; usually as masses showing cleavage. **SG** 2·4–2·5 **Hardness** 6–6½ **Cleavage** One perfect cleavage. **Fracture** Subconchoidal. **Colour and transparency** White, grey or green, sometimes colourless or reddish: transparent to translucent. **Lustre** Vitreous; pearly on cleavage surface. **Distinguishing features** Petalite resembles cleavage masses of feldspar and is often distinguishable only by optical tests although it gives the red flame characteristic of lithium. The perfect cleavage is sometimes distinctive and the name alludes to this, being derived from the Greek word for a leaf. **Occurrence** Petalite occurs typically in lithium-bearing granite pegmatites along with minerals such as spodumene, tourmaline, lepidolite and feldspars.

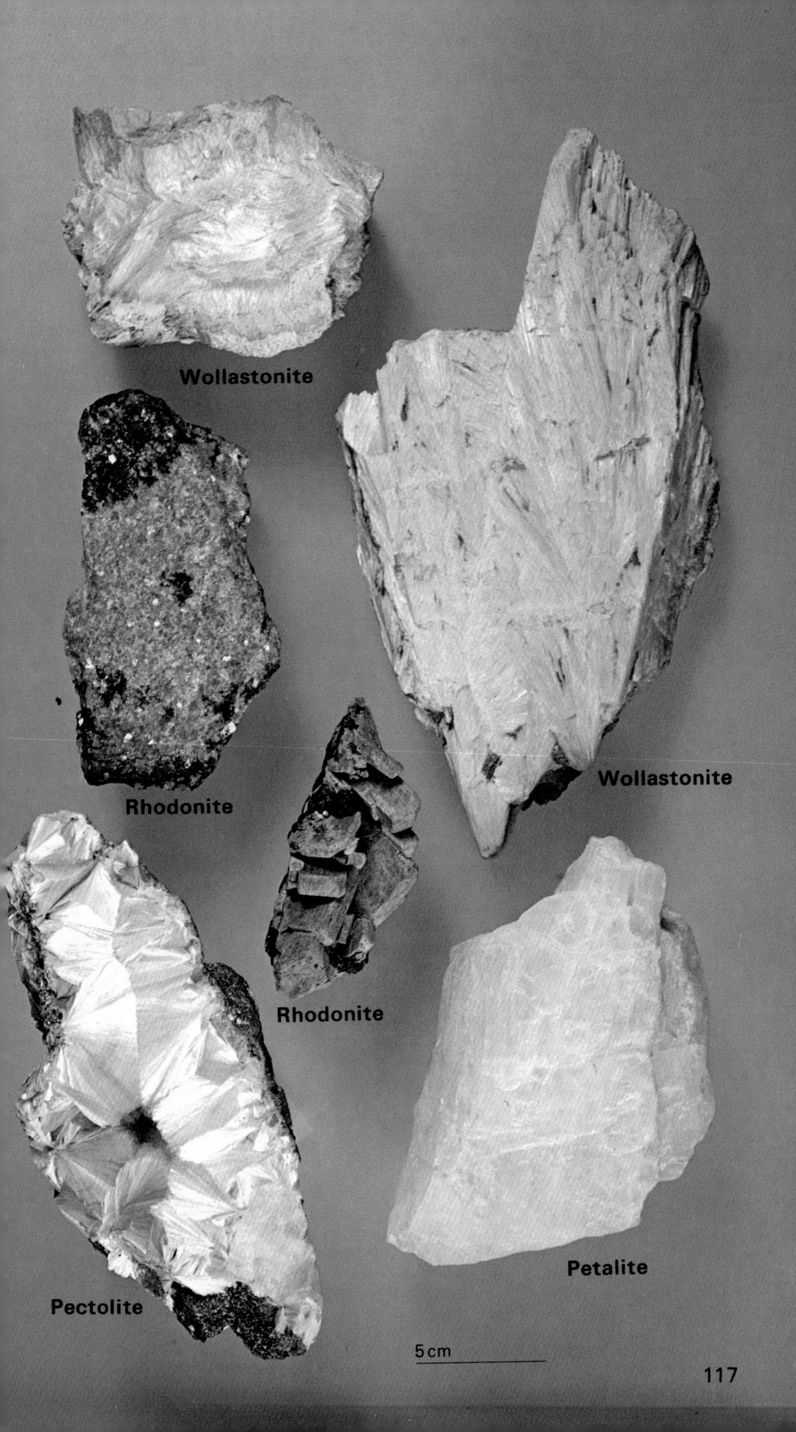
Wollastonite
Wollastonite
Rhodonite
Rhodonite
Pectolite
Petalite
5 cm

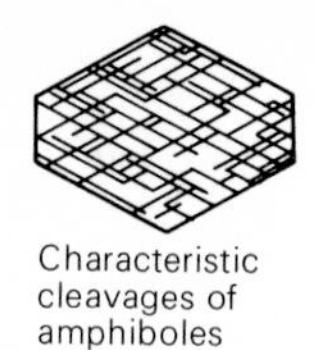

Characteristic cleavages of amphiboles

Amphibole group The amphiboles are an important group of rock-forming silicates that are widely distributed in igneous and metamorphic rocks. The angle between the prism faces and between two cleavages parallel to them is about 120°, and is characteristic of the amphiboles. The amphiboles further differ from the pyroxenes in that they are hydrous silicates, the (OH) group being an essential part of the structure. The chemical formulae of amphiboles are complex because of the extensive atomic substitution that takes place.

Anthophyllite $(Mg,Fe)_7Si_8O_{22}(OH)_2$ Orthorhombic **Cummingtonite-grunerite series** $(Fe,Mg)_7Si_8O_{22}(OH)_2$ **Monoclinic Habit** Individual crystals rare; usually as aggregates of fibrous crystals. **Twinning** Common (cummingtonites). **SG** 2·8–3·4 (anthophyllite): 3·1–3·6 (cummingtonites) (increasing with iron content). **Hardness** 5–6 **Cleavage** Prismatic, perfect. **Colour and transparency** White, grey, green, brown. Brownish colours predominate in the cummingtonite series. Translucent. **Lustre** Vitreous, fibrous varieties silky. **Distinguishing features** Anthophyllite and the cummingtonites are pale in colour. Although generally brown, the cummingtonites are so similar to anthophyllite as to require optical or X-ray tests to distinguish them. Cummingtonite refers to the magnesian members of the series and grunerite to the iron-rich members. Aluminium-rich anthophyllites are called gedrite. **Occurrence** Anthophyllite occurs in medium grade magnesium-rich metamorphic rocks: it does not occur in igneous rocks. Members of the cummingtonite series occur in regionally metamorphosed rocks that are relatively rich in iron and poor in calcium. They occur in contact metamorphic rocks, and cummingtonite occurs in igneous rocks such as certain rhyolites, and as a replacement product of pyroxene in diorites.

Tremolite-actinolite series $Ca_2(Mg,Fe)_5Si_8O_{22}(OH)_2$ **Crystal system** Monoclinic. **Habit** Usually in long bladed or prismatic crystals; sometimes massive, fibrous. **Twinning** Common. **SG** 3·0–3·4 (increasing with iron content) **Hardness** 5–6 **Cleavage** Prismatic, good. **Colour and transparency** White to grey (tremolite), light to dark green (actinolite), (green colour increasing with iron content): transparent to translucent. **Lustre** Vitreous. **Distinguishing features** Slender prismatic habit. Fibrous, radiating tremolite resembles wollastonite, but can be distinguished with the microscope and by lack of reaction with hydrochloric acid. Actinolite is a lighter colour than most hornblendes. **Occurrence** Tremolite is a characteristic mineral of thermally metamorphosed siliceous dolomitic limestones, and it occurs also in some serpentinites. Actinolite generally occurs in schists produced by low to medium grades of metamorphism of basalt and diabase or of pelitic rocks. It is often fibrous and the name asbestos was originally given to this variety. It also occurs in some igneous rocks, usually as an alteration product of pyroxene. Nephrite, a variety of jade, is usually an actinolitic or tremolitic amphibole.

Tremolite
nthophyllite
Tremolite
5 cm
Tremolite
Actinolite
Actinolite
Nephrite
Actinolite

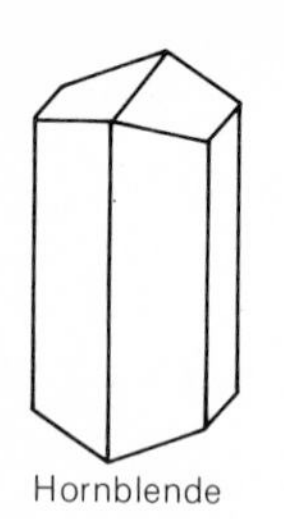

Hornblende

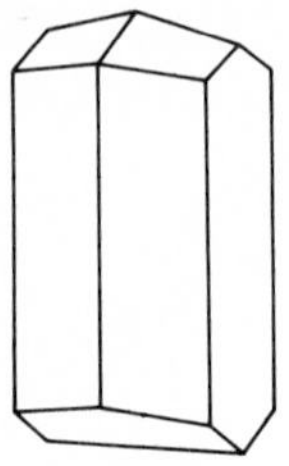

Hornblende: twinned crystal

Hornblende $(Ca,Na)_{2-3}(Mg,Fe,Al)_5(Si,Al)_8O_{22}(OH)_2$ **Crystal system** Monoclinic. **Habit** Crystals usually of long or short prismatic habit; also massive, granular or fibrous. **Twinning** Common. **SG** 3·0–3·5 **Hardness** 5–6 **Cleavage** Prismatic, good. **Fracture** Uneven. **Colour and transparency** Light green, through dark green to nearly black; sometimes with a brownish tinge: translucent to nearly opaque. **Lustre** Vitreous. **Distinguishing features** The 120° cleavage angle distinguishes hornblende (and other amphiboles) from pyroxene. Hornblende is generally darker than other amphiboles and has a wide range of composition. **Occurrence** Hornblende is a very widespread mineral. It occurs in a wide variety of igneous rocks, being a common constituent of granodiorites, diorites, some syenites and some gabbros, and their fine-grained equivalents. It is also a common constituent of many medium grade regionally metamorphosed rocks, and is particularly characteristic of the amphibolites and hornblende schists, and is commonly accompanied by garnet, quartz and calcic plagioclase.

Glaucophane-riebeckite $Na_2(Mg,Fe,Al)_5Si_8O_{22}(OH)_2$ **Crystal system** Monoclinic. **Habit** Good crystals rare; often prismatic or acicular; sometimes fibrous. **SG** 3·0–3·4 (increasing with iron content) **Hardness** 5–6 **Cleavage** Prismatic, good. **Fracture** Uneven. **Colour and transparency** Glaucophane is grey, grey-blue or lavender-blue; riebeckite is dark blue to black. Translucent to subtranslucent. **Lustre** Vitreous; fibrous varieties silky. **Distinguishing features** Colour and association is distinctive for both glaucophane and riebeckite. **Occurrence** Glaucophane occurs typically in sodium-rich schists, derived from geosynclinal sediments which have undergone low-temperature/high-pressure regional metamorphism. It occurs in association with such minerals as jadeite, aragonite, epidote, chlorite, muscovite and garnet. Riebeckite occurs mainly in alkaline igneous rocks such as some granites, syenites and nepheline syenites, and their fine-grained equivalents. Fibrous riebeckite, known as crocidolite or blue asbestos, occurs as veins in bedded ironstones. Riebeckite occurs, though rarely, in schists.

Hornblende
Hornblende
Glaucophane
Riebeckite
Riebeckite (crocidolite)
5 cm

Mica group There are two main groups of micas: dark mica, rich in Fe and Mg; and white mica which is rich in aluminium. In addition there is a series of lithium micas.

Muscovite $KAl_2(AlSi_3O_{10})(OH,F)_2$ **Crystal system** Monoclinic: pseudo-hexagonal. **Habit** Crystals tabular and hexagonal in outline; also as foliated masses and as disseminated flakes. **SG** 2·8–2·9 **Hardness** 2½–3 **Cleavage** Basal, perfect. Cleavage flakes flexible and elastic. **Colour and transparency** Colourless to pale grey, green or brown: transparent to translucent. **Lustre** Vitreous; pearly parallel to cleavage. **Distinguishing features** Perfect cleavage, light colour. **Occurrence** Muscovite is widely distributed. In igneous rocks it is most characteristic of the alkali granites and their pegmatites in which it sometimes forms large masses. It also occurs as a secondary mineral resulting from the decomposition of feldspars. Such fine-grained muscovite is called sericite. It is widely distributed in schists and gneisses and in contact metamorphosed rocks and crystalline limestones. Muscovite survives weathering and transport and is a common constituent of clastic sediments such as sandstones and siltstones.

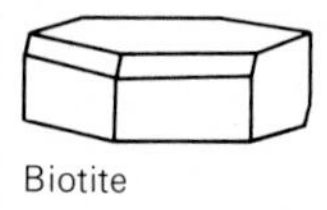
Biotite

Phlogopite-biotite series: Phlogopite $KMg_3AlSi_3O_{10}(OH,F)_2$; **Biotite** $K(Mg,Fe)_3AlSi_3O_{10}(OH,F)_2$ **Crystal system** Monoclinic. **Habit** Crystals tabular or short pseudo-hexagonal prisms; also as lamellar aggregates or disseminated flakes. **SG** 2·7–3·3 (increasing with iron content) **Hardness** 2–3 **Cleavage** Basal, perfect. **Colour and transparency** Phlogopite: yellowish to reddish brown, often with a distinctive coppery appearance; green. Biotite: black, dark brown or greenish black. Transparent to translucent. **Lustre** Vitreous; often submetallic on cleavage surface. **Distinguishing features** Perfect basal cleavage; phlogopite is generally paler in colour than biotite. Biotite is the name commonly given to all dark, iron-rich micas. **Occurrence** Phlogopite most commonly occurs in metamorphosed limestones, and in magnesian igneous rocks and some magnesium-rich pegmatites. It occurs also in kimberlite. Biotite is very widely distributed in granite, syenite and diorite and their fine-grained equivalents; it is characteristic of mica lamprophyre. It is a common constituent of schists and gneisses and of contact metamorphic rocks.

Glauconite Glauconite is a member of the mica family which usually occurs as small, green, rounded aggregates in marine sedimentary rocks. It has a dull lustre and a perfect basal cleavage.

Lepidolite $K(Li,Al)_3(Si,Al)_4O_{10}(OH,F)_2$ **Crystal system** Monoclinic. **Habit** Usually as small disseminated flakes. **SG** 2·8–2·9 **Hardness** 2½–4 **Cleavage** Basal perfect. **Colour and transparency** Pale lilac, also colourless, grey or pale pink: transparent to translucent. **Lustre** Vitreous; pearly on cleavage surfaces. **Distinguishing features** Perfect cleavage, lilac to pink colour. **Occurrence** Lepidolite occurs in granite pegmatites, often in association with lithium-bearing tourmaline and spodumene.

Muscovite
Muscovite
Biotite
Phlogopite
Glauconite
Lepidolite
Lepidolite
5 cm

Chlorite group $(Mg,Fe,Al)_6(Si,Al)_4O_{10}(OH)_8$ **Crystal system** Monoclinic. **Habit** Crystals tabular pseudohexagonal, rarely prismatic; also as scaly aggregates; and massive, earthy. **SG** 2·6–3·3 (increasing with iron content) **Hardness** 2–3 **Cleavage** Basal, perfect. Cleavage flakes are flexible but inelastic. **Colour and transparency** Green; also yellow, brown; translucent to subtranslucent. **Lustre** Vitreous; earthy in fine-grained masses. **Distinguishing features** Green colour and inelastic cleavage fragments distinguish chlorites from micas. **Occurrence** Chlorite occurs in igneous rocks as an alteration product of such minerals as pyroxenes, amphiboles and micas. It also occurs infilling amygdales in lavas. It is characteristic of low grade metamorphic rocks and also occurs in sediments.

Serpentine group $Mg_3Si_2O_5(OH)_4$ **Crystal system** Monoclinic. **Habit** Serpentine occurs mainly as fibrous chrysotile, the most valued type of asbestos, and as lamellar or platy antigorite. **SG** 2·5–2·6 **Hardness** Variable 2½–4 **Cleavage** Basal, perfect (antigorite); none in fibrous chrysotile. **Fracture** Conchoidal, splintery. **Colour and transparency** Various shades of green; also brownish, grey, white or yellow: translucent to opaque. **Lustre** Waxy or greasy; fibrous varieties silky; massive varieties earthy. **Distinguishing features** Green colour, lustre, smooth, rather greasy feel, fibrous habit (antigorite). **Occurrence** Serpentine is a secondary mineral formed from minerals such as olivine and orthopyroxene. It occurs in igneous rocks containing these minerals but typically in serpentinites, which have formed by the alteration of olivine-bearing rocks.

Vermiculite $Mg_3(Al,Si)_4O_{10}(OH)_2.4H_2O$ **Crystal system** Monoclinic. **Habit** Crystals platy. **SG** About 2·3 **Hardness** About 1½ **Cleavage** Basal, perfect. **Colour and transparency** Yellow, brown: translucent. **Lustre** Pearly, sometimes bronzy. **Distinguishing features** Expands greatly, perpendicular to cleavage, on heating. **Occurrence** Vermiculite occurs as an alteration product of magnesian micas, in association with carbonatites.

Kaolinite group $Al_2Si_2O_5(OH)_4$ **Crystal system** Triclinic or monoclinic. **Habit** Microscopic hexagonal plates; usually white earthy masses. **SG** 2·6–2·7 **Hardness** 2–2½ (much less when massive) **Cleavage** Basal, perfect. **Colour** White, sometimes greyish or stained brown or red. **Lustre** Dull, earthy; crystalline plates pearly. **Distinguishing features** Plastic feel. Kaolinite cannot be distinguished from other clay minerals without optical or other tests. **Occurrence** Kaolinite is a secondary mineral produced by the alteration of aluminous silicates, and particularly of alkali feldspars.

Chlorite
Serpentine (antigorite)
Serpentine
(chrysotile)
Serpentine
Vermiculite
Kaolinite
Vermiculite
5 cm

Talc $Mg_3Si_4O_{10}(OH)_2$ **Crystal system** Monoclinic. **Habit** Crystals rare; usually as granular or foliated masses. **SG** 2·6–2·8 **Hardness** 1 **Cleavage** Basal, perfect. **Colour and transparency** White, grey or pale green; often stained reddish: translucent. **Streak** White to very pale green. **Lustre** Dull, pearly on cleavage surface. **Distinguishing features** Extreme softness, soapy feel, greenish white colour. **Occurrence** Talc occurs as a secondary mineral formed as a result of the alteration of olivine, pyroxene and amphibole, and it occurs along faults in magnesium-rich rocks. Talc also occurs in schists produced by low or medium grade metamorphism of magnesian rocks, often in association with actinolite. Massive talc is called steatite or soapstone. It occurs rather less frequently as a result of thermal metamorphism of dolomitic limestones.

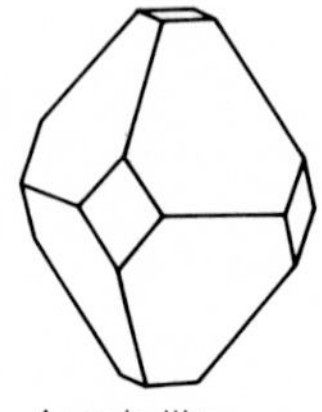

Apophyllite: combination of prism, bipyramid and pinacoid

Apophyllite $KFCa_4Si_8O_{20}.8H_2O$ **Crystal system** Tetragonal. **Habit** Crystals are of varied habit; combinations of prism, bipyramid and pinacoid are most common. **SG** 2·3–2·4 **Hardness** 4½–5 **Cleavage** Basal, perfect; prismatic, poor. **Fracture** Uneven. **Colour and transparency** Colourless, white or greyish; sometimes pinkish or yellowish: transparent to translucent. **Lustre** Pearly parallel to cleavage; elsewhere vitreous. **Distinguishing features** Crystal form, basal cleavage and pearly lustre parallel to it. The basal pinacoid faces are often rough and pitted and contrast with other smooth, bright faces. **Occurrence** Apophyllite occurs in association with zeolites in cavities in basalt and limestone. It also occurs in some hydrothermal mineral veins.

Prehnite $Ca_2Al_2Si_3O_{10}(OH)_2$ **Crystal system** Orthorhombic. **Habit** Crystals rare, tabular; usually in globular and reniform masses with a fibrous structure. **SG** 2·9–3·0 **Hardness** 6–6½ **Cleavage** Basal, good. **Fracture** Uneven. **Colour and transparency** Usually pale watery green; also grey, yellow or white: transparent to translucent. **Lustre** Vitreous. **Distinguishing features** Green colour, habit. **Occurrence** Prehnite occurs most commonly in veins and cavities in igneous rocks, often in association with zeolites. It occurs in very low grade metamorphic rocks, and as a product of the decomposition of plagioclase feldspar. It is named after Col von Prehn, who discovered the mineral at the Cape of Good Hope, South Africa.

Talc
Talc
Apophyllite
Apophyllite
Prehnite
Prehnite
5 cm

Silica group In this group are included those minerals whose composition does not depart significantly from SiO_2. Some varieties are crystalline and include quartz, tridymite and cristobalite; others, generally grouped as chalcedony, are cryptocrystalline. Opal is amorphous.

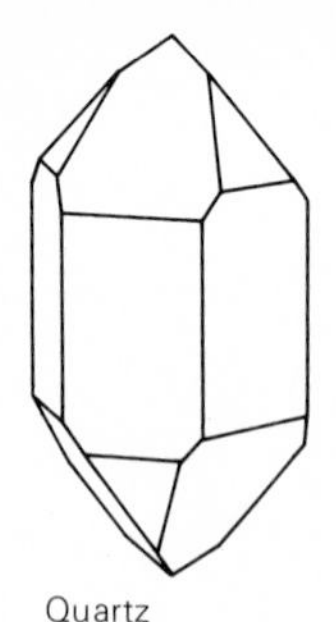
Quartz

Quartz SiO_2 **Crystal system** Trigonal. **Habit** Crystals are usually six-sided prisms and are terminated by six faces. The prism faces are often striated at right-angles to the length of the crystal. Imperfectly developed crystals are common. Right- and left-handed forms can be recognized by the presence of small additional faces. **Twinning** Most quartzes are twinned but twinning is only occasionally observable in crystals. The most common types are Dauphiné twins (double right-handed or double left-handed crystals), Brazil twins (combined right- and left-handed crystals), and Japan twins (contact twins in which the two individuals are nearly at right-angles). **SG** 2·65 **Hardness** 7 **Cleavage** None. **Fracture** Conchoidal. **Colour and transparency** Commonly colourless or white, but the range of colour is very wide (see below): transparent to translucent. **Lustre** Vitreous. **Varieties** Quartz occurs in many varieties. Rock crystal is colourless quartz and sub-varieties include ghost quartz in which growth stages are marked by inclusions, and rutilated quartz (sagenite), which contains hair-like rods of rutile. Amethyst is purple; milk quartz is white; rose quartz is rose-red or pink, and is usually found massive rather than as crystals. Citrine is yellow and transparent and resembles topaz. Smoky quartz (sometimes called cairngorm) is smoky brown to nearly black. Some quartzes contain impurities that not only impart a colour but render them opaque. Ferruginous quartz is an example of this and is commonly brick-red or yellow. **Distinguishing features** Crystal form, conchoidal fracture, vitreous lustre, hardness. **Occurrence** Quartz is one of the most widely distributed minerals. It occurs in many igneous and metamorphic rocks, particularly in granite and gneiss, and it is abundant in clastic sediments. It is virtually the sole constituent of quartzite. Quartz is also a common gangue mineral in mineral veins, and most good crystals are obtained from this type of occurrence. Well formed quartz crystals can be obtained from cavities (geodes), from granite porphyries, and from granite pegmatites.

Quartz: showing striated prism faces

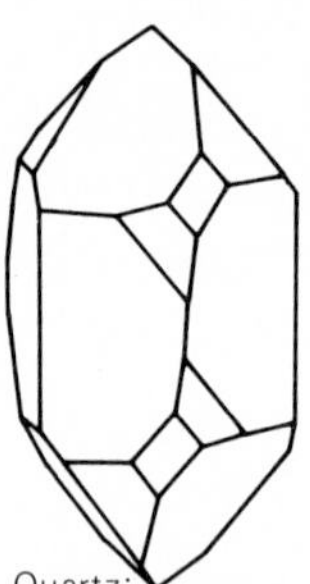
Quartz: right-handed form

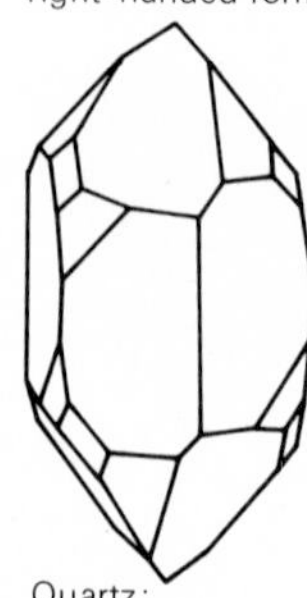
Quartz: left-handed form

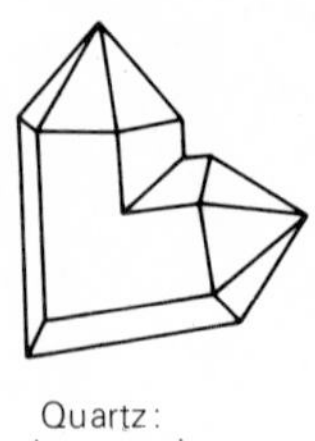
Quartz: Japan twin

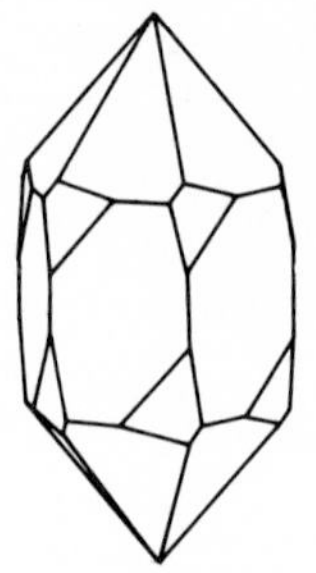
Quartz: Dauphiné twin

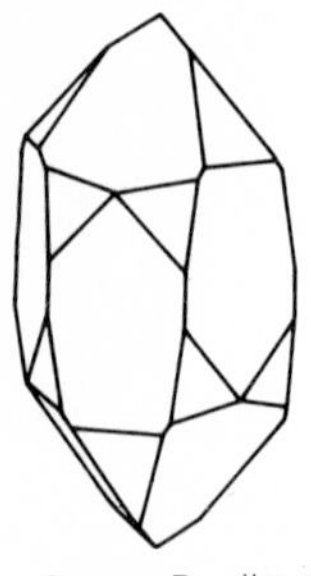
Quartz: Brazil twin

5 cm
Milky quartz
Rock crystal
Rutilated quartz
Citrine
Rose quartz
Amethyst
Smoky quartz

Chalcedony SiO_2 Chalcedony is the name given to compact varieties of silica which comprise minute quartz crystals with sub-microscopic pores. There are two main varieties: chalcedony, which is uniformly coloured, and agate which is characterized by curved bands or zones of differing colour. **Habit** Often mamillated, botryoidal or stalactitic. Chalcedony commonly has a banded structure, which is not always obvious to the naked eye. It often lines cavities in rocks, and is also massive or nodular. **SG** About 2·6 **Hardness** About $6\frac{1}{2}$ **Cleavage** None. **Fracture** Conchoidal. **Colour and transparency** Variable from white, through grey, red, brown, to black (see below): transparent to subtranslucent. **Lustre** Vitreous to waxy. **Distinguishing features** Occurrence and habit, greater density than opal. **Varieties** Various names are applied to the different coloured varieties of chalcedony. Carnelian is red to reddish brown and grades into sard which is light to dark brown; chrysoprase is apple-green and heliotrope, also called blood stone, is green with red spots which resemble spots of blood. Jasper is opaque chalcedony and is generally red but yellow, brown, green and grey-blue varieties occur. Jasper is rarely uniformly coloured, the colour is often distributed in spots or bands. Moss agate consists of a translucent, milky white, bluish white to nearly colourless matrix containing irregularly distributed green, brown, or black moss-like, dendritic impurities of manganese oxide. These often assume attractive figured shapes, and in mocha stone the fern-like branching forms have led to its use in making cameos and other decorative objects. Flint and chert are opaque chalcedony, usually dull grey to black in colour, and which break with a pronounced conchoidal fracture, giving sharp edges. This property was exploited by early man in the fashioning of flint and chert implements. The name chert is used to describe bedded, massive chalcedony, and the name flint is reserved for the black, nodular variety commonly found in chalk. **Occurrence** Chalcedony is precipitated from silica-bearing solutions and hence forms cavity linings, veins and replacive masses in a variety of rocks. Chert and flint may originate either by the deposition of silica on the sea floor, or by the replacement of rocks, notably limestone, by silica from percolating waters.

Chalcedony
Chalcedony
Carnelian
Carnelian
Sard
Chrysoprase
Jasper
Moss agate
Mocha stone
5 cm

Agate SiO_2 Agate is a form of chalcedony which is characterized by bands or zones which differ in colour. **Habit** Agate usually forms concentric or irregular layers usually lining a cavity. **SG** About 2·6 **Hardness** About $6\frac{1}{2}$ **Cleavage** None. **Fracture** Conchoidal. **Colour** The bands are usually variegated in shades of white, milky white or grey; also shades of green, brown, red or black. Commercial agate is often coloured artificially. Onyx is a form of agate with straight, parallel bands that is used particularly for making cameo brooches. **Occurrence** Agate is widely distributed and occurs typically as a cavity filling in lavas. The layering often follows the form of the cavity and gives place inwards to crystals of quartz.

Opal $SiO_2.nH_2O$ **Crystal system** None: amorphous. **Habit** Massive; often as stalactitic, botryoidal and rounded forms; also as veinlets. **SG** Variable, 1·8–2·3 **Hardness** $5\frac{1}{2}$–$6\frac{1}{2}$ **Cleavage** None. **Fracture** Conchoidal. **Colour and transparency** Variable, from colourless, through milky white, grey, red, brown, blue, green to nearly black. Pale coloured forms are common. Transparent to subtranslucent. **Lustre** Vitreous to resinous; sometimes pearly. **Distinguishing features** Form, low density. Opal resembles chalcedony in its occurrence, but is less dense and less hard. **Varieties** Opal is a solidified gel with a variable amount of water, usually about 6–10 per cent. Precious opal has a milky white and sometimes black body colour and exhibits a brilliant play of colours usually in blues, reds and yellows. Fire opal is a variety in which red and yellow colours are dominant and produce flame-like reflections when turned. Hyalite is colourless, botryoidal opal; wood opal is wood that has been replaced in part by opaline silica; common opal is the translucent, pale variety that is variously coloured but lacks the play of colours of precious opal; and hydrophane is a variety which becomes transparent when immersed in water. Siliceous sinter and geyserite are opaline deposits formed around geysers or by precipitation from hot waters. They generally form stalactitic and delicately filamentous forms of various colours. **Occurrence** Opal is deposited at low temperatures from silica-bearing waters. It can occur as a fissure filling in rocks of any kind, but occurs especially in areas of geysers and hot springs. It can also be formed during the weathering and decomposition of rocks. Opal forms the skeletons of organisms such as sponges, radiolaria and diatoms. Diatomite, or diatomaceous earth, is a fine-grained sedimentary rock of friable, chalky appearance that is made up in large part of the skeletons of such organisms. The name opal comes from Sanskrit and means 'gem' or 'precious stone'.

Agate
Agate
Onyx
Agate
Opal
Wood opal
Opal
5 cm

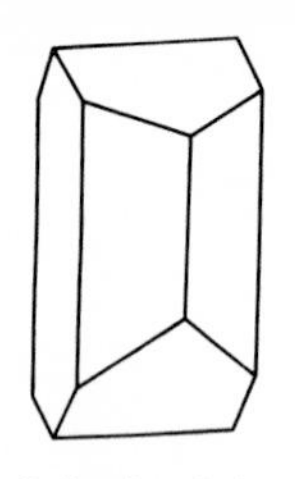

Orthoclase/microcline: prismatic habit

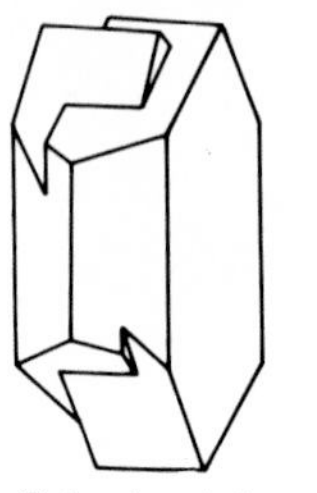

Orthoclase/microcline: Carlsbad twin

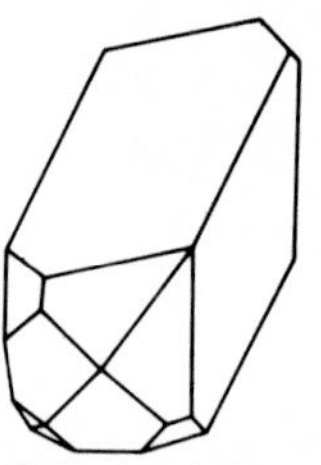

Orthoclase/microcline: Baveno twin

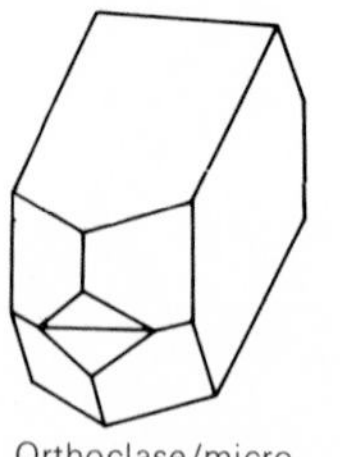

Orthoclase/microcline: Manebach twin

Feldspar group The feldspars are the most abundant of all minerals and are widely distributed in igneous, metamorphic and sedimentary rocks. They have the general formula $X(Al,Si)_4O_8$ in which X is Na, K, Ca or Ba. The feldspars may be grouped into the potassic feldspars and the plagioclase feldspars. In the plagioclases Na and Ca can substitute one for the other. Twinning is common. Carlsbad, Manebach and Baveno twins are simple twins; and albite and pericline twins are repeated twins. In hand specimen twinning shows as a difference in reflectivity of two halves of a crystal in the case of a simple twin, or as a series of parallel striations of different reflectivity in a repeated twin. Albite twinning is common in the plagioclases. **Alteration** Potassic feldspar alters readily to clay minerals, mainly kaolinite, and the plagioclases usually alter to clay minerals or 'sericite'.

Potassic feldspars: Sanidine, Orthoclase and Microcline $KAlSi_3O_8$ **Crystal system** Monoclinic (sanidine and orthoclase); triclinic (microcline). **Habit** Sanidine crystals are usually tabular or prismatic. Orthoclase and microcline are sometimes prismatic, and may be of square cross-section (Baveno habit). **Twinning** Common, on Carlsbad, Baveno or Manebach laws. Microcline also shows repeated twinning on a combination of albite and pericline laws, but this is best seen with a microscope. **SG** 2·5–2·6 **Hardness** 6–6½ **Cleavage** Two perfect cleavages. **Fracture** Conchoidal to uneven. **Colour and transparency** Sanidine is colourless to grey: transparent. Orthoclase is white to flesh-pink, occasionally red. Microcline is similar to orthoclase, but green varieties are called amazonstone. Both are translucent to subtranslucent. **Lustre** Vitreous; rather pearly parallel to cleavage. **Distinguishing features** Orthoclase and microcline are distinguished from other minerals by their colour, cleavages and hardness; they are difficult to distinguish one from the other, though green amazonstone is distinctive. Sanidine is distinguished by its colourless, transparent appearance, tabular habit and occurrence. **Occurrence** Sanidine is the high-temperature form of $KAlSi_3O_8$ and it occurs as phenocrysts in volcanic rocks such as rhyolite and trachyte. It also occurs in rocks that have been thermally metamorphosed at high temperature. Orthoclase is the common potassic feldspar of most igneous and metamorphic rocks, and microcline, the low-temperature variety, occurs in granites, granite pegmatites, hydrothermal veins, and in many schists and gneisses. Like orthoclase, it is found also as grains in sedimentary rocks. Perthite is a potassic feldspar which contains laminae or patches of albite.

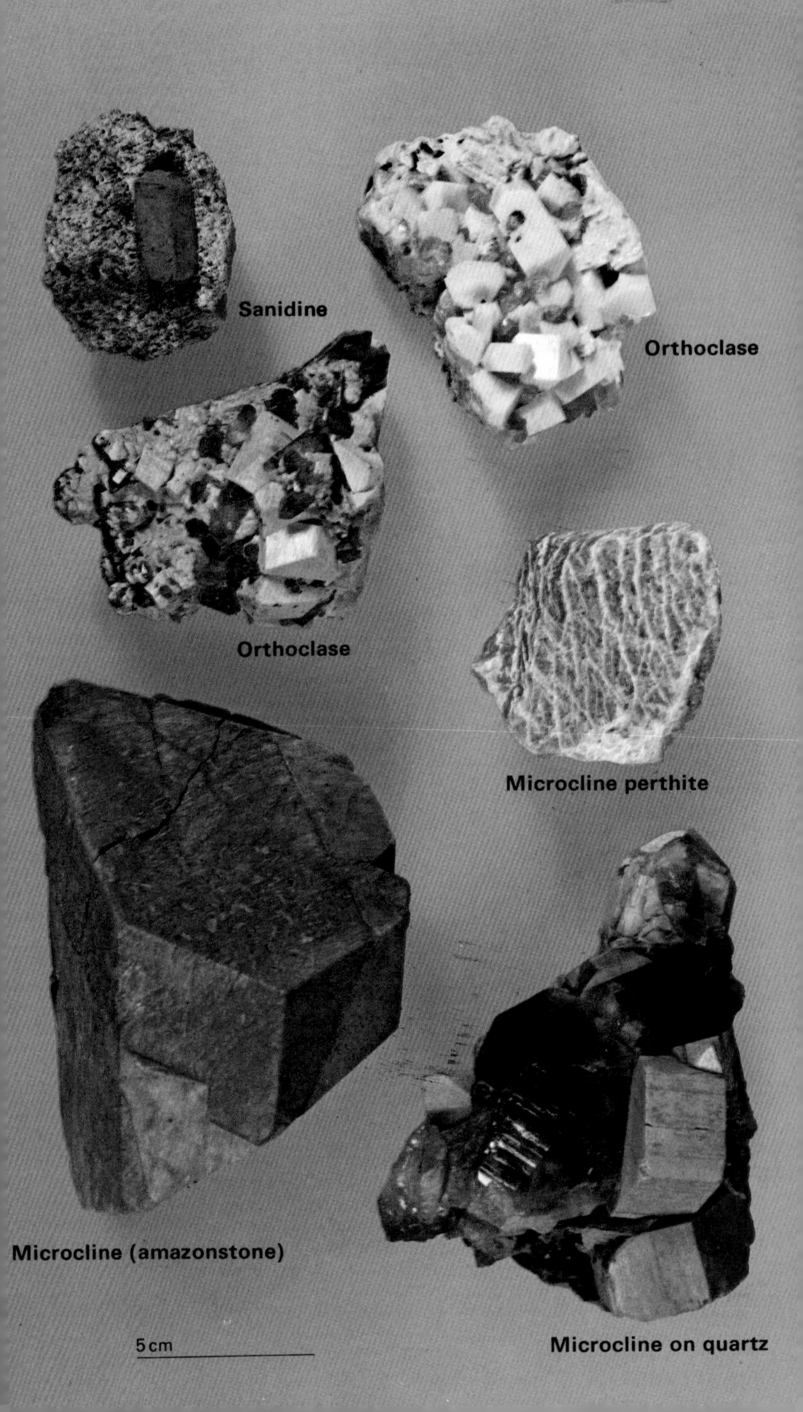
Sanidine
Orthoclase
Orthoclase
Microcline perthite
Microcline (amazonstone)
5 cm
Microcline on quartz

Potassic feldspars: Adularia $KAlSi_3O_8$ **Crystal system** Monoclinic. **Habit** Distinctive simple crystals, usually a combination of prism terminated by two faces. **Twinning** Baveno twins common. **SG** 2·6 **Hardness** 6 **Cleavage** Two perfect cleavages. **Fracture** Conchoidal to uneven. **Colour and transparency** Colourless or milky white, often with a pearly sheen or play of colours (moonstone): transparent to translucent. **Lustre** Vitreous. **Distinguishing features** Simple habit, occurrence. **Occurrence** Adularia is formed at low temperatures and occurs in hydrothermal veins.

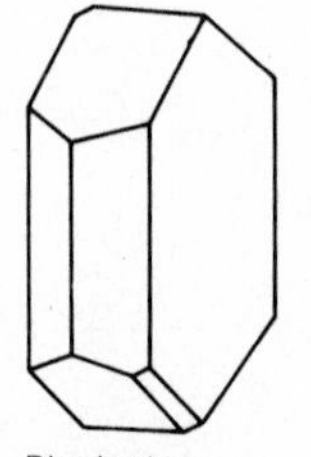
Plagioclase

Plagioclase: repeated albite twinning

Plagioclase $NaAlSi_3O_8$–$CaAl_2Si_2O_8$ **Crystal system** Triclinic. **Habit** Crystals prismatic or tabular; also massive, granular. **Twinning** Repeated twinning is common on albite and pericline laws, as are simple twins on Carlsbad, Baveno and Manebach laws. Both simple and repeated twinning may be shown by one individual. **SG** 2·6–2·8 **Hardness** 6–6½ **Cleavage** Two good cleavages. **Fracture** Uneven. **Colour and transparency** Usually white or off-white; sometimes pink, greenish or brownish: transparent to translucent. **Lustre** Vitreous, sometimes pearly on cleavage surfaces. **Distinguishing features** Plagioclases are most readily distinguished from potassic feldspars by the presence of repeated albite twin lamellae visible on one of the cleavage surfaces. The chemistry changes progressively from albite ($NaAlSi_3O_8$) through oligoclase, andesine, labradorite and bytownite, to anorthite ($CaAl_2Si_2O_8$). Individual members of the plagioclase series are difficult to distinguish without a microscope, but labradorite often shows a spectacular play of colours in blues and greens from its cleavage surfaces. For this reason it is often polished and used as a decorative stone. **Occurrence** The plagioclases are widely distributed minerals. They occur in many igneous rocks and are used as a basis of rock classification. In general the sodic plagioclases are characteristic of granitic igneous rocks and give place to more calcic plagioclases in basalts and gabbros. Between potassic feldspar and albite there exists a continuous series as sodium substitutes for potassium; this series is called the alkali feldspar series. In some layered gabbros, plagioclase (usually labradorite, or bytownite) forms layers that are virtually free from other minerals, and in places there are large masses of oligoclase-andesine rock called anorthosites. Albite is commonly found in pegmatites and in sodic lavas called spilites. Plagioclase is common in metamorphic rocks, in which it often lacks repeated twinning, and it occurs as detrital grains in sedimentary rocks. Calcic plagioclase occurs in meteorites and in lunar rocks.

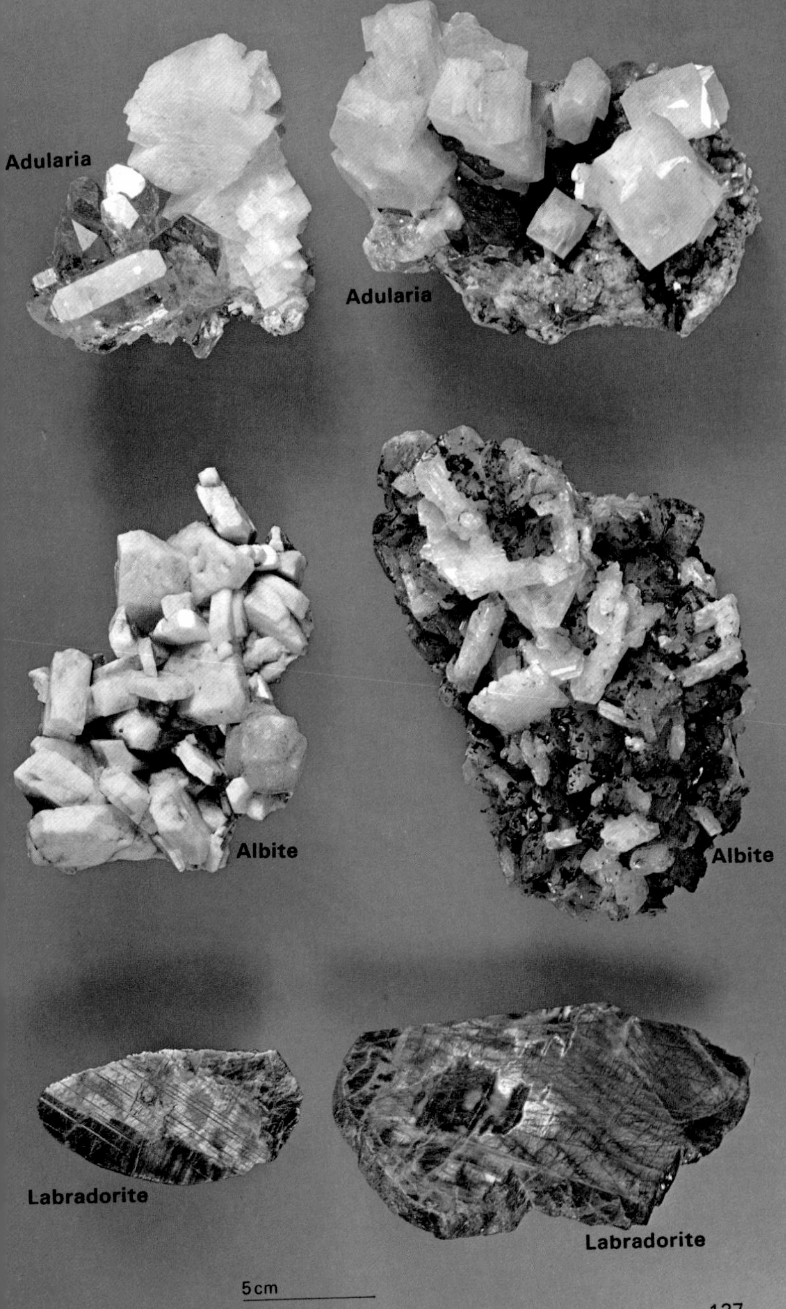
Adularia
Adularia
Albite
Albite
Labradorite
Labradorite
5 cm

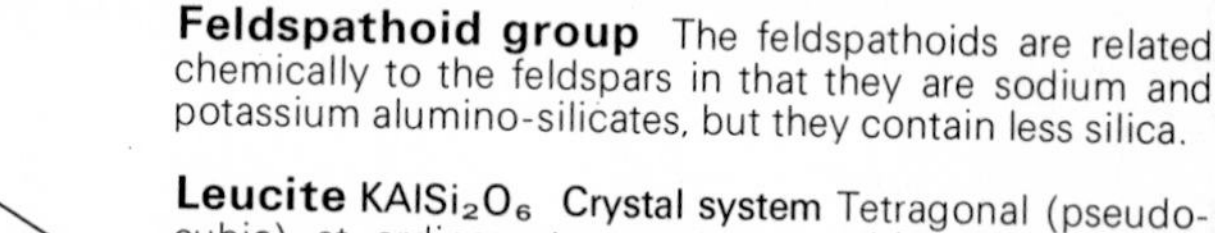

Feldspathoid group The feldspathoids are related chemically to the feldspars in that they are sodium and potassium alumino-silicates, but they contain less silica.

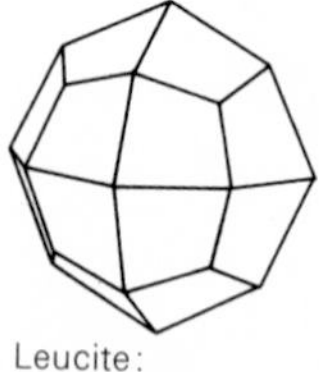

Leucite:
icositetrahedron

Leucite $KAlSi_2O_6$ **Crystal system** Tetragonal (pseudo-cubic) at ordinary temperatures; cubic above 625°C. **Habit** Crystals nearly always icositetrahedra. **SG** 2·5 **Hardness** 5½–6 **Cleavage** Very poor. **Fracture** Conchoidal. **Colour and transparency** Usually white or grey: translucent. **Lustre** Vitreous to dull. **Distinguishing features** Crystal form, occurrence. Analcime and garnet also crystallize as icositetrahedra, but analcime occurs typically in cavities, and garnet is not white or grey. **Alteration** Leucite may alter to pseudoleucite, a mixture of orthoclase and nepheline. **Occurrence** Leucite does not occur in association with quartz, and is unstable at high pressures, and so it has a restricted occurrence. It occurs typically in potassium-rich, silica-poor lavas such as certain trachytes. Fresh leucite does not occur in plutonic igneous rocks, nor in metamorphic rocks. The name comes from a Greek word meaning 'white'.

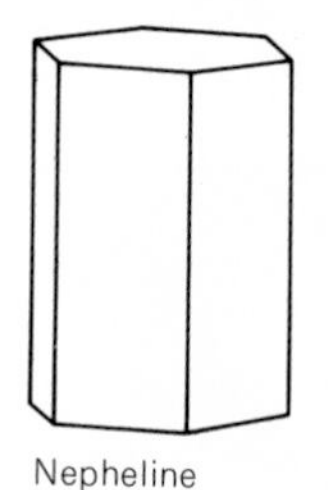

Nepheline

Nepheline $NaAlSiO_4$ **Crystal system** Hexagonal. **Habit** Crystals usually six-sided prisms; also massive and as discrete grains. **SG** 2·6–2·7 **Hardness** 5½–6 **Cleavage** Prismatic, basal, poor. **Fracture** Conchoidal. **Colour and transparency** Usually colourless, white or grey; but also brownish red or greenish: transparent to translucent. **Lustre** Greasy to vitreous. **Distinguishing features** Greasy lustre, gelatinizes in hydrochloric acid. **Occurrence** Nepheline is characteristic of silica-poor alkaline igneous rocks of both plutonic and volcanic associations. It is found, therefore, in nepheline syenites and ijolites and in lavas such as phonolite. The name comes from the Greek word for a cloud, and alludes to its becoming cloudy when placed in acid.

Cancrinite $(Na,Ca)_7Al_6Si_6O_{24}(CO_3,SO_4,Cl)_{1.5-2}\cdot 1-5H_2O$ **Crystal system** Hexagonal. **Habit** Crystals rare but usually prismatic; generally massive or as discrete grains and as veinlets. **SG** 2·4–2·5 **Hardness** 5–6 **Cleavage** Prismatic, perfect. **Colour and transparency** White, grey, yellow, blue: transparent to translucent. **Lustre** Vitreous, inclined to greasy. **Distinguishing features** Colour, occurrence. **Occurrence** Cancrinite is of restricted occurrence, being found typically in nepheline syenites and associated silica-poor alkaline rocks. It occurs in some carbonatites and in some contact metamorphosed limestones. It is named after Count G Cancrin (1774–1845), a Russian Finance Minister.

Leucite
Leucite
Nepheline
Nepheline
Cancrinite
Cancrinite
5 cm

Sodalite $Na_8Al_6Si_6O_{24}Cl_2$ **Crystal system** Cubic. **Habit** Crystals rare, usually rhombdodecahedral; commonly massive, granular. **SG** 2·3 **Hardness** 5½–6 **Cleavage** Rhombdodecahedral, poor. **Fracture** Conchoidal to uneven. **Colour and transparency** Commonly azure-blue; also pink, yellow, green or grey-white: transparent to translucent. **Lustre** Vitreous. **Distinguishing features** Blue colour; distinguished from lazurite by its occurrence, and by absence of associated pyrite. Often shows reddish fluorescence in ultra-violet light. **Occurrence** Sodalite occurs with nepheline and cancrinite in alkaline igneous rocks such as nepheline syenites and also in some silica-poor dyke rocks and lavas.

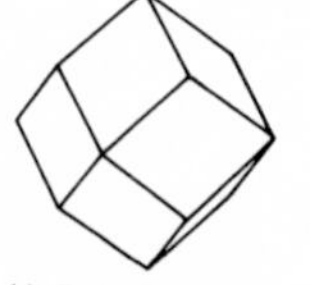

Haüyne/nosean: rhombdodecahedron

Haüyne $(Na,Ca)_{4-8}Al_6Si_6O_{24}(SO_4)_{1-2}$; **Nosean** (Noselite) $Na_8Al_6(SiO_4)_6SO_4$ **Crystal system** Cubic. **Habit** Crystals rhombdodecahedra or octahedra; also as discrete grains. **SG** Haüyne 2·4–2·5: nosean 2·3–2·4 **Hardness** 5½–6 **Cleavage** Rhombdodecahedral, poor. **Fracture** Uneven. **Colour and transparency** Often blue, also grey, brown, yellow-green: transparent to translucent. **Lustre** Vitreous, inclined to greasy. **Distinguishing features** Blue colour, association; haüyne, nosean and sodalite are very similar. **Occurrence** Haüyne and nosean both occur typically in silica-poor lavas such as phonolites. Haüyne is named after RJ Haüy (1743–1822), a French mineralogist. Nosean is named for KW Nose (1753–1835), a German mineralogist.

Lazurite $(Na,Ca)_8(Al,Si)_{12}O_{24}(S,SO_4)$ **Crystal system** Cubic. **Habit** Crystals rare; cubes or octahedra; commonly massive. **SG** 2·4 **Hardness** 5–5½ **Fracture** Uneven. **Colour and transparency** Azure-blue: translucent. **Lustre** Vitreous. **Distinguishing features** Colour, association with pyrite and calcite. **Occurrence** Lazurite is similar in composition to sodalite, nosean and haüyne. Lapis-lazuli is a rock rich in lazurite, and is used for jewellery and as a decorative stone. Lapis-lazuli is a contact metamorphosed limestone. Powdered lazurite was once the source of the pigment ultramarine.

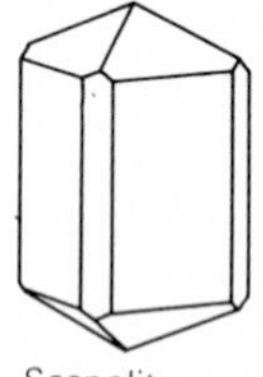

Scapolite

Scapolite group $(Na,Ca,K)_4Al_3(Al,Si)_3Si_6O_{24}(Cl,F,OH,CO_3,SO_4)$ **Crystal system** Tetragonal. **Habit** Crystals prismatic often with uneven faces: usually massive, granular. **SG** 2·5–2·8 (increasing with calcium content) **Hardness** 5–6 **Cleavage** Prismatic, in two sets, good; imparts a splintery appearance to massive scapolite. **Fracture** Subconchoidal. **Colour and transparency** Usually white or bluish grey; also pink, yellow or brownish: transparent to translucent. **Lustre** Vitreous to pearly. **Distinguishing features** Scapolites vary between a sodic end member marialite and a calcic end member meionite. The blocky appearance, pale blue-grey colour and splintery, fibrous cleavage are useful features in identification. **Occurrence** Scapolite occurs in metamorphic rocks, particularly in metamorphosed limestones. The minerals also occur in skarns close to igneous contacts and as a replacement of feldspars in altered igneous rocks.

Haüyne
Sodalite
Lazurite
Lazurite
Scapolite
5cm
Scapolite

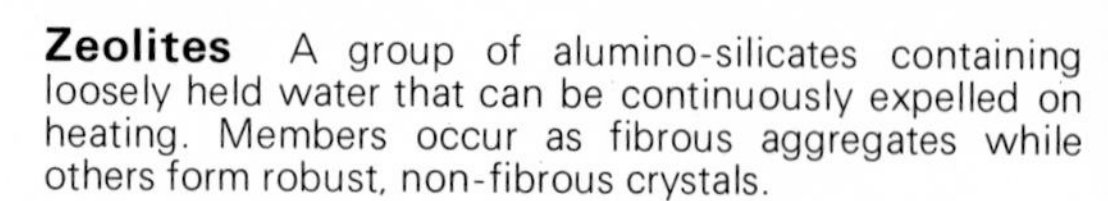

Zeolites A group of alumino-silicates containing loosely held water that can be continuously expelled on heating. Members occur as fibrous aggregates while others form robust, non-fibrous crystals.

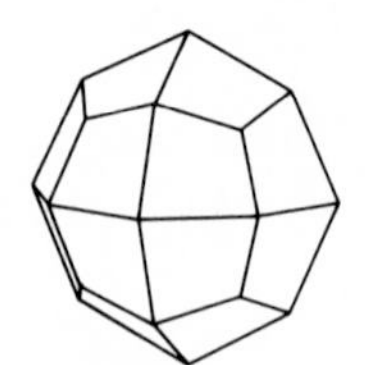

Analcime: icositetrahedron

Analcime (Analcite) $NaAlSi_2O_6.H_2O$ **Crystal system** Cubic. **Habit** Crystals usually icositetrahedral; also massive. **SG** 2·2–2·3 **Hardness** 5½ **Cleavage** Cubic, very poor. **Fracture** Subconchoidal. **Colour and transparency** Colourless, white or grey; sometimes tinged with pink or yellow: transparent to subtranslucent. **Lustre** Vitreous. **Distinguishing features** The crystal form is the same as that of leucite, from which it is distinguished by its occurrence. **Occurrence** Occurs mainly with other zeolites as a secondary mineral in cavities in basaltic rocks and in sedimentary rocks as a secondary mineral.

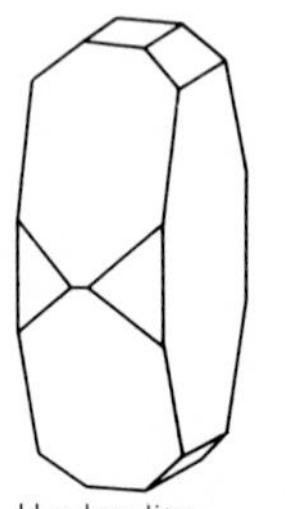

Heulandite

Heulandite $(Na,Ca)_{4-6}Al_6(Al,Si)_4Si_{26}O_{72}24H_2O$ **Crystal system** Triclinic, pseudo-monoclinic. **Habit** Crystals usually tabular and pseudo-orthorhombic. **SG** 2·1–2·2 **Hardness** 3½–4 **Cleavage** One perfect cleavage. **Fracture** Uneven. **Colour and transparency** White, pink, red or brown: transparent to translucent. **Lustre** Vitreous; pearly parallel to cleavage. **Distinguishing features** Tabular, coffin-shaped crystals; pearly lustre parallel to cleavage, vitreous elsewhere. **Occurrence** Occurs with stilbite in cavities in basaltic rocks, and in sedimentary rocks as a secondary mineral.

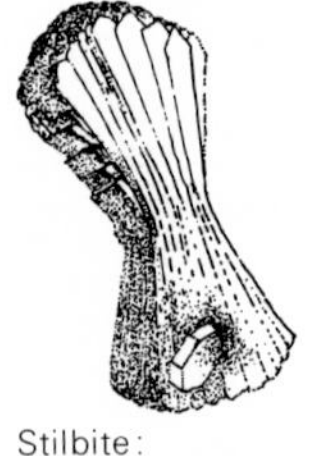

Stilbite: sheaf-like aggregate

Stilbite $NaCa_2(Al_5Si_{13})O_{36}.14H_2O$ **Crystal system** Monoclinic. **Habit** Forms sheaf-like aggregates of twinned crystals. **Twinning** Common, giving cruciform interpenetrant twins. **SG** 2·1–2·2 **Hardness** 3½–4 **Cleavage** One perfect cleavage. **Fracture** Uneven. **Colour and transparency** White; sometimes yellowish or pink; occasionally brick-red: transparent to translucent. **Lustre** Vitreous; pearly on cleavage surfaces. **Distinguishing features** Sheaf-like form, pearly lustre parallel to cleavage, vitreous elsewhere. **Occurrence** In cavities in basalts, often with heulandite.

Harmotome $BaAl_2Si_6O_{16}.6H_2O$ **Crystal system** Monoclinic. **Habit** Crystals usually twins, having a pseudo-orthorhombic or pseudo-tetragonal appearance. **Twinning** Very common, interpenetrant. **SG** 2·4–2·5 **Hardness** 4½ **Cleavage** One good cleavage. **Fracture** Uneven. **Colour and transparency** White; also yellowish or reddish: subtransparent to translucent. **Lustre** Vitreous. **Distinguishing features** Crystal form, occurrence. **Occurrence** In cavities in basalts, often with chabazite.

Chabazite rhombohedral habit

Chabazite $(Ca,Na_2)Al_2Si_4O_{12}.6H_2O$ **Crystal system** Trigonal. **Habit** Crystals rhombohedral but look like cubes. **Twinning** Common: interpenetrant. **SG** 2·0–2·1 **Hardness** 4½ **Cleavage** Rhombohedral, poor. **Fracture** Uneven. **Colour and transparency** Usually white, yellow; often pinkish or red: transparent to translucent. **Lustre** Vitreous. **Distinguishing features** Rhombohedral form. Unlike calcite, does not effervesce with dilute hydrochloric acid. **Occurrence** In cavities in basalts.

Analcime
Heulandite
Stilbite
Stilbite
Chabazite
Chabazite
Harmotome
5cm

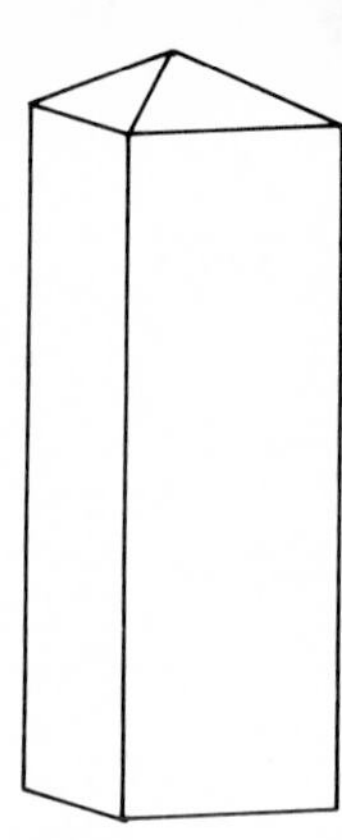
Natrolite

Natrolite $Na_2Al_2Si_3O_{10}.2H_2O$ **Crystal system** Orthorhombic, pseudo-tetragonal. **Habit** Acicular crystals, frequently arranged as divergent or radiating aggregates. **SG** 2·2–2·3 **Hardness** 5 **Fracture** Uneven. **Colour and transparency** Colourless or white: transparent to translucent. **Lustre** Vitreous. **Distinguishing features** Fibrous habit. **Occurrence** Natrolite occurs typically in the cavities of basaltic and other igneous rocks. Mesolite and scolecite are fibrous zeolites of similar composition and occurrence to natrolite. They are monoclinic but mesolite is pseudo-orthorhombic and scolecite is pseudo-tetragonal. It is difficult to distinguish between them in hand specimen.

Thomsonite $NaCa_2(Al,Si)_{10}O_{20}.6H_2O$ **Crystal system** Orthorhombic, pseudo-tetragonal. **Habit** Acicular crystals in radiating or divergent aggregates. **SG** 2·1–2·4 **Hardness** 5–5½ **Cleavage** Two good cleavages. **Fracture** Uneven. **Colour and transparency** White, sometimes tinged with red: transparent to translucent. **Lustre** Vitreous to pearly. **Distinguishing features** Similar to natrolite but usually more coarsely crystalline. Difficult to distinguish in hand specimen from other fibrous zeolites. **Occurrence** Thomsonite, like other zeolites, occurs in cavities in lavas, and as a decomposition product of nepheline.

Laumontite $CaAl_2Si_4O_{12}.4H_2O$ **Crystal system** Monoclinic. **Habit** As small prismatic crystals often with oblique terminations; also massive, or as columnar and radiating aggregates. **SG** 2·2–2·3 **Hardness** 3–3½ **Cleavage** Prismatic, pinacoidal; good. **Fracture** Uneven. **Colour and transparency** White; sometimes reddish: transparent to translucent. **Lustre** Vitreous; pearly on cleavage surfaces. **Distinguishing features** Laumontite loses part of its water on exposure to dry air and becomes powdery, friable and chalky, when it is known as leonhardite. **Occurrence** Laumontite occurs with other zeolites in veins and amygdales in igneous rocks. It is produced as a result of very low grade metamorphism of some sedimentary rocks and tuffs. Laumontite is named after G Laumont, who discovered the mineral.

Natrolite
Natrolite
Mesolite
Scolecite
Thomsonite
Laumontite
5 cm

Rocks

The Earth, and indeed the Moon and planets, are built of the material we call rock. The solid stuff of mountains, the loose sand and gravel of beaches and deserts are all rocks. Rocks are aggregates of minerals, but the petrologist (petrology is the science of rocks), as well as being interested in the mineralogy of rocks, also tries to unravel the record of the geological past which they contain. It is from reading the 'record of the rocks' that so much has been learned about past climates and geography, and about the past and present composition of, and the conditions which prevail within the interior of our planet.

Rocks can be conveniently grouped into *igneous* rocks, *metamorphic* rocks and *sedimentary* rocks. Igneous rocks are formed by the solidification of molten rock material; metamorphic rocks are formed through the alteration of igneous and sedimentary rocks by heat and pressure; while sedimentary rocks are produced by the accumulation of rock waste at the Earth's surface.

Igneous rocks

The so-called *crust* (Fig. 1) of the Earth is about 35 km thick under the

Fig. 1 Section through the Earth's crust

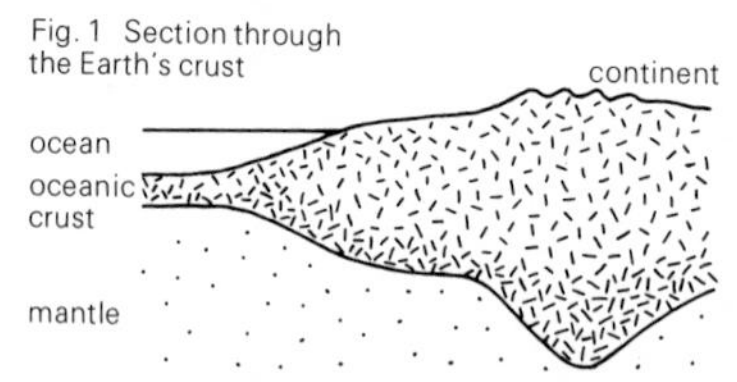

continents but averages only some 7 km beneath the oceans. It is formed mainly of rocks of relatively low density. Beneath the crust there is a layer of denser rock called the *mantle* which extends down to a depth of nearly 3,000 km. Much of the molten rock material which goes to make up the igneous rocks is generated within the upper parts of the mantle. This material, which is called *magma*, migrates upwards into the Earth's crust and forms rock masses which are known as *igneous intrusions*. If magma reaches the Earth's surface and flows out over it, it is called *lava*.

Within some lavas, fragments of dense, green-coloured rocks are sometimes found which consist principally of olivine and pyroxene. These fragments are thought to represent pieces of the mantle, carried upwards by the migrating magma.

The overwhelming majority of lavas consist of the black, rather dense rock called basalt, and most petrologists consider that the primary molten rock material which comes from the mantle has a composition which is near to that of basalt. Although basalt is the most abundant of the lavas, granite is by far the commonest of the intrusive igneous rocks. Granite is mineralogically and chemically different from basalt and for many years geologists have wrestled with the problem of how the two rock types are related. If basalt is assumed to derive from the mantle, is it likely that granite, which is of a quite different composition, could also come from the mantle? Nowadays it is considered that granite may be produced in two ways; either from basalt, or from crustal rocks. When basalt magma starts to crystallize in the upper mantle, or the lower part of the crust, the overall composition of the crystals is not the same as the overall composition of the magma. This means that the liquid part will have a composition different from that of the original magma, and the further the crystallization process goes the greater will be the difference in composition between the liquid and the crystals. If the crystals and the liquid should now be separated by some mechanism, then rocks of two types will result, and each will have a composition different from the original basalt. This process, called differentiation, is capable of producing a great range of rock types, one of which is granite.

The second and perhaps more important way of producing granite is thought to operate within the crust itself. When mountain chains are formed, considerable thicknesses of crustal rocks are squeezed and thickened, and probably the base of the crust bulges down into the mantle. At the same time large volumes of magma move up into the crust. The effect is to heat the base of the crust to temperatures high enough to melt the rocks, so producing more magma. This new magma, which has the composition of granite, is mobile and moves up into the higher levels of the crust where it cools and solidifies as large granite intrusions, which are found in most mountain belts. These two processes account for the majority of igneous rocks.

The recognition and naming of igneous rocks involves an assessment of grain size and the recognition and estimation of the relative amounts of the constituent minerals. Additional information is obtained from colour index, texture, structure, and sometimes from field relations.

Grain size Grain size refers to the size (average diameter) of the mineral grains comprising the rock. Some rocks have large crystals set in a groundmass of smaller grains (see below); in these rocks only the groundmass minerals are taken into account; the large crystals, no matter how obvious, are ignored. Excluding the glassy rocks three broad grain size categories are recognized: *fine-grained*, in which the grains are generally below the limit of resolution of the naked eye (less than about 0·1 mm); *medium-grained*, in which the grains are recognizable with the naked eye, but minerals hard to identify (0·1 to 1–2 mm); and *coarse-grained*, in which the mineral grains can be identified by the naked eye (coarser than 1–2 mm). The coarsest rocks in which the mineral grains have diameters of several centimetres or more are referred to as *pegmatitic*.

Mineralogy This is the most important single feature to be considered when naming igneous rocks. Although magma is a complex silicate melt, most igneous rocks are composed of a few essential minerals belonging to a few mineral groups, namely quartz, the feldspars and feldspathoids (the light coloured or *felsic* minerals), and the pyroxenes, amphiboles, micas and olivines (the dark coloured or *mafic* minerals). Minor constituents are grouped as accessory minerals. All the groups listed are silicates and except for quartz are, within limits, variable in chemical composition. Once the grain size has been decided, rock names are assigned according to the kind and proportions of the constituent minerals. It may be necessary sometimes to know the approximate chemical composition of one particular mineral, but this usually requires at least a microscope and so cannot be determined in the field. The following table summarizes the mineral content and the interrelationships of most of the rocks described in the following pages.

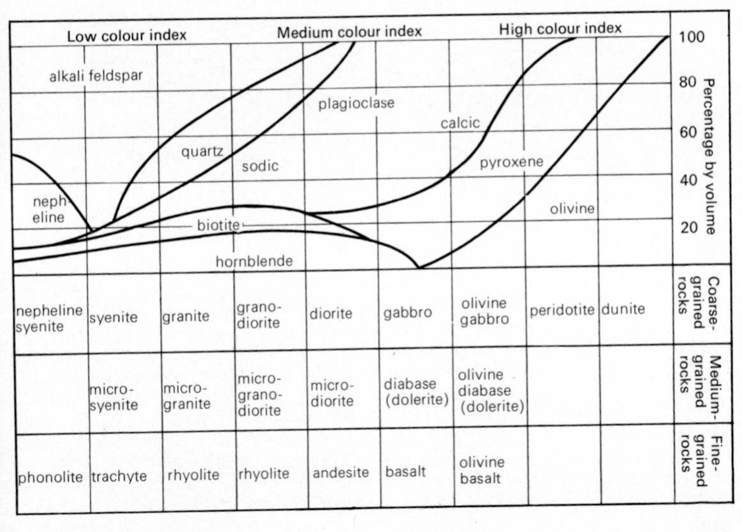

nepheline syenite	syenite	granite	grano-diorite	diorite	gabbro	olivine gabbro	peridotite	dunite	Coarse-grained rocks
	micro-syenite	micro-granite	micro-grano-diorite	micro-diorite	diabase (dolerite)	olivine diabase (dolerite)			Medium-grained rocks
phonolite	trachyte	rhyolite	rhyolite	andesite	basalt	olivine basalt			Fine-grained rocks

Colour index The colour index of a rock is the proportion of dark minerals it contains, on the scale 0 to 100.

Texture Texture refers to the shape, arrangement and distribution of the minerals of the rock. The following descriptive terms are often used.

In a *granular* texture (equigranular) (page 154) all grains are of about the same size and roughly of equant shape. *Poikilitic* texture (page 162) refers to large grains of one mineral enclosing smaller grains of other minerals. If pyroxene encloses plagioclase, as in many gabbros and diabases, the texture is called *ophitic*. In a *porphyritic* texture (page 168) some large grains *(phenocrysts* or *insets)* are set in a finer-grained or glassy matrix *(groundmass)*. 'Porphyritic' is a common adjectival prefix; for example porphyritic granite, porphyritic basalt. *Flow* or *fluidal* texture (page 166) refers to tabular or elongate crystals aligned by flow in the magma, in much the same way as logs in a river. In glassy rocks flow is marked by swirling lines, and often by trains of bubbles.

Structure The structure of rocks refers to the broader features of rock masses rather than those which depend on the interrelationships of the grains.

In a *layered* or *banded* structure (page 160) the rock comprises layers of contrasting mineral composition that appear on a surface as bands differing in colour or texture. A rock with a *vesicular* structure (page 172) contains cavities *(vesicles)* produced by the expansion and escape of gases. Vesicles, which frequently occur in lavas, may be spherical, elliptical or tubular. When vesicles are filled with secondary minerals, the structure is called *amygdaloidal*, and the infilled vesicles *amygdales*. *Xenoliths* (page 158) are fragments of other rocks included in igneous rocks. They may vary greatly in shape and size. *Joints* are cracks or fissures in rocks along which there has been no displacement. Lava flows sometimes show *columnar jointing* (page 170) in which the rock has broken on cooling into parallel hexagonal columns roughly perpendicular to the cooling surface.

Field relations Igneous rocks can be divided conveniently into three major groups: *volcanic (extrusive)* rocks are largely glassy and fine-grained and found in lava flows and tuffs; *hypabyssal* rocks are largely medium-grained and found in minor intrusions (sills, dykes); and *plutonic* rocks are largely coarse-grained and found in major intrusions (batholiths).

Igneous intrusions are described according to their shapes and their relationships with the rocks they intrude (the *country* rocks) (Fig. 2).

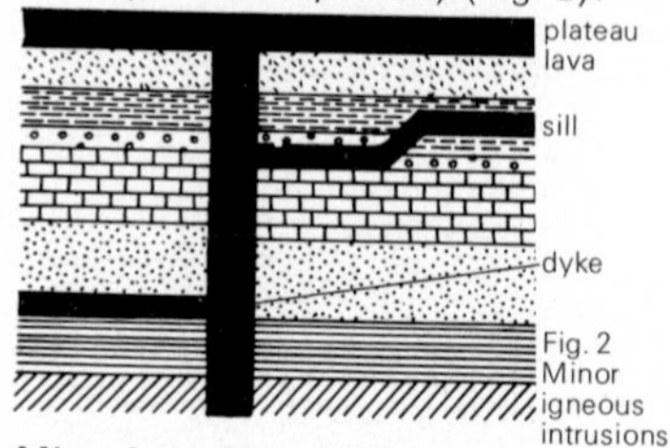

Fig. 2 Minor igneous intrusions

Minor intrusions *Dykes* are sheet-like intrusions which are vertical or nearly so and which cut sharply across bedding (see sedimentary rocks) or foliation (see metamorphic rocks). Dykes range from a few centimetres to hundreds of metres in width. *Sills* are sheet-like intrusions which are essentially horizontal and usually follow bedding or foliation. Like dykes they range from a few centimetres to hundreds of metres in thickness. *Veins* are irregular intrusions which sometimes form a complex network.

Major intrusions *Batholiths* (Fig. 3) are large, cross-cutting intrusions,

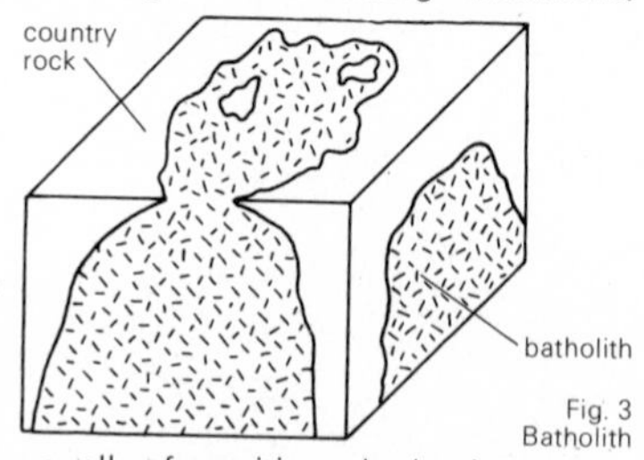

Fig. 3 Batholith

usually of granitic rocks, having steeply dipping contacts and no apparent floor. Exposed batholiths may cover hundreds of thousands of square kilometres. *Stocks* are smaller than batholiths but otherwise similar. They occupy areas of a few square kilometres to tens of square kilometres.

Volcanic rocks *Volcanic cones* (Fig. 4) form when lava and accompanying pyroclasts (lava fragments) are ejected from a vertical pipe-like vent. Lava may, however, flow from a *fissure* from which it may travel for considerable distances forming a *lava plateau*. Lava which flows into water chills rapidly and may give rise to distinctive pillow lavas (page 170).

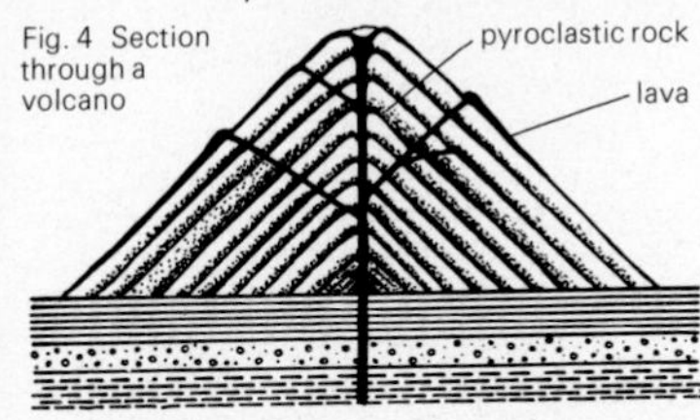

Fig. 4 Section through a volcano

Metamorphic rocks

The Earth's crust as well as being intruded by magma, is from time to time subjected to stresses generated within the crust and mantle which are sufficiently great to cause it to break to form *faults*, and also to bend forming *folds*. These forces are often concentrated along relatively narrow, sinuous belts when the folding, usually combined with intrusion and extrusion of magma, gives rise to mountain chains. The rocks within a mountain chain not only sustain considerable pressures but are also heated both generally and by the large scale intrusion of magma, with the effect that rocks are deformed and recrystallized to varying degrees. Such rocks are called metamorphic rocks.

The term *regional metamorphism* is used to describe the widespread effects of metamorphism that accompanied mountain building, when rocks that now occupy thousands of square kilometres were metamorphosed. Another type of metamorphism, called *contact metamorphism*, is restricted to the vicinity of igneous intrusions. Magma has a high though variable temperature, usually in the range 700 to 1100°C, so that it heats the rocks adjacent to the intrusion, causing recrystallization and growth of new minerals. The area of altered rocks surrounding an igneous intrusion is called a *contact metamorphic aureole* (Fig. 5), the size of which depends on the temperature of the magma and the size of the intrusion.

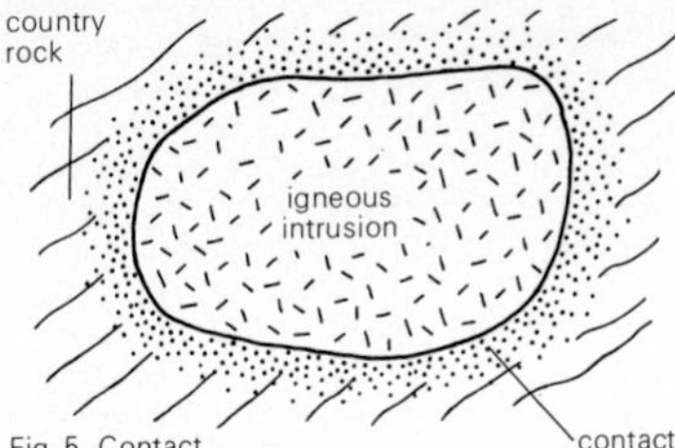

Fig. 5 Contact metamorphic aureole

Metamorphic rocks display a wide range of texture, structure and mineralogy, because of the range of temperatures and pressures to which they may have been subjected, and the wide variety of possible rocks from which they were formed (*parental* rocks). The principal types of metamorphic rock and the nature of the parental rocks are summarized in the following table. The terms 'low', 'medium' and 'high' grade refer to the degree of metamorphism that has affected the rocks.

	Regional metamorphism			**Contact metamorphism**
original rock	low grade	medium grade	high grade	
quartz sandstone	quartz schist	quartzite	quartzite	quartzite
greywacke	schist	schist	gneiss granulite	
limestone–pure	marble	marble	marble	marble
limestone–impure	calcareous schist	calc-silicate rock	gneiss	calcareous hornfels
shale/mudstone	slate/phyllite	schist	gneiss granulite	hornfels
diabase/basalt	greenschist	amphibolite	amphibolite charnockite eclogite	basic hornfels

Texture The minerals of the metamorphic rocks grow in the solid state so that they have to compete for space with the minerals around them, in contrast to minerals of igneous rocks which grow in a fluid (magma). They also often grow under high pressures, and these two factors give to metamorphic rocks their distinctive textures. Most common metamorphic rocks are named accordingly as follows: *slate* is a fine-grained rock with a prominent parting or 'cleavage' along which it can be split into thin sheets (page 178); *phyllite* is a rock somewhat coarser than slate, but still of fine grain size with a lustrous silvery or greenish sheen on the cleavage surfaces; *schist* is a coarse-grained rock with a marked layering, defined by platy or elongate minerals, often finely interleaved with quartz and feldspar (page 180); *gneiss* is a coarse-grained rock composed largely of quartz and feldspar, but with a marked, though often irregular, layered structure (page 190); and *hornfels* is a tough, usually fine-grained, even-textured rock, produced by thermal metamorphism.

Other important textural terms include: *granoblastic*, a texture with mineral grains of the same general size (page 188); *porphyroblastic*, a texture in which large, well-shaped crystals *(porphyroblasts)* are set in a finer-grained matrix (page 174); and *poikiloblastic*, similar to porphyroblastic, but poikiloblasts contain numerous inclusions of another mineral or minerals (page 174).

Structure The structures found in metamorphic rocks are of two types: relict structures inherited from parental sedimentary or igneous rocks which have survived the metamorphism, and new structures produced by the metamorphism itself.

Relict structures inherited from sedimentary rocks include bedding, which may be discernible in a hand specimen, or else as a sequence of different metamorphic rocks that reflect the variation in the original sedimentary succession. Other sedimentary structures such as graded bedding, cross bedding, ripple marks, or even fossils, may be preserved on occasion. Relict structures characteristic of igneous rocks include dykes, fine-grained igneous contacts, pillow structure and amygdales.

The most significant new metamorphic structures are *cleavage* and *folding* (page 178). Cleavage is the structure which allows rocks such as slate to be split along parallel planes. It is a product of pressure or *dynamic metamorphism*, and usually cuts across bedding. In schists which are folded on a minor scale cleavages often cut across the bedding in the folds. Heat and high pressures increase the plasticity of rocks so that they tend to yield by folding rather than rupture *(faulting)*. Folds vary from tiny crumples, common in phyllites, and minor folds measured in centimetres or a few metres, to folds which may be many kilometres across. The style of the folds is also very variable (page 180).

Mineralogy Temperature is probably the most important factor in metamorphism and with increasing temperature, even though the rocks are still solid, chemical reactions take place among the rock components and new minerals are produced. In rocks of a particular composition different minerals are produced at different temperatures, so that the mineralogy often gives a rough guide to the temperature of formation or *grade of metamorphism*. Contact and regional metamorphic rocks have some significant mineralogical differences, however, probably because of the lower pressures involved in contact metamorphism.

The detailed mineralogical variations among the regionally metamorphosed rocks are rather complex, but a few of the more important minerals, and the metamorphic grades at which they appear, are given in the following table. It must be stressed that most rocks will also contain other minerals, while the 'index' minerals will often persist into higher grades.

Parental rock	low grade	medium grade	high grade
pelitic rocks (shale/mudstone)	chlorite biotite	garnet staurolite	kyanite sillimanite

Parental rock	low grade	medium grade	high grade
impure limestone	calcite epidote tremolite	diopside olivine grossular	
diabase/basalt	chlorite	garnet hornblende	

At the highest grades of regional metamorphism the rocks become plastic and the minerals segregate to produce the banded structure characteristic of gneisses. These rocks are approaching their range of melting, and so grade into the igneous rocks of granitic composition.

The most important minerals occurring in the principal types of contact metamorphosed rocks are set out in the following table.

Parental rock	metamorphic equivalent	minerals one or more of:
pelitic rocks	hornfels	biotite, andalusite, cordierite, garnet, hornblende, pyroxene, sillimanite, and feldspar
impure limestone	marble, calcareous schist	calcite, dolomite, tremolite, phlogopite, forsterite, diopside, grossular, idocrase, and wollastonite
diabase/basalt	basic hornfels	chlorite, hornblende, biotite, garnet, and feldspar

Metasomatism Although most of the changes in metamorphic rocks take place in the solid state, there is often a movement of material through the rock, carried by migrating fluids. The process by which a rock is changed by the addition and/or subtraction of material through the agency of such fluids is called *metasomatism*, and many metamorphic rocks owe their origin, in part at least, to this process. Good examples are often found among contact metamorphosed limestones, the addition of material to which produces the rocks known as *skarns*, which are often a good hunting ground for the mineral collector.

Sedimentary rocks

Whereas igneous and metamorphic rocks are produced by internal processes within the Earth, sedimentary rocks are formed by processes which are active at the Earth's surface. The surface of the land is continually being attacked by agents of weathering and erosion, such as rain and rivers, wind and moving ice. These physical agents are helped by chemical decay from percolating waters, and together they break up even the toughest rocks and produce rock waste. This is transported, mainly by rivers but also by wind, and in higher latitudes by ice. Eventually this material, now referred to as *sediment*, is deposited at river mouths, in lakes, or in the sea. It is the accumulation of this material, often in deposits many kilometres thick, which goes to make up the sedimentary rocks.

Sedimentary rocks of a different kind are produced from the huge quantities of material carried to lakes and the sea, not as rock or mineral fragments, but dissolved in river water. When the water of lakes or seas become saturated with salts, often as a result of evaporation in arid climates, various salts are precipitated and form sedimentary rocks known as *evaporites*.

Yet another group is produced by the gradual accumulation of animal skeletons, such as shells and corals, which are composed essentially of calcium carbonate and go to form the rocks called limestones.

The main features to be considered when studying sedimentary rocks in the field are texture (including grain size), structure, mineralogy, field relationships and, to some extent, colour. In the classification used in this guide the sedimentary rocks are first divided into three major divisions according to their mode of origin: *mechanical origin*, sediments which have been transported as solid particles by water, wind or ice (they are further subdivided according to grain size); *chemical origin*, sediments formed by precipitation from solution of dissolved salts, and sometimes by chemical replacement of one mineral by another (they are further subdivided according to chemical composition); and *organic origin*, sediments formed by the accumulation of organic matter, whether animal or plant (they are further subdivided according to chemical composition).

Mechanical origin	
Coarse	conglomerate, breccia, tillite
Medium	sandstone, arkose, orthoquartzite, grit
Fine	siltstone, greywacke, mudstone, shale

Chemical origin	
Calcareous	calcareous mudstone, oolitic and pisolitic limestone (in part), dolomite, travertine
Siliceous	flint, chert
Ferruginous	ironstone
Saline	rock salt, gypsum rock
Phosphatic	phosphate rock (in part)

Organic origin	
Calcareous	biochemical limestone, oolitic and pisolitic limestone (in part)
Carbonaceous	coal
Phosphatic	phosphate rock (in part)

Texture This refers to the size, arrangement and shape of the individual grains of the rock. Size categories are: *coarse*, >2 mm (gravel); *medium*, 2 to $\frac{1}{16}$ mm (sand); and *fine*, within which the range $\frac{1}{16}$ to $\frac{1}{256}$ mm is called silt; and $<\frac{1}{256}$ mm, is called clay.

These terms can be applied only loosely in the field, and the distinction between silt (siltstone) and clay (mudstone) is one that requires a microscope, although with practice, it can be made by eye. The shape of individual grains of the medium- and coarse-grained sediments reveals something of their origin. Angular grains, for instance, are unlikely to have been transported very far, whereas rounded grains suggest considerable transport. Sediments with nearly spherical, polished grains are often deposited by wind. The range of grain sizes within a single sediment is also significant. If the range is small, that is, if all grains are about the same size, the sediment is a *mature* one, having been well worked and 'sorted' by currents; while a sediment containing a wide range of grain sizes is an *immature* one, and has probably been deposited rapidly, or as in the case of greywacke (page 194) and till (page 192), by a particular mechanism. Other textural features are referred to in the rock descriptions.

Structure Structure refers to the large scale features which are best seen in the field. These are often particularly informative as to the nature of the environment in which the sediments were accumulated. The most commonly observed structures are described below.

Bedding is of almost universal occurrence in sedimentary rocks, and is a layering expressed by variations of texture and mineralogy. Individual layers were deposited at about the same time, and in an approximately horizontal attitude, although the rocks may have been folded subsequently. Layers having a thickness greater than one centimetre are referred to as *strata*, while layers less than one centimetre thick are *laminae* or *laminations*. The surface separating one stratum or lamination from the next is a *bedding plane*. A collection of strata of one rock type, which can be delimited on a geological map, comprises a *formation*.

Current bedding is a type of bedding in which strata are inclined so that they wedge out at one end and are truncated at the other. The truncated surface is produced by local, contemporaneous erosion (see unconformity). Individual current-bedded units may be a few centimetres or many metres in thickness. Current bedding forms when sediment, commonly sand, is transported by wind or water, and accumulates on a sloping surface, which may be the sea bed at the mouth of a river, or the lee side of a sand-dune. Individual strata wedge out and are inclined in the direction of the current, so that the original direction of flow of wind or water may be determined (page 194).

In *graded bedding* the size of grains varies from coarse at the bottom of a stratum to fine at the top. This type of bedding forms by the rapid settling of large quantities of sediment so that larger particles, settling more rapidly, accumulate at the bottom, and the slower settling finer material remains at the top (see greywacke page 194). In strongly folded rocks this feature can be used to determine the 'right way up'.

Slump bedding is a folded or contorted structure produced by the sliding or slumping of wet, recently deposited sediment down a slope on the sea floor (page 196). It is often associated with graded bedding.

Ripple marks are commonly seen on bedding planes and are produced by the movement of water or wind over the sediment, thus shaping it into parallel ridges. Cross-sections of individual ripples usually reveal small scale current bedding (page 194).

An *unconformity* in a succession of rocks represents a time interval during which there was erosion, followed later by further deposition. The time gap represented may have been of short duration so that the older and newer series of rocks are parallel (sometimes called a *disconformity*); sometimes, however, the time span was sufficiently long for there to have been either considerable erosion or for the older rocks to have been tilted before the newer rocks were deposited. Such unconformities can sometimes be traced for considerable distances, and may be of importance in correlating one sequence of rocks with another. They are indicative of relative movements of land and sea (Fig 6).

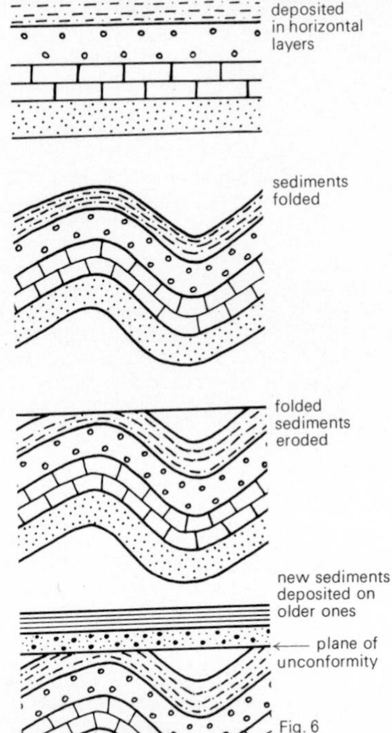

Fig. 6 Development of an unconformity

For details of organic structures (fossils) see pages 210 to 309 and for details of concretions and nodules see page 204.

Mineralogy Although there is a wide variation in the chemical composition and mineralogy of sedimentary rocks, these are of less importance in field identification and classification than in the igneous and metamorphic rocks. For help with mineral identification the reader is referred to the mineral section (pages 6 to 145).

Field relations It is helpful when studying rocks in the field always to examine the relationships between associated rock groups because this will help to elucidate the nature of the environment in which they were formed, and how it changed with time. This in turn may reveal something of the geography of the area at the time of the formation of the sediment (palaeogeography).

Igneous rocks

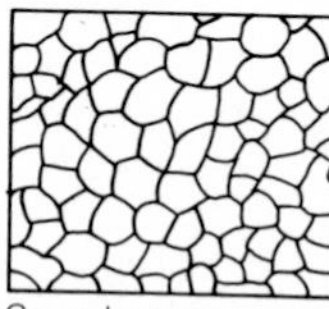

Granular texture

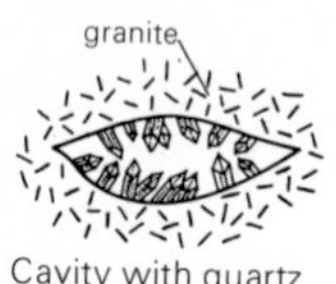

Cavity with quartz crystals

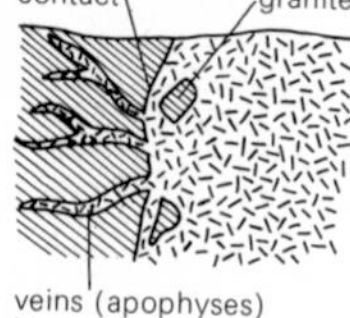

Contact of granite with country rock

Granite **Colour** Most commonly shades of white, grey, pink and red, but usually mottled in almost any combination of these. **Colour index** 0 to 30 **Grain size** Coarse to very coarse; usually constant over large areas. **Texture** Normally granular; commonly porphyritic. Phenocrysts are invariably of feldspar and usually develop good crystal shapes; they often attain sizes of 8 to 10 cm, and they may be aligned owing to flow of the granite magma. Granites may be foliated, due to the parallel arrangement of minerals such as mica and hornblende. **Structure** Granites are typically homogeneous but they may have a banded aspect (see gneiss on page 190). Xenoliths are common. Drusy structure is not unusual. Druses are irregular cavities in the rock into which well formed crystals of quartz, feldspar and other minerals project. Dykes and veins of microgranite, quartz porphyry and pegmatite commonly cut granites. **Mineralogy** Alkali feldspar with or without sodic plagioclase (usually oligoclase) plus quartz, which must comprise 10 per cent or more of the volume of the rock. The feldspar is white or pink; if both a white and a pink feldspar are present the former tends to be plagioclase and the latter orthoclase. If plagioclase is the dominant feldspar then the rock is granodiorite. Biotite (biotite granite) and/or muscovite (biotite-muscovite or muscovite granite) are usually present, and hornblende may occur. Apatite, sphene, zircon and magnetite are common accessories, but a very wide range of other minerals has been recorded from granites. **Field relations** Batholiths, stocks, bosses (a stock of circular cross-section), sills and dykes. Most of these bodies are clearly intrusive, having sharp, cross-cutting relationships with the country rocks which they metamorphose. In some of the more deeply eroded areas of metamorphic rocks, however, the granites often grade imperceptibly into foliated granite and granite gneiss, and may be partly metamorphic in origin. When sharply cross-cutting the granite is often finer grained near the contacts owing to more rapid cooling. Irregular veins or *apophyses* commonly extend into the country rocks. Granite often breaks down to various clay minerals (such as kaolin) and quartz sand, but it an also be particularly resistant to breakdown and form rugged outcrops and hills.

Granodiorite **Colour** Greys predominate. **Colour index** Often a little higher than for granite. **Grain size** and **texture** As for granite. **Mineralogy** Plagioclase (oligoclase to andesine) more abundant than orthoclase; otherwise as for granite. **Field relations** As for granite. Granodiorites are quantitatively the commonest of the granite family, and indeed, are probably the most voluminous of all the plutonic igneous rocks.

Muscovite granite

Granite with xenolith

5 cm

Biotite granite

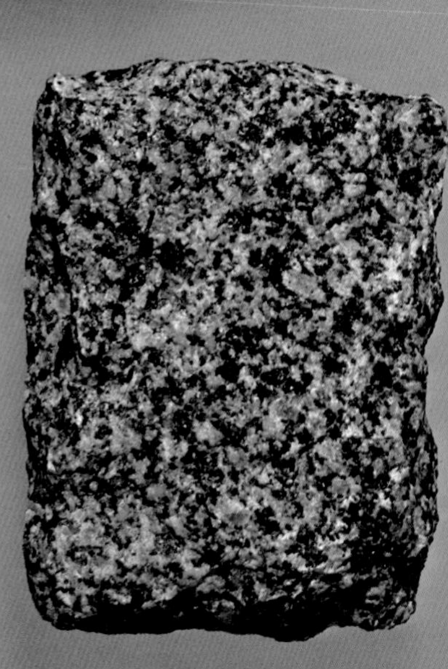

Hornblende-biotite granite

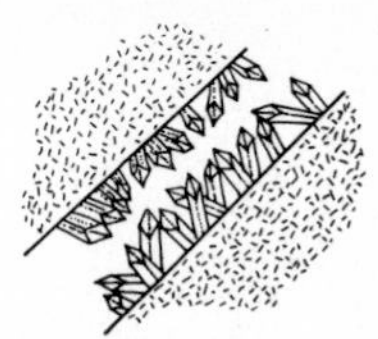

Pegmatite with large crystals growing inwards from the walls

Granite pegmatite **Colour** White, pink and red, but unevenly coloured because of the large size of individual crystals. **Grain size** Very coarse to giant-grained (by definition). Giant crystals have been reported, for example a spodumene crystal 14 m long from the Black Hills, South Dakota, and beryl crystals as much as 6 m long from the same locality. **Texture** The coarseness of grain is striking, but this is usually variable and parts are often finer grained. The crystals often grow parallel or subparallel to each other, and commonly perpendicular to the walls of the intrusion. Graphic or runic texture is common (graphic granite) in which large alkali feldspar crystals contain numerous elongated, regularly spaced, sharply angular quartzes which resemble the characters in certain ancient forms of writing. **Mineralogy** Alkali feldspar and quartz, usually with muscovite. Accessory minerals include all those found in granites and comprise a very varied assemblage. A few of these are beryl, biotite, chalcopyrite, corundum, fluorite, galena, magnetite, oligoclase, allanite, pyrite, pyrrhotine, rutile, sphene, spodumene, topaz and zircon. **Field relations** Dykes, veins and irregular segregations, which are usually of rather limited extent. Pegmatites tend to be concentrated in the marginal parts of granite intrusions, or in the country rocks in the immediate vicinity of the granite. Pegmatites are produced by the crystallization of the last residual fluids after the bulk of the granite has solidified, and many of the rarer elements tend to be concentrated in them. They are, therefore, a very important source of many ore minerals for which they are often worked. A wide range of minerals with crystals attaining large sizes occur in pegmatites making them the principal single source of the choicest mineral specimens.

Peralkaline granite **Colour** White, grey and pink. **Colour index** 0 to 30 **Grain size** Coarse. **Texture** Similar to granite but rarely foliated. **Structure** Similar to granite but xenoliths less common. **Mineralogy** Quartz (greater than 10 per cent) plus alkali feldspar. Characterized by presence of alkali-rich pyroxene (aegirine) and/or alkali-rich amphibole (for example riebeckite). Biotite may also occur. **Field relations** Stocks, bosses, sills and dykes. Peralkaline granites are much less common than granites and granodiorites and do not form such large intrusions. They tend not to occur in areas of mountain building, as do the normal granites, but are found in more stable areas of the continents.

Graphic granite
Tourmaline-bearing pegmatite
Riebeckite granite
5cm
Granite pegmatite

Syenite **Colour** Red, pink, grey or white. **Colour index** 0 to 40. **Grain size** Coarse; can be pegmatitic. **Texture** Tends to be equigranular, but also porphyritic and/or fluidal. Essentially similar to granite. **Structure** Drusy cavities common. **Mineralogy** Principally alkali feldspar and/or sodic plagioclase (albite or oligoclase), usually with biotite, amphibole or pyroxene. Up to 10 per cent quartz may occur (quartz syenite), when it grades into granite; or nepheline may be present, when it grades into nepheline syenite. **Field relations** Stocks, dykes and sills. May be associated with, or grade into, granites, or form individual intrusions which are rarely more than a few kilometres in diameter. Syenites are not very common rocks.

Nepheline syenite **Colour** Usually grey but tones of green, pink and yellow. **Colour index** Usually 0 to 30; but occasionally higher. **Grain size** Coarse; can be pegmatitic. **Texture** Equigranular, porphyritic and/or fluidal. Phenocrysts, when present, are usually of feldspar or nepheline. **Mineralogy** Alkali feldspar and/or sodic plagioclase (albite or oligoclase), and often alkali pyroxene or amphibole and/or biotite. Common accessory minerals are cancrinite and sodalite but a great range of rarer minerals has been recorded. **Field relations** Stocks, dykes and sills. Tend to be associated with other highly alkaline rocks (that is rocks which contain a high proportion of minerals rich in sodium and potassium) such as syenites, and peralkaline granites. Nepheline syenites are relatively rare rocks.

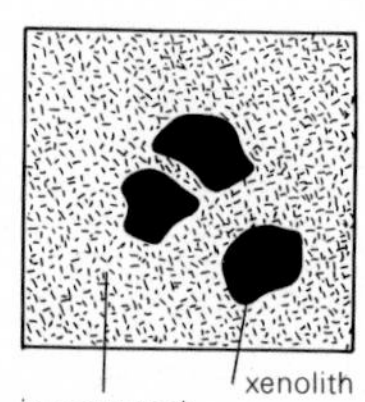

Xenoliths

Diorite **Colour** Speckled black and white in hand specimen; occasionally shades of dark green or pink. The dark minerals are more noticeable than in gabbro. **Colour index** 40 to 90, but very variable, often over short distances. **Grain size** Coarse; may be pegmatitic. **Texture** Equigranular or porphyritic. In porphyritic varieties the feldspar or hornblende may form phenocrysts. Diorites often vary rapidly in texture; an equigranular variety may grade into a porphyritic one within a few centimetres. They are sometimes foliated due to the roughly parallel arrangement of the minerals. **Structure** Xenoliths are common. **Mineralogy** Essentially plagioclase (oligoclase or andesine) and hornblende; biotite and/or pyroxene may occur. Alkali feldspar and quartz (quartz diorites) may be present, when diorite grades into granodiorite. Common accessory minerals are apatite, sphene and iron oxides. **Field relations** Forms independent stocks, bosses and dykes, but also comprises local variants of masses of granite, and sometimes gabbro, into which they merge imperceptibly.

Syenite

Nepheline syenite

5 cm

Diorite

Diorite

pyroxene encloses plagioclase crystals
Ophitic texture

top of layer is mostly light minerals
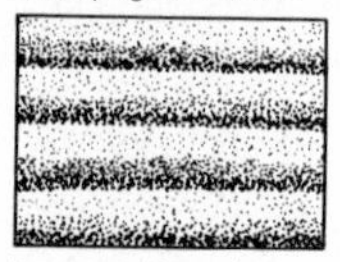
bottom of layer is mostly dark minerals
Layered structure

Gabbro **Colour** Grey, dark grey, black; may have a bluish or greenish tone. **Colour index** 30 to 90; with decrease of coloured minerals gabbro grades into anorthosite and with an increase it grades into pyroxenite and peridotite. **Grain size** Coarse; can be pegmatitic. **Texture** Granular; porphyritic texture rare. Ophitic texture is common. **Structure** Layering, defined by alternating layers of light and dark coloured minerals, often occurs. Individual layers vary from several metres to a centimetre or two in thickness. Often contain pegmatitic veins or segregations. **Mineralogy** Essentially plagioclase (labradorite or bytownite) and pyroxene; quartz (quartz gabbro), olivine (olivine gabbro) or hornblende may occur, and iron oxides, chromite and serpentine are common accessories. **Field relations** Stocks, sills and dykes. Individual intrusions may be of considerable size (a few kilometres is usual). Rare, very large sheet-like intrusions, called *lopoliths*, have diameters of hundreds of kilometres, but within these intrusions other rocks such as pyroxenite and anorthosite are also important. Layering is common and field observation indicates that the layers are often arranged like a stack of saucers. If the intrusion has been affected by earth movements such as folding or faulting the layering may be steeply inclined.

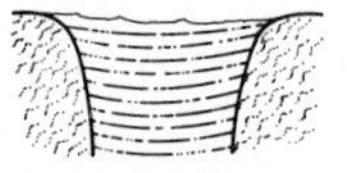
Layered gabbro intrusion, as seen in cross-section

Anorthosite **Colour** Grey to white. **Colour index** Less than 10 **Grain size** Medium to coarse. **Texture** Granular. Elongate crystals sometimes occur in parallel alignment and this may be emphasized by streaks and patches of dark minerals. **Structure** May have a layered structure (see gabbro). **Mineralogy** Comprises at least 90 per cent plagioclase (oligoclase/andesine to bytownite). Accessory minerals include pyroxene, olivine and iron oxides. **Field relations** Stocks, dykes, batholiths. In smaller intrusions anorthosites are usually associated with gabbros comprising part of a layered sequence but anorthosites also form huge masses, sometimes covering thousands of square kilometres, within areas of metamorphic rocks. These masses are variable in composition, grading into gabbro with increase in dark minerals.

Troctolite **Colour** Grey, studded with black, brown or reddish spots (hence the popular name troutstone). **Colour index** 30 to 90 **Grain size** Coarse. **Texture** Granular. **Structure** May have a layered structure, or be part of a layered sequence. **Mineralogy** Plagioclase (labradorite to anorthite) and olivine. The olivine is usually altered to greenish serpentine. **Field relations** Usually associated with gabbro or layered anorthosite.

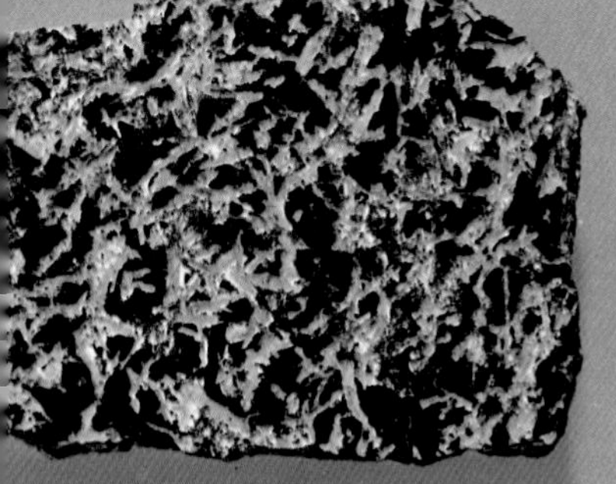

Olivine gabbro

Anorthosite

Layered gabbro

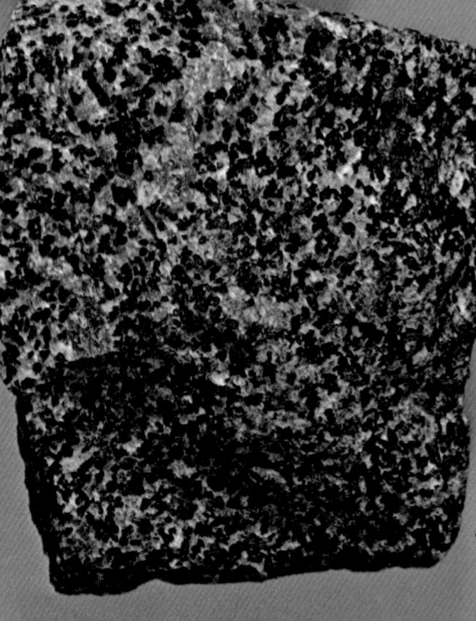

Troctolite

small crystals enclosed in larger ones
Poikilitic texture

Peridotite **Colour** Dull green to black. Dunites (see below) are light to dark green or shades of yellow and brown. **Colour index** Greater than 90 **Grain size** Medium to coarse. **Texture** Granular. Dunite usually has a sugary texture. Poikilitic texture is common and can be seen by careful inspection of cleavage surfaces. Porphyritic texture is rare. **Structure** Layering may occur. **Mineralogy** Comprises dark minerals only; feldspar is negligible or absent. Olivine is essential (pure olivine rock is called dunite); pyroxene and/or hornblende are usually present. Biotite (mica peridotite), chromite and garnet sometimes occur. **Field relations** Independent dykes, sills and small stocks, but also forms parts of large layered gabbroic intrusions together with pyroxenite and anorthosite (see gabbro). It is unlikely that a magma of peridotite composition exists but peridotite is thought to be an *accumulate* formed by the crystallization, settling and accumulation of olivine crystals from a gabbro magma. Peridotite occurs also as xenoliths in basalt which may represent fragments brought up from deep layered intrusions, or from the Earth's mantle.

Pyroxenite **Colour** Green, dark green to black. **Colour index** Greater than 90 **Grain size** Medium to coarse. **Texture** Granular. **Structure** May be layered. **Mineralogy** Predominantly clino- or orthopyroxene. Olivine, hornblende, iron oxides, chromite or biotite may be present. Feldspar is minor in amount, or absent. **Field relations** Small independent intrusions—stocks or dykes—and as individual bands in layered gabbro (see gabbro).

Serpentinite **Colour** Greyish-green, green to black. Often banded, streaked or blotched in bright greens or reds. **Grain size** Medium to coarse. **Texture** Compact, dull waxy, with a smooth to splintery fracture. **Structure** Often banded; commonly criss-crossed by veins of fibrous chrysotile serpentine. **Colour index** Greater than 90 **Mineralogy** Principally serpentine minerals. Olivine, pyroxene, hornblende, mica, garnet and iron oxides may be present. **Field relations** Stocks, dykes and lenses. Serpentinites are secondary rocks, having formed by the serpentinization of other rocks, principally peridotite. They commonly occur as pods or lenses in folded metamorphic rocks, probably representing altered olivine-rich intrusions.

Kimberlite **Colour** Bluish, greenish or black. **Grain size** Coarse. **Texture** Usually porphyritic, with a range of minerals, which tend to be rounded and perhaps broken, forming phenocrysts. **Structure** Xenoliths are common. **Mineralogy** Principally serpentinized olivine, phlogopite and pyrope; together with orthopyroxene and chrome diopside. Accessory minerals include ilmenite, chromite and often diamond. Calcite may be abundant. **Field relations** Steeply dipping circular to oval pipes, which are rarely more than a few hundred metres in diameter, and occasionally as dykes. Kimberlite pipes are the only primary source of diamonds.

Dunite

Serpentinite

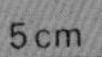

Kimberlite

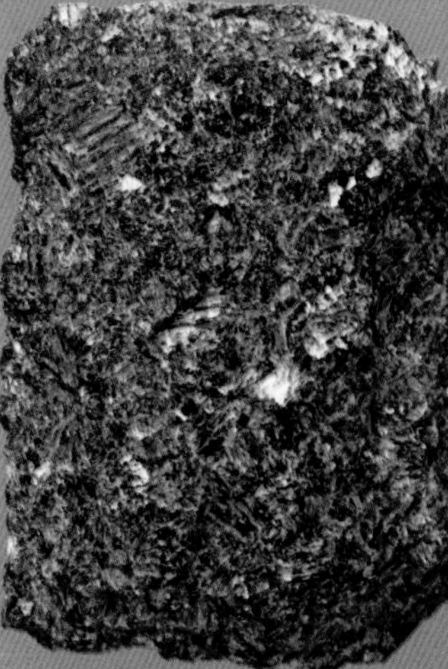
Pyroxenite

Porphyritic microgranite **Colour** Light to dark grey; yellowish or reddish. **Colour index** 0 to 30 **Grain size** Groundmass medium-grained. **Texture** Porphyritic; phenocrysts commonly have good crystal shape, and may be aligned due to flow. **Mineralogy** Essentially the same as granite. Phenocrysts are quartz and feldspar (white, grey or reddish) with, more rarely, hornblende or biotite. The groundmass comprises the same minerals but is generally too fine-grained for them to be distinguished. **Field relations** Dykes, sills, veins. Commonly intruded into granite and the surrounding rocks.

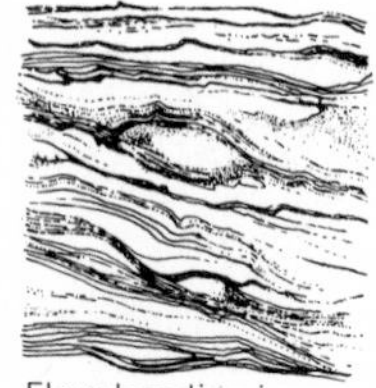

Flow banding in rhyolite

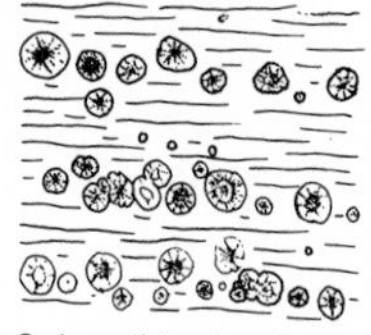

Spherulitic rhyolite

Rhyolite **Colour** Usually light coloured; white, grey, greenish, reddish or brownish. The colour may be even, or in bands of differing shades. **Grain size** Fine to very fine. **Texture** Frequently shows alternating layers that differ slightly in granularity or colour. Phenocrysts not uncommon (porphyritic rhyolite). *Flow banding* is sometimes evident, defined by swirling layers of differing colour or granularity, and by aligned phenocrysts. **Structure** Vesicles or amygdales may be present. (Pumice is a highly vesicular variety of rhyolite.) May contain *spherulites* which are spherical bodies, often coalescing, comprising radial aggregates of needles, usually of quartz or feldspar. Spherulites are generally less than 0·5 cm in diameter, but they may reach a metre or more across. They form by very rapid growth in quickly cooling magma, and by the crystallization of glass. **Mineralogy** As for granite, but rapid cooling results in minute crystals. Phenocrysts of quartz, feldspar, hornblende or mica occur. **Field relations** Flows, dykes and plugs. Rhyolite (or granite) magma is highly viscous and so flows only very slowly, so that if it is extruded it forms very short, thick flows or is confined as a plug in the throat of a volcano.

Obsidian – conchoidal fracture

Obsidian and pitchstone **Colour** Shiny black, also brown or grey. Pitchstones have a dull rather than a shiny lustre. **Grain size** None; the rock is glassy. **Texture** Glassy, but obsidian may contain rare phenocrysts; pitchstones contain numerous phenocrysts. **Structure** May be spotted or flow banded and spherulites (see rhyolite) are common. Being a siliceous glass it breaks with a conchoidal fracture and may be fashioned to a sharp cutting edge. It was used for making cutting tools by primitive peoples. **Mineralogy** Essentially a glass. Rare phenocrysts (abundant in pitchstones) of quartz and feldspar. **Field relations** Dykes and flows. Commonly associated with rhyolites to which they are chemically equivalent.

Pitchstone
Rhyolite
Obsidian
Porphyritic microgranite
Pumice
5 cm

Microsyenite **Colour** Grey, reddish, pinkish or brownish. **Colour index** 0 to 40 **Grain size** Medium. **Texture** Granular; commonly porphyritic. A fluidal texture of closely packed prisms of alkali feldspar is often developed. **Mineralogy** Essentially alkali feldspar; a little biotite, hornblende, pyroxene or quartz is usual. Phenocrysts are usually alkali feldspar; rarely biotite or hornblende. In the type of microsyenite known as rhomb porphyry the feldspars have characteristic rhomb-shaped cross-sections. **Field relations** Dykes and sills; occasionally lava flows. Microsyenites are associated with intrusions of syenite or nepheline syenite, and with trachyte and phonolite.

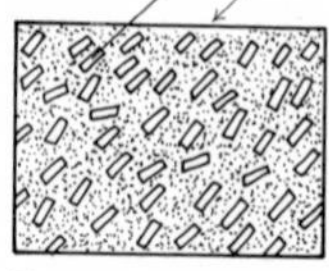

Flow texture

Trachyte **Colour** Usually grey, may be white, pink or yellowish. **Colour index** 0 to 40 **Grain size** Fine. **Texture** Almost invariably porphyritic. The rectangular phenocrysts are sanidine. A fluidal texture (known in these rocks as *trachytic*) is characteristically developed, but the fine grain size usually precludes this being seen with the naked eye. **Mineralogy** Dominantly alkali feldspar both in the groundmass and as phenocrysts. A little quartz (less than 10 per cent) or oligoclase may be present. Dark minerals are typically alkaline types such as aegirine or alkali amphibole, and are present only in small amounts so that trachytes are light in colour and weight. **Field relations** Lava flows and as narrow dykes and sills. Trachyte lavas occur in association with basalts in basaltic volcanoes, but are usually subsidiary to basalt. Occasionally trachyte forms flows of considerable extent.

Phonolite **Colour** Dark green to grey. **Colour index** 0 to 30 **Grain size** Fine. **Texture** A rather dense (compact) texture; usually porphyritic. Has a rather greasy lustre. **Structure** Often has a platy structure so that it breaks into flat slabs. Reputed to ring when struck with a hammer. **Mineralogy** Alkali feldspar, usually sanidine, nepheline and aegirine or alkali amphibole (such as riebeckite). The phenocrysts are usually rectangular feldspars or nephelines. **Field relations** Lava flows, sills and dykes. Sometimes associated with trachyte; commonly found in the vicinity of nepheline syenite.

Leucitophyre **Colour** Grey to dark grey, but may contain white spots in a grey or black groundmass. **Colour index** 20 to 70 **Grain size** Medium to fine. **Texture** Invariably porphyritic. The characteristic eight-sided or rounded phenocrysts of leucite are distinctive. **Mineralogy** Alkali feldspar, leucite and pyroxene; nepheline, phlogopite and alkali amphibole may be present. **Field relations** Dykes and lava flows. Relatively rare rocks, and associated with other rocks containing feldspathoids.

Microsyenite
(rhomb porphyry)

Phonolite

5 cm

Leucitophyre

Trachyte

Porphyritic texture

Microdiorite **Colour** Grey to dark grey; occasionally greenish or pinkish. **Colour index** 40 to 90 **Grain size** Medium. **Texture** Usually porphyritic; hence these rocks are sometimes called *porphyrites*. **Mineralogy** As for diorite. Phenocrysts usually hornblende or biotite but may be augite. **Field relations** Dykes and sills, often forming dyke swarms (see page 170) in the vicinity of diorite or granite intrusions.

Andesite **Colour** Shades of grey, purple, brown, green or almost black. **Grain size** Fine; less commonly partly glassy. **Texture** Frequently porphyritic. **Structure** Flow structures may be evident; can be vesicular or amygdaloidal. **Mineralogy** Only the phenocrysts are recognizable in hand specimen; these are white tabular plagioclase feldspars, plates of biotite mica or prisms of hornblende or augite. The microscope shows the groundmass to consist of plagioclase (oligoclase-andesine) with one or more of the minerals hornblende, biotite and orthorhombic or monoclinic pyroxene. **Field relations** As lava flows but may also form dykes. Andesite lavas are second in abundance only to basalt. They are usually associated with basalt and rhyolite, and are commonest in areas of mountain building.

Mica lamprophyre **Colour** Grey to black; often weathers to brownish shades. **Colour index** 30 to 70 **Grain size** Medium to fine. **Texture** Porphyritic; rarely granular. Biotite phenocrysts are characteristic and are abundant and of large size, giving the rock a very distinctive appearance. Biotite-rich varieties are noticeably 'soft' when hammered. **Mineralogy** The phenocrysts of biotite are readily identified in hand specimen, while more rarely reddish orthoclase and prisms of hornblende may occur. The groundmass, as seen with the microscope, essentially comprises either orthoclase or sodic plagioclase, biotite, and pyroxene or amphibole. Carbonate is often present, when the rock effervesces with dilute hydrochloric acid. **Field relations** Dykes, sills and small plugs which are usually found in association with granite, syenite or diorite.

Hornblende lamprophyre **Colour** Greenish, greyish or black when fresh; tends to weather to shades of red or brown. **Colour index** 30 to 70 **Grain size** Medium to fine. **Texture** Porphyritic; hornblende phenocrysts usually form long, slender prisms, and are often aligned. Less commonly granular. **Mineralogy** Phenocrysts of hornblende, set in a groundmass consisting of hornblende together with orthoclase or sodic plagioclase. **Field relations** Dykes, sills and small plugs which occur in the vicinity of granites, syenites and diorites.

Pyroxene lamprophyre

Andesite

Porphyritic microdiorite

Hornblende lamprophyre

Mica lamprophyre

Diabase (Dolerite) **Colour** When fresh it is black, dark-grey or green; may be mottled black and white. **Grain size** Medium. **Texture** Occasionally ophitic texture (see gabbro) can be distinguished in hand specimen. May be porphyritic. **Structure** Vesicles and amygdales occur. Sometimes has segregations of coarser rock enriched in feldspar. **Mineralogy** Phenocrysts comprise olivine (olivine diabase) and/or pyroxene or plagioclase. The groundmass comprises the same minerals with iron oxide, and sometimes with some quartz, hornblende, or biotite. **Field relations** Dykes and sills. These may form swarms of hundreds or perhaps thousands of individual dykes or sills which often radiate from a single volcanic centre.

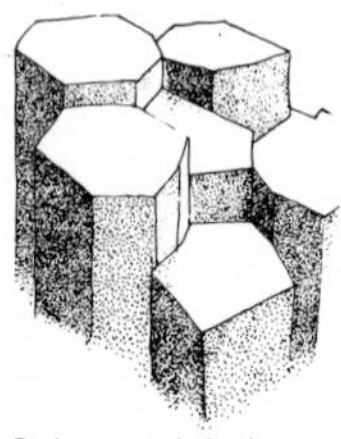

Columnar jointing in basalt

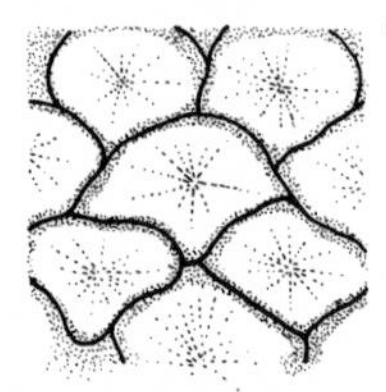

Pillow lava, as seen in cross-section

Basalt **Colour** When fresh it is black or greyish black; often weathers to a reddish or greenish crust. **Grain size** Fine. **Texture** Usually dense with no minerals identifiable in hand specimen; a freshly broken surface is dull in appearance. May be porphyritic. **Structure** Often vesicular and/or amygdaloidal. Xenoliths are relatively common and usually consist of olivine and pyroxene; they have a green colour. Columnar jointing is common and often spectacular, as at the Giant's Causeway, Northern Ireland; individual columns tend to be hexagonal, and up to 0·5 m across. *Spheroidal* weathering structures are sometimes developed in which successive layers break away like the skins of an onion leaving a rounded core. **Mineralogy** Phenocrysts are usually olivine (green, glassy), pyroxene (black, shiny) or plagioclase (white-grey, tabular). If olivine is present the rock is called olivine basalt. Microscopic examination shows the groundmass to consist of plagioclase (usually labradorite), pyroxene, olivine and magnetite, with a wide range of accessory minerals. Amygdales may be filled, or partly filled with zeolites, carbonates or silica, usually in the form of chalcedony or agate. **Field relations** Lava flows and narrow dykes and sills. The edges of dykes or sills are often finer grained than the centres or even glassy, due to rapid cooling on intrusion. Most basalts occur as lava flows either in volcanoes or as extensive sheets building up a *lava plateau,* which may cover hundreds of thousands of square kilometres, and may be fed by numerous fissures. The surface forms of lavas are of two principal types; smooth or ropy (the surface looks like rope) which is known by the Hawaiian term of *pahoehoe,* and scoriaceous which is rough and clinkery and has the Hawaiian name *aa*. Another common form is *pillow lava* which consists of pillows or balloon-like masses of basalt —usually with a very fine-grained or glassy outer layer. They are formed by the eruption of lava into water.

5 cm
Vesicular basalt
Olivine basalt
Diabase
Ropy basalt lava
Amygdaloidal basalt

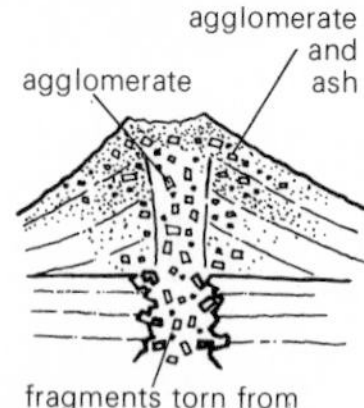

Section through a volcanic cone

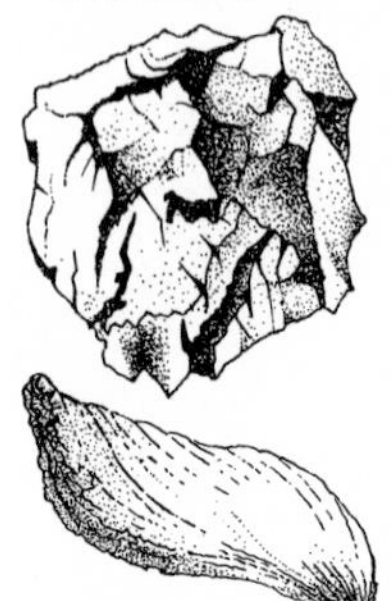
Typical shapes of volcanic bombs

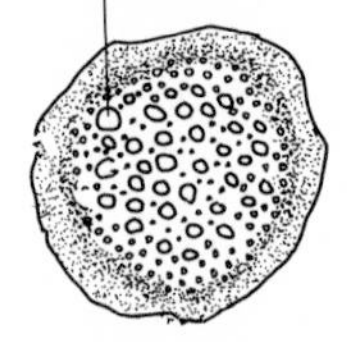
Typical vesicular interior of a volcanic bomb

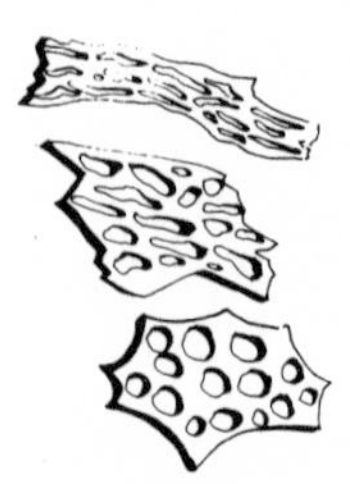
Ignimbrite: pumice fragments (fiamme)

Pyroclastic rocks

Agglomerate **Structure** Angular to sub-rounded fragments, more than 64 mm in diameter, in a finer-grained matrix. Most fragments are irregular and highly vesicular; others are rounded, ellipsoidal or spindle-shaped, usually with vesicular interiors, and are called bombs. Bombs are thrown from volcanoes as molten lava clots and acquire their shapes in flight. In modern volcanoes agglomerate is unconsolidated but becomes consolidated with time. **Composition** Bombs have the composition of the lava erupted by the volcano, such as basalt or andesite, but blocks may also be torn from the sides of the crater or from the rocks beneath the volcano. **Field relations** Agglomerate generally accumulates in the crater of a volcano or on the flanks close to the crater. Associated with tuffs and lavas.

Ash and tuff **Structure** Ash comprises unconsolidated volcanic fragments less than 2 mm in diameter; when consolidated it is called tuff. Ash and tuff commonly contain fragments (up to 64 mm in diameter) which are called lapilli (lapilli ash; lapilli tuff). Lapilli may be angular but are commonly spherical or ellipsoidal having been ejected in a molten state. Ashes and tuffs are usually layered like sedimentary rocks, and often graded within one layer so that larger fragments occur near the base. **Composition** Fragments are essentially of three types: crystalline rock (*lithic* ash, lithic tuff), for example rhyolite, trachyte or andesite; glassy fragments (*vitric* ash, vitric tuff) formed by ejection of liquid lava (these are commonly pumice fragments comprising glass with abundant vesicles); and individual crystals (*crystal* ash, crystal tuff) such as feldspar, augite and hornblende. Most ashes and tuffs are mixed including lithic, vitric and crystal fractions. **Field relations** Ashes are blown from volcanoes during eruptions and with lava flows and agglomerates build up volcanic cones. Generally the larger fragments are found near the crater, while finer-grained material may be carried by wind to greater distances, sometimes hundreds of kilometres from the volcano. They are associated with lavas, agglomerates and sedimentary rocks.

Ignimbrite **Structure** Essentially pieces of pumice, that is highly vesicular glass, usually less than one centimetre across, in a finer-grained matrix of glass fragments. Pumice fragments are often flattened particularly towards the base of flows when they are called *fiamme*. Layering commonly occurs, and columnar jointing is often developed. **Composition** Glass, usually having the composition of rhyolite or trachyte. Some phenocrysts may occur. **Field relations** Ignimbrites are considered to be deposited from ash flows which are ejected rapidly from volcanoes as great volumes of hot expanding gases and incandescent glass fragments. They are controlled by gravity and rush at great speed down the flanks of the volcano.

Volcanic bomb
Crystal tuff
Lithic tuff
Volcanic bomb
Volcanic bomb
Ignimbrite
5 cm

Metamorphic rocks

Contact metamorphism

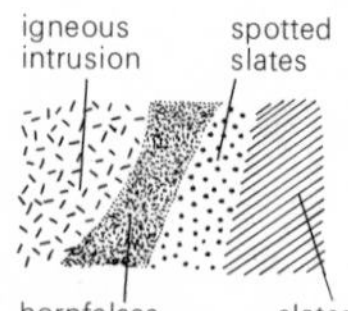

Contact metamorphic aureole showing relationship of hornfelses and spotted slates to igneous intrusion

Spotted slate **Colour** Black, purple, greenish or grey with darker spots. **Texture** Fine-grained, homogeneous; same texture as slate. **Structure** Same structures as slate, that is good cleavage, possibly also sedimentary structures. Characterized by the presence of spots, usually of a spherical or ovoid shape but may be rectangular, up to 3 or 4 mm in diameter. The spots are usually rather vague and filled with inclusions; they may pass gradually into hornfels in which spots are identifiable as cordierite or andalusite. **Mineralogy** Usually too fine-grained for identification in hand specimen, but sometimes the rectangular shapes of andalusite crystals can be identified. **Field relations** In the outer parts of contact metamorphic aureoles involving the metamorphism of pelitic rocks. They usually grade into hornfelses of higher grade towards the igneous intrusion.

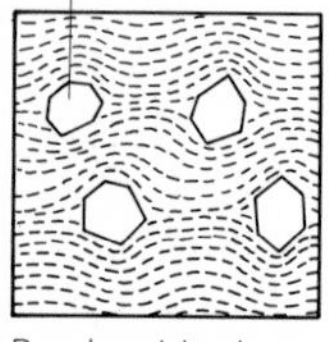

Porphyroblastic texture

Andalusite (Chiastolite) **cordierite hornfels** **Colour** Black, bluish, greyish, greenish. Often speckled with darker coloured porphyroblasts. **Texture** Matrix homogeneous, fine-grained, though a little coarser than slate; contains porphyroblasts or poikiloblasts of andalusite or cordierite which exceptionally may be several centimetres in diameter. The equigranular texture makes the rock tough and splintery. **Structure** Relict structures from the parental sedimentary rocks are usually obliterated by the metamorphic recrystallization, but occasionally primary bedding may be preserved. **Mineralogy** The matrix is too fine-grained to be distinguished, though tiny flakes of mica may be detected with a lens. Andalusite forms rectangular prisms of square section of a black colour, but if of large size may be reddish. Andalusite is sometimes present as poikiloblasts in which the very fine-grained dark inclusions assume a characteristic cross shape in cross-section, or a thin dark band along the centre of longitudinal sections; these crystals are known as chiastolite. Cordierite is rarely as well shaped as andalusite, occurring usually as rounded grains. When well formed it is hexagonal in section. **Field relations** Contact metamorphic aureoles; grades outwards into lower grade rocks such as spotted slate. Andalusite hornfelses may be directly in contact with the igneous intrusion which caused the metamorphism, or hornfelses of higher grade, such as pyroxene or sillimanite hornfels, may be interposed between them.

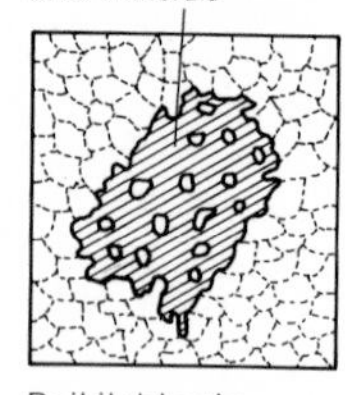

Poikiloblastic texture

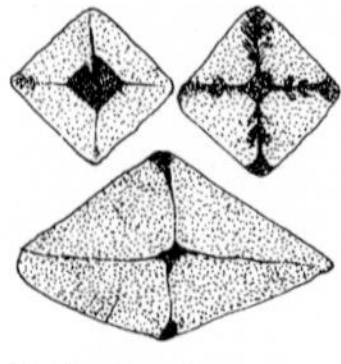
Inclusions in chiastolite

Pyroxene hornfels **Colour** Similar to andalusite-cordierite hornfels. **Texture** Homogeneous, fine- to medium-grained; often with porphyroblasts. **Structure** All primary sedimentary structures destroyed by recrystallization. **Mineralogy** Porphyroblasts of pyroxene, cordierite or andalusite, where present, are usually the only recognizable minerals. **Field relations** The innermost part (that is highest temperatures) of contact metamorphic aureoles.

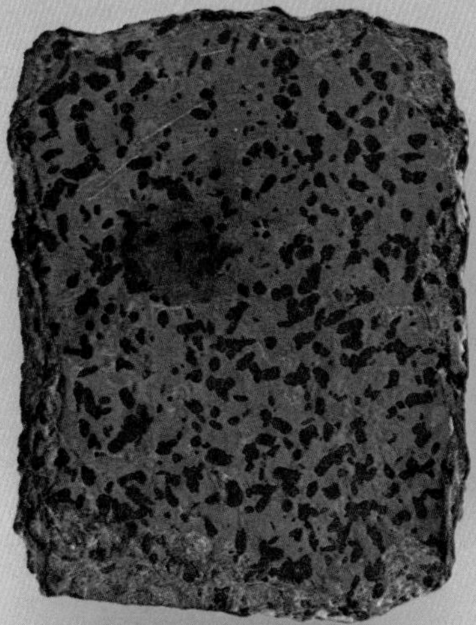

Spotted slate

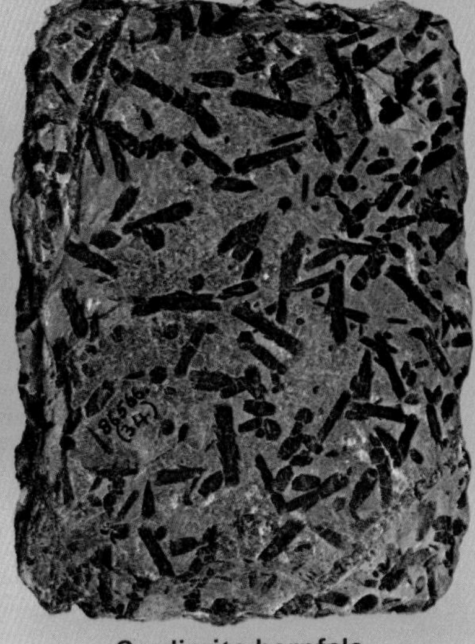

Cordierite hornfels

Chiastolite hornfels

Pyroxene hornfels

Chiastolite hornfels

5 cm

Marble **Colour** White or grey but a wide range of black, red and green occurs, often in streaks and patches. **Texture** Medium- to coarse-grained; granular; often of a sugary appearance. **Structure** Sedimentary structures such as bedding may be preserved. Fossils may be abundant at low metamorphic grades, but with increasing recrystallization they are destroyed. **Mineralogy** Essentially calcite, but may contain greater or lesser amounts of dolomite. Some brucite, olivine, serpentine, tremolite, phlogopite etc, may be present when it grades into calc-silicate rock and skarn. Marbles are readily scratched with a knife, which serves to distinguish them from the much harder white quartzites. **Field relations** Marbles are produced by the metamorphism of limestone around igneous intrusions. They are thus found in the vicinity of such intrusions but can usually be traced into unmetamorphosed limestone. They are associated with rocks such as hornfelses.

Calc-silicate rock **Colour** Similar colour range to marble but rarely pure white. **Texture** Medium- to coarse-grained; may contain porphyroblasts of calcium silicate minerals, sometimes of large size. **Structure** Original sedimentary bedding may be preserved. **Mineralogy** Calcite together with a wide range of possible minerals including periclase, olivine, serpentine, tremolite, diopside, woilastonite, idocrase and grossular garnet, which are commonly concentrated into bands, patches or nodules. **Field relations** Similar to marble except that calc-silicate rocks are formed by the metamorphism of 'impure' limestone containing some sandy or shaly material, which contributes the silicon, aluminium etc, necessary for the formation of calcium-silicate minerals.

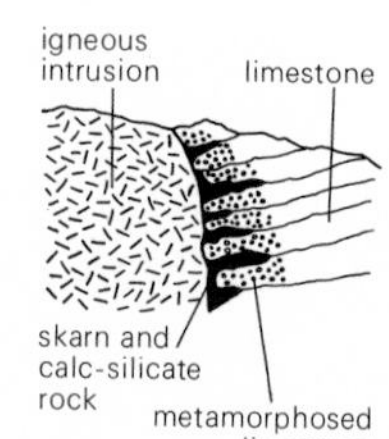

Skarn at contact of igneous intrusion, as seen in cross-section

Skarn **Colour** Brown, black, grey but commonly variable even on scale of a hand specimen. **Texture** Fine-, medium- or coarse-grained. **Structure** Minerals often concentrated into layers, nodules, lenses or radiating masses. **Mineralogy** As for calc-silicate rocks but with iron-rich pyroxene and garnet in addition. Sulphides of iron, zinc, lead and copper are associated with these silicate minerals. **Field relations** Skarns are produced at the contacts of granite, and sometimes of syenite and diorite, with limestone. Silicon, magnesium and iron from the magma migrates into, and reacts with, the limestone producing the silicate minerals, and sometimes forming ore deposits. Many good mineral specimens are obtained from skarns.

Halleflinta **Colour** Grey, buff; may be pink, green or brown. **Texture** Fine, even-grained; gives the rock a splintery fracture. **Structure** A layering, after bedding, may be apparent. **Mineralogy** Too fine-grained to distinguish with the naked eye. **Field relations** Represents metamorphosed tuffs which have been impregnated with secondary silica; often found in metamorphic aureoles.

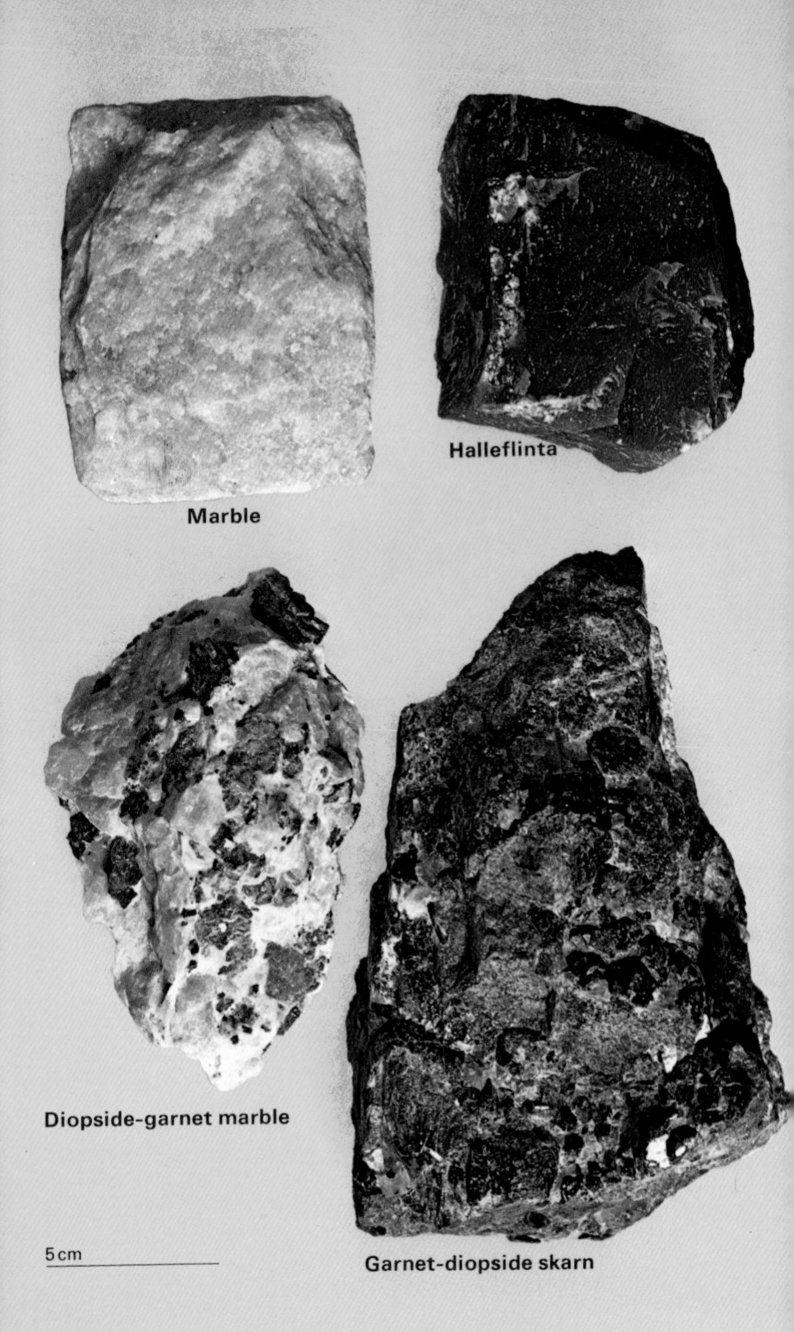
Marble
Halleflinta
Diopside-garnet marble
Garnet-diopside skarn
5 cm

Regional metamorphism

bedding cut across by cleavage
cleavage

Slate: showing relationship of cleavage to bedding

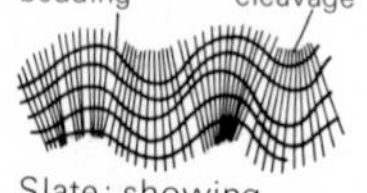

Slate: showing relationship of cleavage to large-scale folds

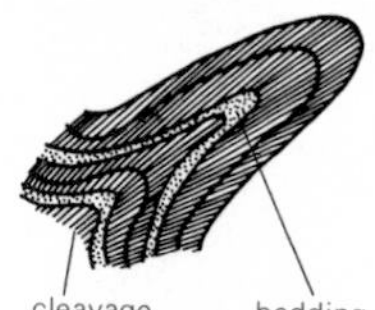

Cleavage developed across a small-scale fold

Slate **Colour** Black and shades of blue, green, brown and buff. **Texture** Fine-grained. **Structure** By definition, slates are characterized by a single, perfect cleavage (slaty cleavage), enabling it to be split into parallel-sided slabs. On the cleavage surfaces sedimentary structures such as bedding and graded bedding can often be seen. Fossils may be preserved but are invariably distorted. Folds are often apparent in the field. **Mineralogy** Too fine-grained to distinguish with the naked eye. Pyrite porphyroblasts often occur, usually as cubes. **Field relations** Slates are produced by low-grade regional metamorphism of pelitic sediments (shales, mudstones) or fine-grained tuffs. They may be associated with other metamorphosed sedimentary or volcanic rocks.

Phyllite **Colour** Usually greenish or greyish but with a characteristic silvery sheen. **Texture** Fine- to medium-grained. A well-developed schistosity is caused by the parallel arrangement of flaky minerals enabling the rock to be split easily into slabs, and producing the characteristic sheen. **Structure** Minor folds or corrugations are often present. **Mineralogy** Chlorite and/or muscovite are essential constituents, and give to the phyllites their green or grey colour. **Field relations** Phyllites are produced from pelitic rocks under conditions of low grade metamorphism. They usually grade into mica schists.

Chlorite schist **Colour** Usually green, may be greyish. **Texture** Fine- to medium-grained. Schistosity well developed. **Structure** Commonly folded on a large or small scale. Larger sedimentary features such as bedding may still be preserved. **Mineralogy** Chlorite, an essential constituent, usually develops as tiny subparallel or parallel flakes indistinguishable to the naked eye, but occasionally forms coarser knots and patches. Porphyroblasts of albite or chloritoid may occur. **Field relations** Chlorite schists form from pelitic rocks under the same grade of metamorphism as phyllites, with which they are associated.

Glaucophane schist **Colour** Dark coloured with a characteristic purplish or bluish tinge. **Texture** Fine- to medium-grained. Has a schistosity but not usually particularly well developed. Amphibole needles are often in parallel alignment. **Structure** Folding may be apparent. **Mineralogy** Characterized by the presence of glaucophane (a blue alkali amphibole); quartz, albite, jadeite, garnet and chlorite may also occur. **Field relations** Not common rocks, and probably represent about the same grade of metamorphism as chlorite schists. They are usually formed by high-pressure metamorphism of igneous rocks such as basalt and diabase, but may also be derived from sediments.

Slate

Glaucophane schist

Chlorite schist

Phyllite

5 cm

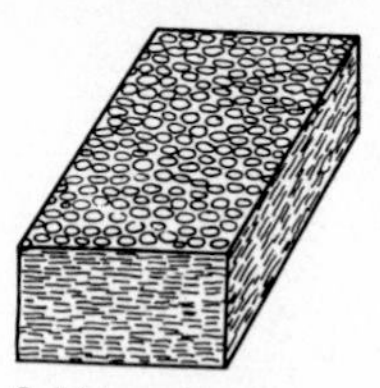

Schistose texture

Sericite schist and **muscovite schist** **Colour** Usually grey or white, and in coarser varieties the individual mica flakes are brightly reflecting. **Texture** Fine-, medium- or coarse-grained. Invariably has a well-developed schistosity. **Structure** Small scale corrugations may occur; sometimes comprises alternate mica-rich and mica-poor layers which may follow the original bedding. **Mineralogy** The dominant mineral is a white mica; if it is very fine-grained it is known as sericite, and the rock is called sericite schist; if the mica is coarser it is referred to by the name muscovite, and the rock is a muscovite schist. The muscovite flakes commonly reach 2 or 3 mm across and are easily prized free with a knife. Quartz is usually present and a little chlorite or garnet may be recognizable. **Field relations** Sericite and muscovite schists represent a moderate grade of metamorphism and tend to be associated with phyllites, chlorite schists and garnet-bearing schists. They are formed from pelitic sediments, while sandstones containing some argillaceous material give rise to quartz-muscovite schists.

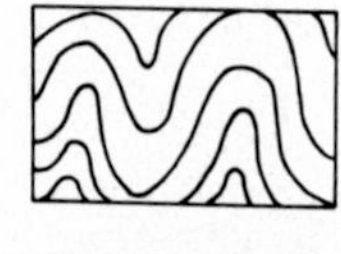

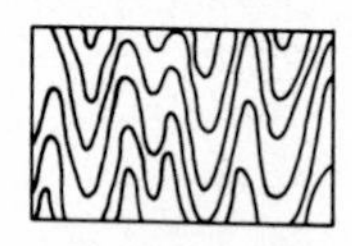

Some different styles of folding

Biotite schist **Colour** Usually brownish or black; individual mica flakes may reflect light brightly. **Texture** Coarse-, medium- or fine-grained. A well developed schistosity is invariably present, produced by parallel or subparallel mica flakes. **Structure** Many have a banded or striped appearance, caused by variation in the amount of mica present, which may represent original bedding or may be due to *metamorphic segregation*, that is the migration of material within the substance of the rock to form layers of differing chemical and mineral composition. Minor folding common. **Mineralogy** Biotite is the diagnostic mineral of these rocks. The green or brown elastic flakes are easily recognizable, and may be readily prized from the rock with a knife. Biotite is developed partly at the expense of chlorite and/or muscovite, but these minerals may still be present; indeed all proportions of biotite, muscovite and chlorite may occur. Quartz is usually concentrated into mica-poor layers. Feldspar may be present, commonly in the form of conspicuous white porphyroblasts. **Field relations** Biotite is one of the index minerals of regional metamorphism and its presence indicates a grade of metamorphism higher than that of the chlorite schists. With decreasing metamorphism biotite schists grade into sericite and chlorite schists and phyllites, while with increasing metamorphism they pass into garnet-bearing schists. Biotite schists are abundant rocks in most regional metamorphic terrains, and represent metamorphosed pelitic sediments.

Folded sericite schist

Quartz-muscovite-biotite schist

Biotite schist

5 cm

Folded biotite schist

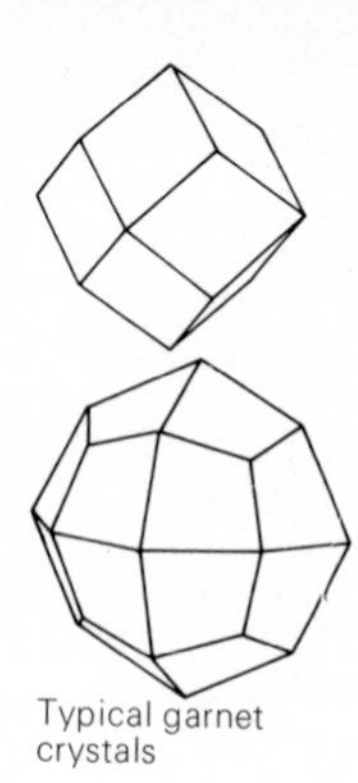

Typical garnet crystals

Garnet-mica schist **Colour** Black, brown, reddish or pinkish. **Texture** Medium- to coarse-grained. A schistosity is invariably well developed. Garnet porphyroblasts are common and may reach a centimetre or more across; the schistosity often bends around the porphyroblasts giving an eye-like appearance. **Structure** Folds of various sizes may occur, and a banded appearance due to variations in the amount of mica or concentration of garnet in particular layers is usual. **Mineralogy** Together with the essential garnet, biotite, muscovite and quartz may be found. The garnet is usually the reddish variety called almandine, and it often forms well shaped crystals that can sometimes be freed from the rock. **Field relations** Garnet-bearing schists are widespread in terrains of regional metamorphism. Garnet is a metamorphic index mineral and indicates a degree of metamorphism higher than that of the biotite schists. Garnet-biotite schists are usually derived from pelitic sediments.

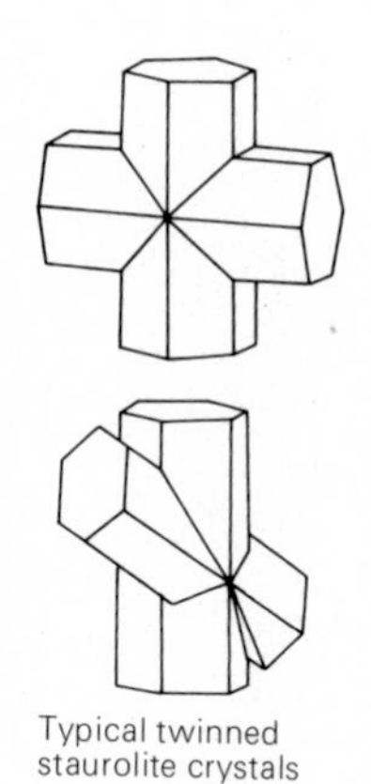

Typical twinned staurolite crystals

Staurolite schist **Colour** Black, brown, reddish. **Texture** Medium- to coarse-grained. The schistosity is usually good, but may be interrupted by porphyroblasts of staurolite and garnet. The staurolites are often randomly orientated, cutting across the foliation. These rocks are often coarse and banded owing to metamorphic segregation (see biotite schist) when they become gneisses. **Structure** May be folded. **Mineralogy** Staurolite forms porphyroblasts with the characteristic prismatic habit and often as twins in the form of a cross. The most commonly associated minerals are garnet, biotite, muscovite, feldspar and quartz. **Field relations** Staurolite-bearing schists are rocks of comparatively high metamorphic grade, and tend to be associated with other high grade rocks such as kyanite and sillimanite schists in the central parts of metamorphic belts. They are the product of metamorphism of pelitic sediments.

Albite schist **Colour** Grey, greenish, brownish. **Texture** Fine-, medium- or coarse-grained. May be phyllitic, schistose or massive. Porphyroblasts of albite are usually conspicuous. **Structure** May be folded. **Mineralogy** White or grey porphyroblasts of albite may be associated with chlorite, epidote, biotite, muscovite or garnet. **Field relations** Rocks containing conspicuous porphyroblasts of albite may be found amongst phyllites and schists of the chlorite, biotite or garnet metamorphic zones. Albite schists probably result from the metamorphism of rocks which originally contained a high proportion of albite, perhaps feldspathic grits or sandstones; or the albite may be a product of metasomatism, having migrated into the schists.

5 cm

Albite schist

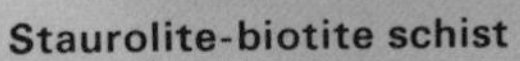

Staurolite-biotite schist

Garnet-muscovite-biotite schist

Garnet-biotite schist

Bladed kyanite crystal

Kyanite schist **Colour** Grey, brown, reddish, with distinctive blue crystals. **Texture** Medium- to coarse-grained. Schistose; may also be gneissose. Kyanite often forms porphyroblasts, sometimes of large size. **Structure** May be folded. **Mineralogy** Kyanite forms sky blue porphyroblasts of a simple bladed habit which lie parallel to the foliation, or may be concentrated into clusters of crystals; it may occur also in veins with quartz. Other minerals which may be present include garnet, staurolite (either may form porphyroblasts), biotite, muscovite, quartz and feldspar. **Field relations** Found in the central, high grade part of metamorphic belts, in association with sillimanite and staurolite schists.

Sillimanite schist **Colour** Brown, grey, reddish brown. **Texture** Medium- to coarse-grained. Schistosity is not usually striking; may be gneissose. **Structure** May be folded. **Mineralogy** The sillimanite forms needles, which tend to be extremely fine, and slender prisms. Commonly associated minerals include biotite, muscovite, garnet, feldspar and quartz, and these may be concentrated in contrasting layers. **Field relations** At the highest grades of metamorphism sillimanite replaces kyanite as an index mineral, so that sillimanite schists and gneisses are found in the central parts of metamorphic belts in association with rocks such as kyanite and staurolite schists, and perhaps granites and migmatites.

Pyroxene granulite **Colour** Dark colours–grey, brown. **Texture** Medium- to coarse-grained. Tough, massive rocks which may be layered or banded, but not usually schistose. **Mineralogy** Pyroxene, either hypersthene or diopside, is characteristic. Garnet, kyanite, sillimanite, biotite, hornblende, quartz or feldspar may also be present. **Field relations** The granulites are considered to be formed at very high temperatures and pressures, which imply great depths in the crust. They are, therefore, found in the very old continental shield areas, which have undergone considerable erosion to reveal rocks formed at great depths.

Eclogite **Colour** Greenish, reddish; sometimes pale to dark green with brownish red spots. **Texture** Medium- to coarse-grained. Massive, banded. Large porphyroblasts of garnet or pyroxene may occur. **Mineralogy** Dominantly composed of pyroxene of a green variety called omphacite, and a reddish garnet. Kyanite and even diamond sometimes occur. The garnet–pyroxene mineralogy is unique and diagnostic. **Field relations** Restricted occurrence being found as lenses and blocks in masses of metamorphic and igneous rocks; particularly in association with peridotites and serpentinites in regions with major faults, and as xenoliths in serpentinite and kimberlite. This suggests that they have been transported from considerable depths, probably from the base of the crust or the mantle. Their very high density–they are among the densest known silicate rocks–supports this hypothesis.

5 cm

Kyanite-biotite schist

Garnet-pyroxene granulite

Sillimanite-garnet schist

Eclogite

Actinolite-chlorite schist **Colour** Green. **Texture** Fine- to medium-grained; schistosity well developed. **Structure** Primary features of igneous origin such as amygdales may be recognizable. **Mineralogy** Flaky green chlorite and fine green needles of actinolite, which sometimes form radiating clusters or are concentrated into small patches, are characteristic. Feldspar, epidote and calcite may also be identified. **Field relations** Produced by the metamorphism of basic igneous rocks such as diabase and basalt. They occur in phyllites and chlorite schists as cross-cutting layers which trace the position of the original basic dykes, sills or lava flows.

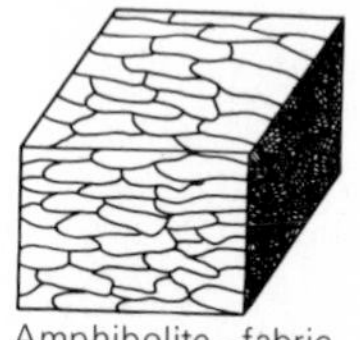

Amphibolite—fabric of subparallel prisms

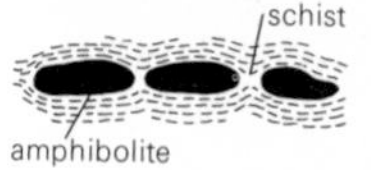

Disrupted layer of amphibolite within schist

Amphibolite and hornblende schist **Colour** Black, dark green, green; may be streaked or flecked with white, grey or red. **Texture** Medium- to coarse-grained. A well-developed foliation or schistosity may be present, but this is due to stout prisms of hornblende lying in one plane and often aligned within it, rather than to flaky minerals such as micas. The rock does not split as easily as schist. Some varieties are massive, the minerals having no preferred orientation; such rocks have an 'igneous' appearance and are conveniently called *epidiorite*. Porphyroblasts, particularly of garnet, may be present. **Structure** A fine banding or layering of alternating dark and light coloured layers may be developed. These rocks are relatively massive and unyielding so that small scale folds are rarely developed. **Mineralogy** Amphibole, commonly hornblende but sometimes actinolite or tremolite, comprises some 50 per cent and often as much as 100 per cent of the rock. Actinolite and tremolite tend to form fine needles or long prisms whereas hornblende forms short, stubby, black shining prisms. Other commonly associated minerals are feldspar (particularly in hornblende schist), chlorite, epidote, pyroxene and garnet; the latter often forming dark red porphyroblasts. A massive rock composed wholly or largely of amphibole is generally called amphibolite, whereas one with a schistosity, and in which there is a large proportion of another mineral, such as feldspar, is called hornblende schist. **Field relations** Result mostly from the metamorphism of igneous rocks such as diabase and basalt, but are of a higher metamorphic grade than actinolite-chlorite schists. As the primary igneous rocks were often sills or dykes cutting sedimentary rocks, so amphibolites and hornblende schists are found as conformable and cross-cutting layers in metamorphosed sediments such as schists, marbles and quartzites. Amphibolites are tough rocks which tend to resist shearing and so they often occur as disrupted lenses and fragments among schists and gneisses. They are relatively common rocks in most metamorphic terrains.

Epidiorite
Actinolite-chlorite schist
5 cm
Garnet-hornblende schist
Hornblende schist
Amphibolite

Marble **Colour** White (the best statuary marble), yellow, red, black or green; either uniform or blotched, banded or veined in differing shades. **Texture** Medium- to coarse-grained; tends to be evenly granular, often sugary in appearance. **Structure** Commonly massive but may have a layering or banding which is usually a primary bedding structure. Schistosity or cleavage is rarely present in pure marbles, but being relatively plastic they may flow readily under high pressures so that folds of a highly contorted type may be developed. **Mineralogy** Calcite and/or dolomite are the major constituents, so that these rocks are relatively soft (they can be scratched easily with a knife which distinguishes them from the harder, white quartzites), and calcite marbles effervesce with dilute hydrochloric acid. If the original limestone contained sand, silt or clay, then the resulting marble contains such minerals as phlogopite, diopside, tremolite, grossular garnet, olivine, serpentine (after olivine) and many others. It is the presence of these minerals that produces the attractive range of colours and structures found in marbles. **Field relations** Marbles are formed by the metamorphism of sedimentary limestones, so they are found in metamorphic terrains in association with other metamorphosed sediments such as quartzite, phyllite and many types of schist. Note that marble can also be formed by contact metamorphism of limestones (see page 176).

Granoblastic texture

Quartzite **Colour** White, grey, reddish. **Texture** Medium-grained; usually of a granoblastic texture. **Structure** Usually massive but primary sedimentary features may be preserved, such as bedding, graded bedding or current bedding. **Mineralogy** Essentially composed of tightly interlocking grains of quartz. A little feldspar or mica may also be evident. White varieties are distinguished from marble by their greater hardness. **Field relations** Quartzites are metamorphosed quartz sandstones and are found in association with other metamorphosed sedimentary rocks such as phyllite, schist and marble.

Quartzo-feldspathic schist **Colour** White, grey, reddish, brownish. **Texture** Medium- to coarse-grained. A poorly developed schistosity is usually apparent, causing the rock to break into slabs. **Structure** May be folded. **Mineralogy** Quartz is a major constituent, together with feldspar. Micas, both muscovite and biotite, usually occur and tend to concentrate in particular layers. **Field relations** Grades, with an increase of quartz, into quartzite, and by a decrease in quartz into schist and phyllite. Represents metamorphosed grit and sandstone which originally contained a high proportion of feldspar, and perhaps also some mica or clay. Occurs in association with other metamorphosed sedimentary rocks such as quartzite, schist and marble.

Serpentine marble

Quartzo-feldspathic schist

Quartzite

Marble

5 cm

Gneissose texture

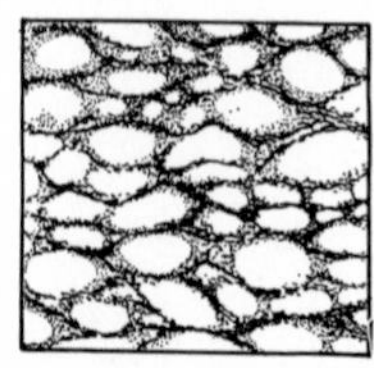

Augen texture

Layering in gneiss

Ghost structure in migmatite

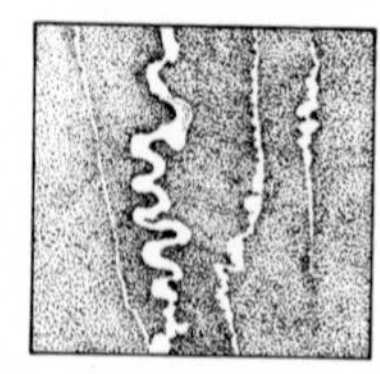

Ptygmatic vein in migmatite

Gneiss and **augen gneiss** **Colour** Grey or pink but with dark streaks and layers. **Texture** Medium- to coarse-grained. Characterized by discontinuous, alternating light and dark layers, the former usually having a coarsely granular texture while the latter, which often contains mica, may be foliated. Augen gneiss contains large porphyroblasts of feldspar, or aggregates of feldspar and quartz, often a centimetre or more across, which are eye-shaped, hence the term augen. **Structure** In addition to the gneissose texture described above, gneisses tend to be banded on a large scale with layers and streaks of darker and lighter coloured gneiss. Granite and quartz veins and pegmatites are common. May be folded. **Mineralogy** Feldspar is abundant and, together with quartz, forms the granular, lighter coloured layers. Muscovite, biotite and hornblende are commonly present, while any of the minerals characteristic of the higher grades of regional metamorphism may occur. **Field relations** At the highest grades of metamorphism rocks may approach melting temperatures when they are able to recrystallize freely and so produce the textures characteristic of gneisses. Thus gneisses occur, in association with migmatites and granites, in the central parts of metamorphic belts.

Migmatite **Colour** A mixture of darker coloured host rocks and lighter coloured (white, pink, grey) granitic rocks. **Texture** Medium- to coarse-grained. The various components of migmatites may display schistosity or gneissose or augen textures. **Structure** Migmatites are mixed rocks comprising a host, usually of schist or gneiss, and a granitic component which may form layers, pods or veins, or be more evenly distributed through the rock as porphyroblasts of feldspar, or clusters of feldspar and quartz grains (granitization). If the granitization is very extensive the rock as a whole may approach granite in composition, but original structures such as layering, folding etc are usually still discernible, and are called *ghost structures*. Quartzite, amphibolite and marble tend to resist granitization, and may form layers and isolated masses in the migmatite. Evidence that these rocks were 'plastic' is often provided by swirling folds, and complexly folded granite veins, known as *ptygmatic veins*. **Mineralogy** The host rocks have a mineralogy appropriate to schists or gneisses of high metamorphic grade. The granite fraction comprises essentially alkali feldspar and quartz. **Field relations** Found locally in the inner part of the contact aureoles of large granite intrusions but on a regional scale in terrains of medium and high grade metamorphism, particularly in the old Archaean continental shields which, because of the considerable erosion to which they have been subjected, reveal rocks that were formed at considerable depths.

Folded granite gneiss

Gneiss

5 cm

Augen gneiss

Granite gneiss with ptygmatic folds

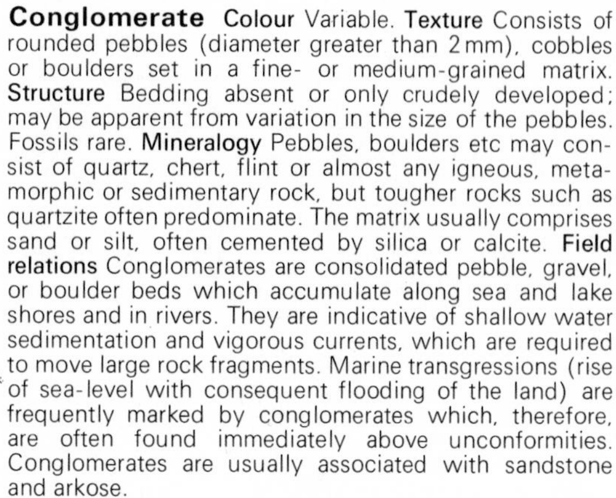

Sedimentary rocks

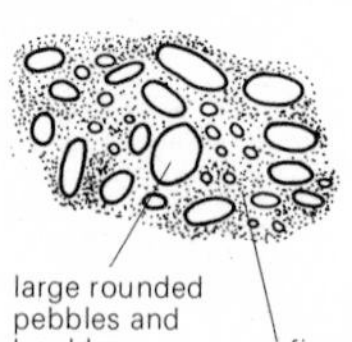

Conglomerate

Conglomerate **Colour** Variable. **Texture** Consists of rounded pebbles (diameter greater than 2 mm), cobbles or boulders set in a fine- or medium-grained matrix. **Structure** Bedding absent or only crudely developed; may be apparent from variation in the size of the pebbles. Fossils rare. **Mineralogy** Pebbles, boulders etc may consist of quartz, chert, flint or almost any igneous, metamorphic or sedimentary rock, but tougher rocks such as quartzite often predominate. The matrix usually comprises sand or silt, often cemented by silica or calcite. **Field relations** Conglomerates are consolidated pebble, gravel, or boulder beds which accumulate along sea and lake shores and in rivers. They are indicative of shallow water sedimentation and vigorous currents, which are required to move large rock fragments. Marine transgressions (rise of sea-level with consequent flooding of the land) are frequently marked by conglomerates which, therefore, are often found immediately above unconformities. Conglomerates are usually associated with sandstone and arkose.

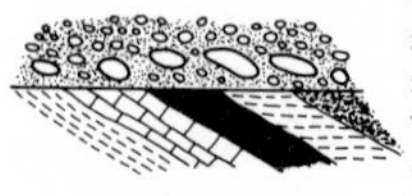

Conglomerate above unconformity

Breccia

Breccia **Colour** Variable. **Texture** Consists of angular rock fragments (2 mm to many metres in diameter) set in a fine- to medium-grained matrix. In some breccias the fragments can be seen to match along their opposed sides, indicating only modest disturbance. **Structure** Bedding not usual, though in some types of breccia bedding is apparent in the matrix. Fossils rare. **Mineralogy** The fragments may be of any type of igneous, metamorphic or sedimentary rock. The matrix usually consists of silt or sand cemented by calcite or silica. **Field relations** Many breccias represent consolidated talus or scree material, that is accumulations of rock fragments formerly lying on steep hill slopes, or at the foot of cliffs. They are often found above unconformities, and associated with conglomerate, arkose and sandstone. Other breccias are produced by the fragmentation of rocks during faulting.

Ice-scratched pebble

Till and tillite **Colour** Tills are reddish brown or grey; tillites, dark grey to greenish black. **Texture** Angular and some rounded pebbles, cobbles and boulders, characteristically unsorted, that is, of a wide range of sizes, set in a fine- to medium-grained unconsolidated (till) or consolidated (tillite) matrix. Pebbles are often scratched as a result of abrasion during glacial transport. **Structure** Unbedded. **Mineralogy** The rock fragments may be of any type, and in till are in a clayey or sandy matrix, while in tillites the matrix is consolidated into shale or even slate. **Field relations** Till, also known as boulder clay, is deposited by glaciers, and forms widespread surface deposits in the higher latitudes of the northern continents. The recognition of tillites, which are fossil tills, in sedimentary and metamorphic rocks of considerable antiquity (for instance Precambrian, Palaeozoic), indicates that there have been a number of ice ages.

Conglomerate

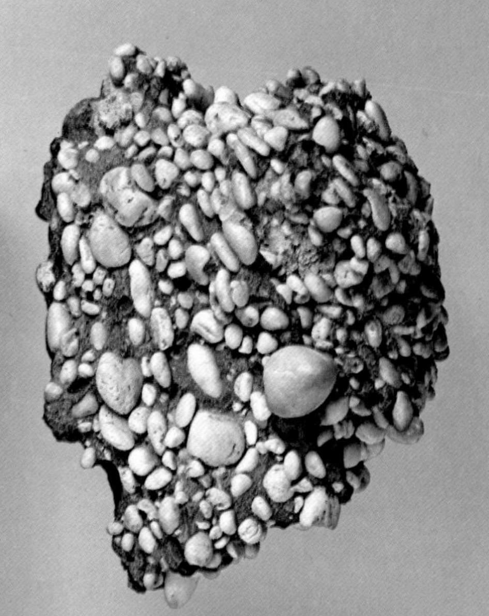

Conglomerate

Tillite

Breccia

5 cm

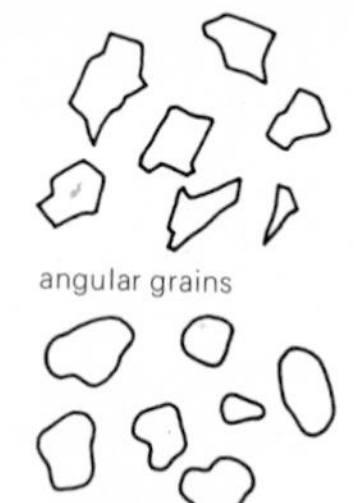

Sand grains

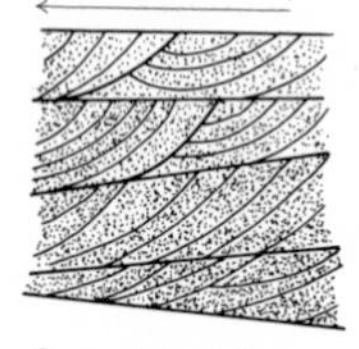

Current bedding

Sandstone, grit and orthoquartzite **Colour** Very variable; frequently red, brown, greenish, yellow, grey, white. **Texture** Medium-grained. Usually well sorted, that is grains all about the same size; grains sharply angular (grit), or subangular to rounded (sandstone). **Structure** Bedding usually apparent; current bedding and ripple marks common; graded bedding may occur. Concretions and fossils may be found. **Mineralogy** Quartz is the main component but is often accompanied by feldspar, mica or other minerals. The grains may be cemented by silica, calcite or iron oxides. Rocks composed almost wholly of quartz grains with a silica cement are known as 'pure' sandstones or orthoquartzites. Green glauconite occurs in the variety known as greensand. Sands and sandstones rich in olivine, rutile, magnetite and other minerals, are found locally. **Field relations** Sandstones are associated with most other sedimentary rocks. Most sands accumulated either in water, usually the sea, or as wind-blown deposits in arid continental areas. Desert sandstones tend to be red, and the individual sand grains are often almost spherical and polished.

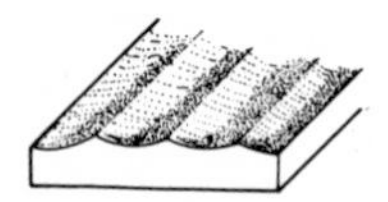
Ripple marking

Arkose **Colour** Red, pink or grey. **Texture** Medium-grained, but usually nearer to the coarse end of the scale; grains angular. **Structure** Bedding may be obscure or well developed; often current bedded. Fossils rare. **Mineralogy** Contains 25 per cent or more of feldspar, rarely more than 50 per cent; the rest is mainly quartz, but some biotite and muscovite may occur. The cement is usually calcite or iron oxides. **Field relations** Arkoses are derived from the disintegration of granite and granite gneisses, and because they are composed of quartz and feldspar they resemble granites, but the angular, fragmental nature of the grains serves to distinguish arkose from the closely interlocking igneous texture of granite. Arkoses occur above unconformities in the immediate vicinity of granitic terrains, or in thick deposits associated with conglomerates (containing granite boulders) derived from granites or gneisses.

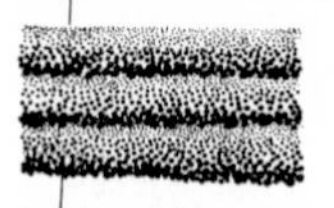

Graded bedding in greywacke (three units)

Greywacke **Colour** Grey to black, sometimes greenish; usually dark in colour. **Texture** Typically contains sharply angular grains (up to 2 mm) in a finer-grained matrix. **Structure** Frequently massive; graded bedding is typical. Individual graded beds are coarse-grained (sandy, perhaps with a few small pebbles) at the base, and pass upwards into silt or clay at the top. Fossils rare. Slump structures common. **Mineralogy** Coarser grains consist of quartz, feldspar and rock fragments; matrix too fine-grained to be distinguished by the naked eye. The green colour is due to the presence of chlorite. **Field relations** Greywackes, the typical sediments of *geosynclines*, that is rapidly subsiding marine basins of deposition, are thought to have been deposited by *turbidity currents;* these are masses of sediment-charged water which flow down slopes on the sea bed, and deposit their load of sediment in deep water. Individual units are probably deposited very rapidly, hence their ill-sorted nature.

Coarse sandstone (grit)
Arkose
Sandstone
Orthoquartzite
Greywacke
5 cm

Siltstone **Colour** Grey to black, brown, buff, yellow. **Texture** Grain size $\frac{1}{16}$ to $\frac{1}{256}$mm; individual grains can often just be distinguished with the naked eye. Compact, even-textured; but may be earthy. **Structure** Often shows finely laminated bedding, sometimes picked out by contrasting colouring; may be homogeneous or massive. Fine scale current bedding and ripple marking also occur. Fossils often abundant, as are nodules and concretions. **Mineralogy** Too fine-grained for minerals to be distinguished except for rare larger grains of quartz and feldspar. On some bedding surfaces the glint of micas may be seen. **Field relations** Siltstones form by the compaction of sediment of silt grade which may have accumulated in the sea, in lakes, or be amongst the residual materials of glacial action.

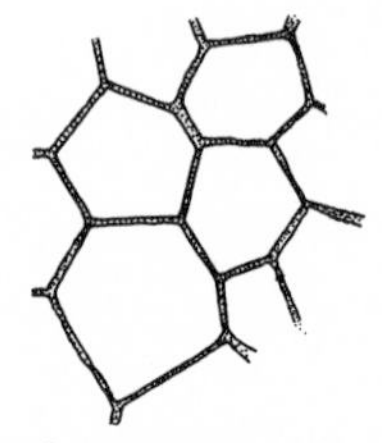
Sun cracks in mudstone

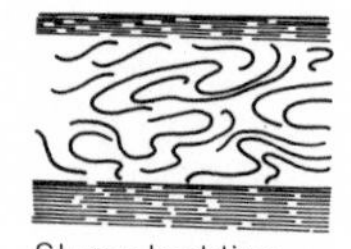
Slump bedding

Mudstone, shale and clay **Colour** Black, grey, white, brown, red, dark green or blue. **Texture** Grain size less than $\frac{1}{256}$mm; individual grains are too small to be distinguished with the naked eye. Mudstone and shale feel smooth, and a pure clay is not gritty when smeared between the fingers. Clays are plastic and often sticky when wet. **Structure** When consolidated and relatively massive it is known as mudstone (or claystone); if finely bedded so that it splits readily into thin layers it is called shale. When soft and uncompacted it is termed clay. Sun cracks, rain prints etc sometimes occur on bedding surfaces; and fossils and concretions are common. **Mineralogy** Too fine-grained for minerals to be distinguished with the naked eye, or usually even with the microscope. Clays consist of a mixture of clay minerals together with detrital quartz, feldspar and mica. Iron oxides are usually abundant and contribute the red and yellow colours. Black shales are rich in carbonaceous matter, and pyrite and gypsum commonly occur in them, sometimes as well shaped crystals. **Field relations** Clays tend only to occur in the younger geological formations, being consolidated into mudstones and shales with time. Being very fine-grained, clay is easily transported by water into the sea and lakes, where it accumulates with silt, sand and calcareous organisms to form typical sequences of shales, siltstones, sandstones and limestones. Some clays are *residual*, having formed *in situ* as soils; such are the bauxitic clays (see mineral section).

Aeolian clay (loess) **Colour** Yellow, brown, buff, grey. **Texture** Fine-grained. Easily powdered in the fingers; compact, earthy, porous. **Structure** Bedding poor. Fossils infrequent. **Mineralogy** Too fine-grained for individual minerals to be seen with the naked eye. **Field relations** Aeolian clays are deposited by wind and the material is probably ultimately of glacial origin. Great thicknesses cover China (loess) and it is widespread elsewhere.

Mudstone
Shale
Shale
Siltstone
Loess
5 cm

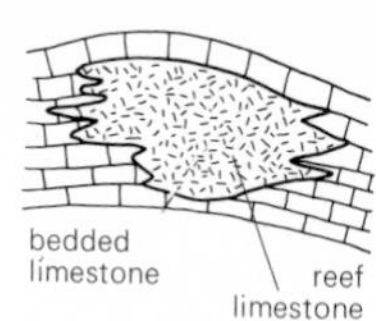

Reef limestone, as seen in cross-section

Limestone (biochemical) **Colour** White, grey, cream or yellow when pure; red, brown, black when impure. **Texture** Highly variable from very fine-grained, and porcellaneous, to coarsely crystalline and of sugary appearance. If fossils are present their abundance and nature partly determine the texture. **Structure** Bedding usually developed. A wide variety of fossils occur and it is rare not to find some evidence of organic remains. The fossils may be complete, fragmental, or partly destroyed by recrystallization. In richly fossiliferous types the rock usually comprises an assortment of fossil fragments amongst an interstitial finer-grained limestone matrix. Large scale structures such as fossil coral reefs with the corals in their original attitudes may sometimes be observed in large field exposures. Limestones are often criss-crossed by calcite and mineralized veins. **Mineralogy** Essentially comprises finely divided calcite (calcareous mud) or larger crystals which may be derived from animal skeletons such as crinoid plates or by recrystallization, particularly along veins. Finely crystalline silica in the form of chert, and forming bedded or nodular masses, sometimes occurs. Quartz, silt or muddy sediment may be present, and with an increase of these constituents limestones pass into calcareous sandstones and shales and mudstones. **Field relations** Biochemical limestones are formed principally of the accumulations of the calcareous skeletons of organisms, and they are widely distributed. They form in three principal ways: as reefs, which comprise corals, algal colonies, etc, together with the remains of animals living in and on the reefs such as crinoids and brachiopods; as widespread bedded limestones consisting of the skeletons of bottom living (benthonic) organisms including many types of gastropods, lamellibranchs and brachiopods; and as accumulations of the skeletons of floating (pelagic) organisms. The first two types are characteristic of relatively shallow water, while the pelagic limestones may form in deeper water. Some limestones, which can usually be distinguished by the nature of their fossils, are formed in fresh water.

Chalk **Colour** White, yellow, grey. **Texture** Fine-grained; porous; compact or friable. **Structure** Bedding not usually apparent on small scale. Flint and marcasite nodules common. Fossils usually present. **Mineralogy** Chalk is a very pure limestone formed of calcite, containing only small amounts of silt or mud. Secondary silica (flint) and marcasite are common. **Field relations** Chalk is a pelagic limestone consisting mainly of the tests of coccoliths, foraminifera and other free-swimming micro-organisms embedded in a fine-grained calcareous mud. It formed in open seas in which there was little or no deposition of other sediments. Most chalks are Cretaceous in age, but similar deposits are accumulating in some parts of the oceans at the present time.

Crinoidal limestone

Chalk

Shelly limestone

Fossiliferous freshwater limestone

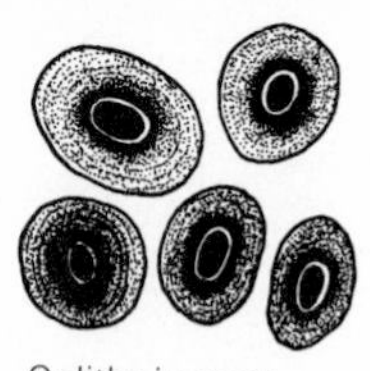

Ooliths in cross-section, as seen with a microscope

Oolitic and pisolitic limestone **Colour** White, yellow, brown, red. **Texture** Ooliths are spheroidal or ellipsoidal structures built up of concentric layers, and measuring up to 2 mm in diameter (commonest size about 1 mm). Larger, more irregular structures, up to pea size, are called pisoliths. Rock composed essentially of closely packed ooliths is called oolite and resembles fish roe. The ooliths may also be dispersed through a finer-grained matrix. **Structure** Often current bedded. Fossil fragments occur. **Mineralogy** Usually composed of calcite but ooliths composed of dolomite, silica and hematite (see page 202) also occur. The matrix is also calcite but a few grains of quartz and other detrital minerals are usually present. **Field relations** Ooliths are forming at the present time in certain warm, shallow, strongly agitated parts of the sea such as the Bahamas Banks. They form by the precipitation and accretion of carbonate around quartz grains and shell fragments rolled along the bottom. Oolitic limestones tend to grade into other limestones and sandstones.

Calcareous mudstone **Colour** White, grey, yellowish. **Texture** Fine-grained, exhibiting subconchoidal fracture; homogeneous. **Structure** May be bedded; fossils rare. **Mineralogy** Calcite. Some detrital material may be present but it is very fine-grained. **Field relations** Probably formed in relatively deep water, partly as accumulations of the skeletons of free-swimming micro-organisms, and partly as chemical precipitates. The lack of fossils may be due to their recrystallization and breakdown.

Dolomite **Colour** White, cream or grey, but often weathers brown or pinkish. **Texture** Coarse, medium or fine; compact, sometimes earthy. **Structure** Bedding tends to be large scale. May be massive or contain complex concretions and nodular growths. Conspicuously jointed. Organic remains usually destroyed by recrystallization. **Mineralogy** Contains a high proportion of dolomite (the mineral and rock have the same name). Detrital minerals and secondary silica (chert) may be present. **Field relations** Dolomites are usually interbedded with other limestones and are commonly associated with salt and gypsum deposits. Most dolomites are thought to be of secondary (replacement) origin; the calcite of the original limestone having been replaced *in situ* by dolomite, probably by percolating watery solutions.

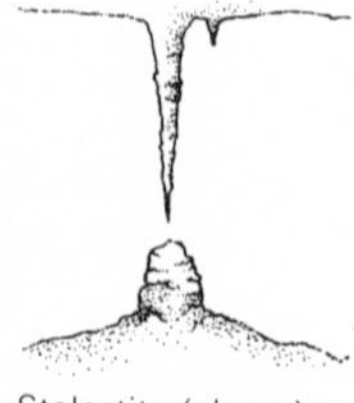

Stalactite (above); stalagmite (below)

Travertine and tufa **Colour** White, yellow, red, brown. **Texture** Compact to earthy; friable. **Structure** Tufa is a porous or spongy rock, while travertine is more dense and often banded. Stalactites are pendant growths from cave roofs, and stalagmites the corresponding floor accumulations; internally they display concentric growth rings. **Mineralogy** Principally calcite; impurities of iron oxides are responsible for yellow and red colours. **Field relations** These rocks are produced by the precipitation of calcite through water evaporation around springs or in caves, where they form thin deposits of no great extent. They are also deposited from water around geysers and hot springs.

5 cm

Pisolitic limestone

Calcareous mudstone

Travertine

Oolitic limestone

Dolomite

Ironstone **Colour** Brown, red, green, yellow. **Texture** Fine, medium or coarse; sometimes oolitic. **Structure** Finely and coarsely bedded; also current bedded. Commonly nodular. Organic remains common but usually fragmentary. **Mineralogy** Characterized by the presence of a high proportion of iron-bearing minerals (at least 15 per cent iron) the commonest of which are siderite, hematite, magnetite, pyrite, limonite, chamosite and glauconite. Detrital minerals usually present, while calcite and dolomite are common cementing agents. **Field relations** They are usually interbedded with cherts, limestones and sandstones, and may be classified as mudstones, oolites, limestones, sandstones, etc with suitable mineralogical prefixes. Most ironstones are thought to be chemical deposits, the iron having been precipitated from solution.

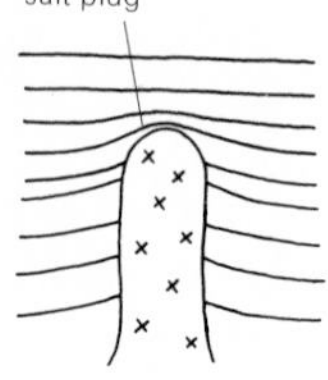

Form of salt plug, as seen in cross-section

Evaporites

Rock salt **Colour** Colourless, white, orange, red, yellow or rarely purple. **Texture** Massive, coarsely crystalline, glassy or sugary; or as distinct cubic crystals (halite). **Structure** In thick, structureless, massive beds, commonly with partings of shale. Often strongly distorted owing to flow. Fossils rare. **Mineralogy** Essentially halite; easily detected by strong saline taste. Impurities include associated salts (carbonates, sulphates), clay minerals and iron oxides. **Field relations** Formed by the evaporation of saline waters in lagoons, seas and inland lakes—hence the name evaporites. Particularly associated with shales and dolomites; and with *red beds* (marls and sandstones) which are indicative of formation under desert conditions. Often forms *salt plugs* due to the upward intrusion of the low density rock salt into overlying sediments.

Rock gypsum **Colour** White, pink, red, green or brown. **Texture** Coarse to fine; massive, saccharoidal (sugary) or fibrous; earthy; friable. **Structure** May show bedding, which is often strongly distorted. Usually interbedded with sandstone, mudstone or limestone in which large gypsum crystals occur. Fossils rare. **Mineralogy** Gypsum is commonly associated with anhydrite, halite, calcite, dolomite, clay minerals and iron oxides. **Field relations** Most gypsum deposits are thought to have been formed by the hydration (addition of water) of anhydrite, which forms in similar environments to rock salt.

Phosphate rock **Colour** Black, brown, yellow, white. **Texture** Bedded phosphate is fine- to coarse-grained; compact, earthy or granular (sometimes oolitic). Guano is friable; earthy. **Structure** Bedded phosphate rocks are usually nodular and contain organic remains, often replaced by phosphate minerals. Guano is usually bedded. **Mineralogy** Essentially phosphates of calcium, iron and aluminium, but very complex; associated with common detrital minerals. **Field relations** The most extensive phosphate rock deposits are associated with marine sediments, typically glauconite-bearing sandstones (greensand), limestones and shales. Guano is an accumulation of the excrement of sea birds and is found principally on oceanic islands.

Chamositic ironstone
Rock salt
Oolitic ironstone
osphate rock
Rock gypsum
m

Nodules and concretions

Pyrite nodules **Colour** Bronzy yellow (when freshly broken) but weather to brown, yellow or black. **Texture** When broken open usually reveal radiating, acicular crystals. **Structure** May be spherical, nodular, botryoidal or cylindrical. The surface may be smooth, rough or covered with wart-like knobs. **Mineralogy** Pyrite. **Field relations** Pyrite nodules are widespread in a broad range of sedimentary rocks, but particularly in pelitic rocks, especially when these are of a black or blackish grey colour. They also occur in limestone.

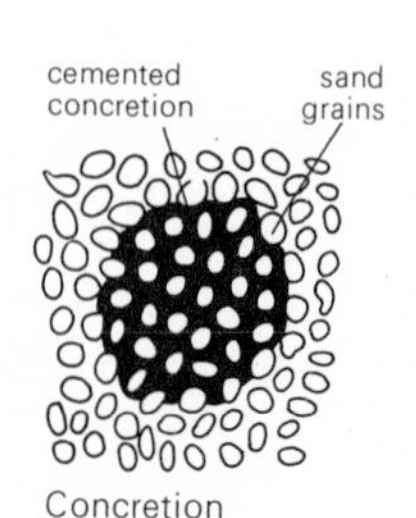

Concretion

Flint and chert nodules **Colour** Blue-grey, grey, to nearly black when fresh, but weather to a whitish, powdery crust (patina). **Texture** Very fine-grained and smooth; conchoidal fracture. Rough on weathered surfaces. **Structure** Flint and chert form rounded nodules of widely differing forms, but chert also forms massive beds. Flint nodules are often hollow and may contain a fossil, such as a sponge or echinoid. **Mineralogy** Composed of silica, mainly the variety chalcedony. Some authors distinguish flint and chert compositionally but the differences, if any, are slight. **Field relations** Flint and chert nodules occur typically in limestone and chalk. They are usually patchily distributed but often concentrated along one bedding plane. Their origin is not fully understood; some appear to be secondary replacements of the host rock, whilst others may represent primary deposition on the sea bed of colloidal silica.

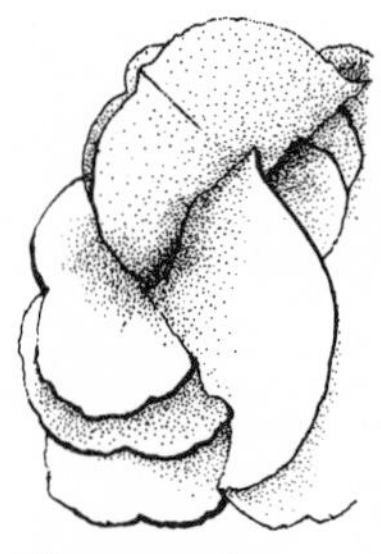

'Desert rose' cemented by gypsum

Concretions **Colour** Similar to host rock. **Texture** Similar to host rock. **Structure** Spherical, ellipsoidal, disc-shaped etc; with sand crystals the shape is determined by the crystallographic habit of the cementing mineral. Bedding is unbroken from the host rock through the concretion. Concretions vary in size from a few millimetres to several metres across. **Mineralogy** Concretions comprise essentially the same material as the host sediment but are cemented (concreted) together, usually by silica, carbonate or iron oxides. The cement gives the concretion added strength so that it is resistant to weathering and can readily be detached, as a discrete unit, from the surrounding rock. **Field relations** Found in a wide variety of rocks but particularly in shales, siltstones and sandstones. Concretions form by the local deposition within the sediment, probably by percolating waters, of the cementing mineral. Sand crystals are formed by the crystallization in loose sand of crystals of minerals such as baryte, calcite and gypsum (see desert rose on page 74).

'Desert rose' cemented by baryte

Septaria **Colour** Black to dark brown or yellow. **Texture** Similar to host sediment. **Structure** Spheroidal to ellipsoidal. Distinguished by a radiating and polygonal pattern of veins, more easily seen when cut open, and which in weathered specimens may stand up as a series of wall-like ridges. **Mineralogy** Pelitic sediment cemented by carbonate; the veins are usually calcite. **Field relations** Found in pelitic sediments. The mechanism of formation is complex and is not fully understood.

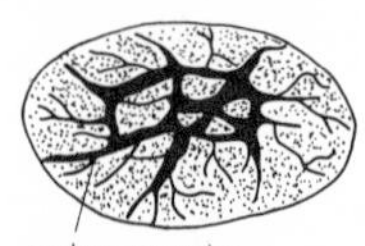

Septaria, as seen in cross-section

5 cm
Pyrite nodule
Pyrite nodule
Pyrite nodule
Flint nodule
Sandstone concretion
Mudstone concretion
Septarian concretion
Mudstone concretion

Meteorites

Every day thousands of solid bodies, which have originated in space, enter the Earth's atmosphere. Most of these are burned up but it is estimated that each year something like 500 survive the passage through the atmosphere and land on the surface of the Earth. These bodies are known as meteorites. As few as about ten meteorites which are seen to fall are recovered each year, and only some 2,000 authentic meteorites are recorded. The importance of meteorites is two-fold; firstly, they provide the best evidence we have for the composition and early history of the solid matter of the solar system; and secondly, it is considered probable that the composition of some meteorites approaches that of the interior of our planet, about which we have very little direct information.

Mineralogy The commonest minerals that occur in meteorites are of two kinds: the silicates, consisting principally of olivine, pyroxene and plagioclase; and the nickel-iron alloys, kamacite (Fe, Ni with 4–7 per cent Ni) and taenite (Fe, Ni with 30–60 per cent Ni). The presence of kamacite and taenite is the most outstanding difference between the mineralogy of meteorites and terrestrial rocks, in which they occur only very rarely. The iron sulphide troilite (FeS) is also common in meteorites but the majority of the sixty or so other minerals recorded are found only in small amounts.

Classification Meteorites are classified, principally on their mineralogy, into three groups, namely, irons, stony-irons and stones. The stones are further subdivided according to the presence or absence of small spherical bodies called *chondrules*. Stones containing chondrules are called chondrites; those without are called achondrites.

Irons The largest meteorites which have been found are irons, and the biggest of these is estimated to weigh 61,000 kg. They are usually irregular in shape and may have deep cavities in them, or protuberances from the surface, which may have been caused by collisions in space, fragmentation in flight, weathering, or impact with the Earth's surface. The surface may be smooth, furrowed or covered with shallow depressions. A freshly fallen iron will have a black *fusion crust*, owing to melting by atmospheric friction in flight. This crust is very thin (less than a millimetre), and is usually confined to one surface. Most irons have a brownish colour due to oxidation of the iron during weathering.

The irons consist mainly of nickel-iron alloys. Many irons when cut and etched with acid reveal a complex intergrowth of kamacite and taenite known as *Widmanstätten structure*, which is found only in meteorites.

Stony-irons The meteorites of this group are composed of nickel-iron and silicate minerals in about equal proportions, and usually consist either of well shaped crystals of olivine in a continuous matrix of nickel-iron, or of plagioclase and pyroxene set in a discontinuous nickel-iron matrix. The stony-irons comprise only about 4 per cent of the known meteorites. The surface features and weathering described for the irons apply also to the stony-irons, but in the latter, particularly the olivine-bearing ones, the silicate minerals may weather out preferentially giving the surface a rough, pitted aspect.

Stones About 90 per cent of all meteorites which have been observed to fall are stones; more than 90 per cent or more of these are chondrites. Most chondrites have a composition of about 30 per cent pyroxene, 40 per cent olivine, 10 per cent plagioclase, 5–20 per cent nickel-iron and 6 per cent troilite. The chondrules are composed mainly of olivine or orthopyroxene and may be abundant or sparse. The achondrites are coarser-grained than the chondrites, and resemble some terrestrial rocks in texture and mineralogy. They are rather variable in composition but consist principally of one or more of the minerals plagioclase, pyroxene and olivine.

Individual stones tend to be equidimensional in shape, though some may be angular owing to fragmentation by collision or terrestrial impact. Sometimes they are conical or dome-shaped (like the Apollo command modules) owing to a constant orientation when travelling through the atmosphere, so that one side suffers considerable heating with consequent loss of material. The fusion crust of

Stony meteorites

hondrite showing ablation crust

Chondrite showing chondrules

recciated achondrite

Weathered stony meteorite

stones is thicker than that of irons, often black in colour, and may be dull or shiny. It may have a fluted or furrowed form caused by flow of molten material from the front to the back of the meteorite during flight through the atmosphere.

The interior of stones is usually grey or dark grey in colour, granular in texture and round chondrules may be apparent. Some nickel-iron metal may be evenly disseminated throughout or occur in occasional patches. They are difficult to recognize in the field because of the similarity in texture to terrestrial rocks.

Summary of features for recognizing meteorites Meteorite falls are so infrequent there is no point in going especially to search for them, unless one has been reported to have fallen in a particular area. If you think that you have found a meteorite, however, the main features to look for in hand specimen are: firstly, the presence of chondrules; secondly, the presence of fusion crust; and thirdly, the presence of nickel-iron alloy, either comprising the whole meteorite, or as finely disseminated, shining grains or patches on a freshly broken surface. If you have recovered an object that was seen to fall from the sky, then it may be a meteorite, and it should be submitted to a museum or university for expert examination.

Tektites

Tektites are small glassy objects which, unlike meteorites, are found only in certain, rather limited areas of the Earth's surface. They are named according to the area in which they are found and the principal types are: *australites* from the southern part of Australia, Tasmania and coastal islands; *philippinites* from the Philippine Islands and southern China; *javaites* from Java; *malaysianites* from Malaysia; *indochinites* from Thailand and Indochina; *Ivory Coast tektites* from the Ivory Coast, West Africa; *bediasites* from Texas, United States; *Georgia tektites* from Georgia, United States; and *moldavites* from western Czechoslovakia.

It has been estimated that something like 650,000 tektites have been collected, of which the philippinites account for some 500,000.

Tektites are usually small, the majority being less than 300 gm in weight, and about 1 to 3 cm across, but some examples up to 12 kg are recorded. The shape of tektites is very variable but discoid, lensoid, button-shaped, tear-drop, dumb-bell, spherical and boat shapes commonly occur.

Some tektites are smooth and shiny but others have a rough, strongly etched and abraded surface, often with a system of grooves which reflect flow patterns within the glass. Most tektites are jet-black in colour but thin flakes are transparent or translucent in shades of brown. The moldavites, however, are dark green and in thin flakes are transparent and bottle-green in colour.

Chemically, tektites comprise a silica-rich glass which is also rich in alumina, potash and lime, and can be matched by a few igneous and sedimentary rocks. This has led to theories for the origin of tektites by the melting of terrestrial rocks through the impact of large meteorites or comets with the Earth. Other theories proposed an extra-terrestrial origin but a terrestrial origin is now favoured.

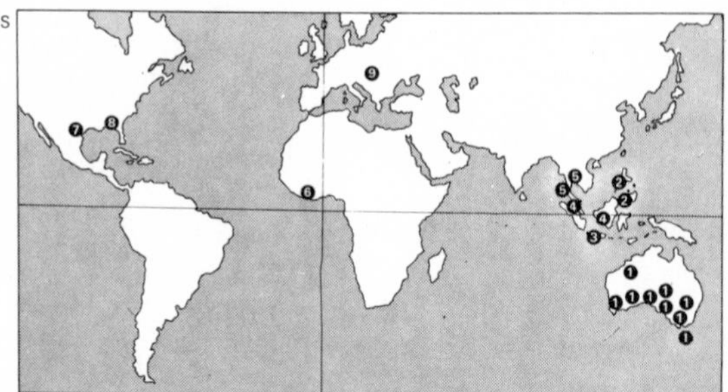

World finds of tektites

Key
1 australites
2 philippinites
3 javaites
4 malaysianites
5 indochinites
6 Ivory Coast tektites
7 bediasites
8 Georgia tektites
9 moldavites

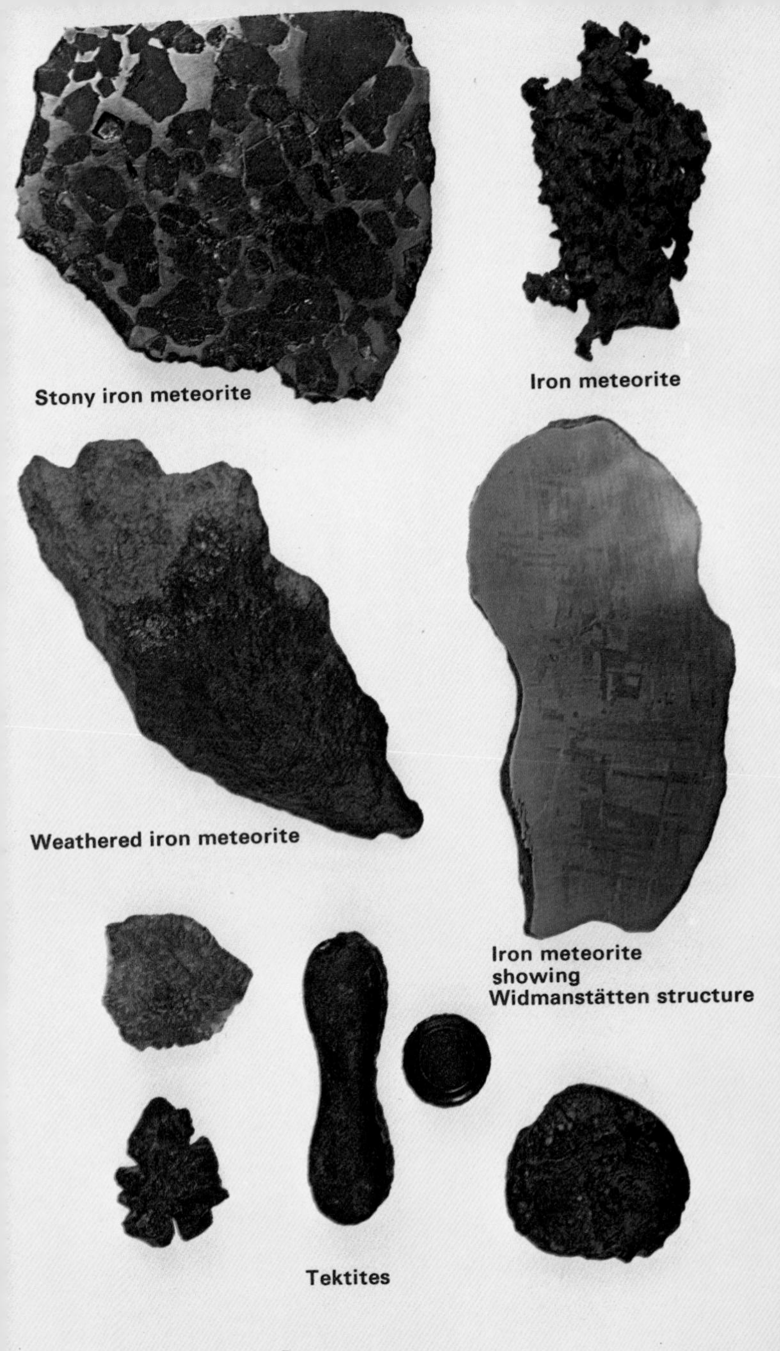

Stony iron meteorite

Iron meteorite

Weathered iron meteorite

Iron meteorite
showing
Widmanstätten structure

Tektites

Fossils

Fossils are the remains of animals or plants that are preserved in the rocks. It is very unusual for complete organisms to be preserved and fossils usually represent the hard parts such as bones, shells or tests of animals and the leaves, seeds or woody parts of plants. Fossils may be internal moulds (for example most of the ammonites shown here on pages 250 to 259), external moulds (for example brachiopods on pages 270 to 279) or the original material impregnated with chemicals from the surrounding rocks (for example vertebrates on pages 302 to 309). Alternatively they may be indirect evidence of life in the past such as footprints, burrows or borings. These are called *trace fossils* and with the exception of the borings of *Teredo* (page 262) they are not dealt with in this book. Most trace fossils, however, are readily identifiable by reference to similar phenomena in modern environments.

Fossils are found in most sedimentary rocks and are particularly common in limestones and some shales. The majority of fossils represent aquatic animals, as conditions for preservation are usually better in aquatic environments than on land. In many cases terrestrial animals and plants are preserved only in aquatic sediments, either in the sea, rivers, lakes or estuaries. For example, fossil land mammals are often found in the same deposits as fish, crocodile and turtle remains, which indicates aquatic conditions at the time of deposition.

How fossils are formed Remains of soft-bodied organisms are known, and even jellyfish may be fossilized in very special conditions. The vast majority of fossils, however, consist only of the hard parts of the organism, and in general the possession of such hard parts may be regarded as an essential prerequisite for fossilization. Even the hardest parts of an organism will be broken down or dispersed if they are exposed to scavenging animals, bacterial action or the weather, and for fossilization to occur it is essential that these factors be excluded; rapid burial shortly after the death of the organism usually ensures this. The medium in which the remains are buried may vary widely but the commonest materials are muds, sands and volcanic ash. Some of the more spectacular fossils result from preservation in special conditions, for example insects preserved in amber (page 292), which is itself fossil resin, the complete mammoths preserved in the permanently frozen ground of Siberia, or the hundreds of thousands of mammal bones preserved in the tar pits of California.

Many remains of Caenozoic invertebrates and vertebrates represent the original material of the animal. It is often unchanged chemically or may simply be impregnated with minerals, such as silica or calcite, that enter in solution from the surrounding sediments. This process tends to increase the weight and the hardness of the fossil and is known as *petrifaction*. Alternatively chemical changes may occur in the material of the remains leading to recrystallization. This rarely affects the appearance of the fossil but may totally alter the fine structure. In many cases, especially in Palaeozoic and Mesozoic fossils, the original material of the organism may have been entirely dissolved away. This tends to leave a space in the consolidated sediments that may become occupied by minerals from the surrounding rocks—a process known as *replacement*. If this occurs gradually as the original molecules of the remains are dissolved away, then the fine structure of the organism will be preserved. If the solution of the original material is rapid, however, and replacement is not immediate, then the original structure will be lost although the original external appearance may be retained. Some replaced fossils are very beautiful especially if the replacing material is silica or some of the iron compounds such as pyrite (iron sulphide). In some cases the tissues of the organism may be converted into a film of carbon—a process known as *carbonization*. The results of this process are well demonstrated here by some of the plants (page 214 to 221) and the graptolites (page 280).

Names Almost all living animals and plants have popular names which may be well known and in some cases are relatively precise, referring only to a single kind of animal or plant over a large area. On the other hand, fossils are rarely familiar enough to have acquired popular names and those few that are well known are generally vague, usually referring to large groups, for example ammonites and dinosaurs. Popular names are not used scientifically as they have three important disadvantages. Firstly, they are imprecise and may refer to quite different animals or plants in different areas; secondly, they are not international but differ in other languages; and thirdly, many rarer animals and plants and the majority of fossils have never received popular names. To overcome this each organism, either modern or fossil, has a scientific name which is in Latin or is a Latinized form of a word from another language. The disadvantage of scientific names is that many of them are long and difficult to pronounce or remember. This is a minor objection, however, in the face of their many advantages. Scientific names are international and are therefore the same to scientists of every nationality; they are precise and define exactly the kind of organism referred to; and there is a name for every known organism. If a new form is discovered a name is created by the worker who first publishes its description.

Animals and plants are grouped in several categories which indicate their degree of relationship, one to another, and some of these categories are used or mentioned in this section of the book. Each animal or plant belongs to a species. This is a group of very similar individuals that have the potential to interbreed freely to produce fertile offspring, but are unable to breed successfully with members of other species. There are a few examples of animals and more examples of plants in which the species boundaries seem to break down, allowing successful breeding between members of different species. These are exceptions, however, and successful breeding is not possible between members of the large majority of different species. The scientific name for a species consists of two parts, i.e. it is a binomial, with a *generic* name followed by a *trivial* or *specific* name, for example *Homo sapiens*. The specific name is always used in conjunction with the generic name and is meaningless when used alone. The generic name refers to the *genus* which is a group of species that are generally similar and are fairly closely related. For example, the genus *Equus* (horses) includes several species such as the domestic horse *Equus caballus*, the wild ass *Equus asinus* and the zebra *Equus zebra*. Members of these three species are clearly very similar in general appearance, anatomy, way of life and behaviour, but differ in minor features such as their colour and details of their anatomy. Note that the generic and specific names are always printed in italics (or underlined when writing and typing). The generic name always begins with a capital letter and may be used alone to refer to the genus.

Genera are grouped into families which are major groups of generally similar organisms, for example the Felidae, which includes all the cat-like animals such as the domestic cat, lynx, tiger, lion, mountain lion, cheetah and jaguar. Family names are not printed in italics but may be easily identified as they always end with the letters 'ae'. Families are grouped into orders the members of which differ in many important features from members of other orders. For example, the order Carnivora includes the flesh-eating mammals such as the cats, weasels, hyaenas, dogs and raccoons. Members of the Carnivora differ in many obvious features from members of a major order of plant-eating mammals such as the Artiodactyla which includes the pigs, deer, giraffes and antelopes. The names of orders are not printed in italics and are not always easily identifiable. They always begin with a capital letter and most of them end with the letter 'a', but this is by no means always the case and several names of higher categories also have the same ending. Specimens are grouped on the basis of orders in several parts of this section of the book. For example, in the bryozoans (pages 232 to 237) the four orders Cryptostomata, Trepostomata, Cyclostomata and Cheilostomata are used. Orders are grouped into classes and several of these are referred to. For example, four classes of molluscs are described: these are the Gastropoda, Cephalopoda, Pelecypoda (referred to as bivalve molluscs) and Scaphopoda. In many cases the names of orders and classes have been converted into popular names by slight alteration of the ending, for example gastropods. Members of different classes differ from each other in many ways but their differences are less fundamental than those distinguishing members of the different phyla (singular phylum), which are the major divisions within the animal kingdom and form the basis on which the major parts of this section of the book are arranged. Phyla described here include the Arthropoda, Mollusca, Bryozoa and Echinodermata.

There are well over a million species of animals and plants so it is clearly impractical to attempt a treatment at the species level, and even at the generic or family level it is impossible to be comprehensive. An attempt has therefore been made to show common, widespread genera of the larger forms within the phyla that are well represented as fossils. Formal names for the phyla are not used if a more popular name is considered equally good, as confusion is unlikely to occur at this level. Animals are much more important than plants as fossils and the greater part of this section is therefore devoted to fossil animals. Several important groups of animals are omitted, either because they are relatively rare as fossils (for example Annelida), or because their members are usually very small (for example Foraminifera) and require specialized collecting techniques and specialist knowledge for their identification. The majority of specimens are referred to by their generic names and are grouped in the sections either in orders or classes. The molluscs are the most important larger fossil animals and consequently they are given far more space than any other group (pages 238 to 269). Some groups, such as the insects and fishes, are extremely difficult to identify and in these cases a few examples are given with no further attempt to explain the groups to which they belong.

Why study fossils? Palaeontology, or the study of fossils, is an important branch of geology. The findings of palaeontologists also have a large and increasing importance in zoological and botanical studies as well as having direct application in the search for minerals. Fossils provide the only direct evidence of life in the past and are used in the interpretation of anatomical features and for studies of the relationships of living organisms. Detailed evolutionary histories of many animals and plants have been discovered by the study of fossils and some of these are among the most interesting and popular stories known from any of the natural sciences.

Fossils may be used to provide indications of the age of the rocks in which they occur, and to provide correlations between rocks of the same age over wide geographical areas. Indeed, until fairly recently all estimates of the age of rocks were relative and were based upon the interpretation of their contained fossils. This use of fossils is still of prime importance and fossils are widely used in the oil industry and in the search for other fossil fuels such as coal and natural gas. Estimates of the ages of rocks from their contained fossils depend on comparisons of their faunas and floras with similar assemblages in rocks of known age. It is therefore necessary to identify the fossils accurately and the identification and interpretation of fossils is the work of many geologists in museums, geological surveys, universities and in industry. This is not to say that the amateur cannot identify his specimens; indeed the identification of most fossils collected is well within the ability of any good collector.

During this century new methods have been developed which allow the age of some rocks to be assessed on an absolute time scale, that is giving the age in millions of years. These methods are based upon the rate of change of radioactive isotopes which are trapped in some rocks when they are formed.

For Caenozoic rocks the most useful of these methods relies on the change of radioactive potassium into argon and is known as the potassium-argon or K/Ar method. In older rocks different radioactive changes are used, such as the conversion of rubidium to strontium, while in relatively recent remains the age of bone and wood can be calculated from the amount of a carbon isotope (C^{14}) that they contain. These techniques have not superseded the use of fossils in geological work but they have allowed the geological time scale to be worked out with greater accuracy. Fossils are used in conjunction with these new techniques; they are valuable for dating almost all sedimentary rocks whereas the new techniques are more restricted in their application. An interesting example of the use of the two techniques is presented by the rich deposits of Lower Miocene mammals known from East Africa. Many of the remains are preserved either in volcanic sediments or in beds sandwiched between layers of volcanic ash. The volcanic material can be accurately dated by the K/Ar method, thus providing accurate dates for the fossil mammals. These faunas may then be compared with faunas of fossil mammals from the rest of Africa and from Europe and Asia, and where close agreement is found the date for these other deposits may be determined, even though they contain no volcanic rocks.

Fossils suitable for dating rocks must have wide geographical ranges and must change relatively quickly with time. Ammonites and other molluscs, brachiopods, trilobites and echinoderms fit these two requirements and are particularly useful for dating many rocks. The ammonites are particularly important in this respect and here they have been divided into fairly narrow time ranges. The amateur collector, however, is more likely to encounter the situation in reverse. He will probably know the age of a deposit, either from his geological map or from geological books on the area, and knowing the age, he will be more easily able to identify his specimens from the smaller number of genera that occur in a given period of geological time. In this section the time range of each genus is given in broad terms and the geographical range is then added with abbreviations as follows: NA North America; SA South America; E Europe; Af Africa; Aust Australasia; Worldwide indicates probable occurrence in all these regions plus Asia. (In this book, the term Recent is used for those specimens occurring during the last few thousand years.)

Fossil plants

These are relatively less important than fossil animals in geology, although pollen and algae are important for dating some rocks. Representatives of the major groups of larger land plants are included here.

Psilopsids Silurian–Devonian: NA E Af Asia Aust The earliest known land plants. Very simple and lacking leaves, roots or seeds.

Psilophyton Devonian: NA Forking growth form typical, stems covered with spines and having shoots with coiled tips in young forms.

Coal Measure plants

The Coal Measures are the major source of fossil plants and tip heaps from coal mines are the best places to collect. Many groups of advanced plants were already in existence by coal measure times.

Lycopsids Devonian–Recent: Worldwide Living forms club mosses; small and unimportant members of tropical floras; about 900 living species; reproduction by spores. In Coal Measures they were very important and some had trunks reaching up to 30 m high before branching. Trunks and branches covered with leaves having a spiral arrangement.

Lepidodendron Carboniferous: E The end of a branch showing short, very narrow, leaves and a single branch. Branching is by repeated forks. Scars left by leaf bases in older parts of stem are oval and arranged spirally.

Sphenopsids Devonian–Recent: Worldwide Living form *Equisetum* (horsetail). Important in Coal Measures, some forms over 40 m tall. Stem jointed and internal casts show vertical ridges. Shoots and branches arise at joints of stem.

Calamites Internal cast of stem showing vertical ridges and joints.

Annularia These are the detached leaves of a form similar to, or identical with *Calamites*.

Fern-like plants Devonian–Recent: Worldwide Some Carboniferous fern-like plants are probably true ferns, but some reproduced by seeds rather than spores and are known as seed-ferns.

Pecopteris This is probably a true fern and carries numerous small leaflets which have a distinct mid-rib, and are attached to the stem by their whole bases.

Ptychocarpus Generally similar to *Pecopteris* and common in Europe and North America.

Neuropteris This is a seed-fern and not closely related to the above two genera. The growth form of the true ferns and the seed-ferns was almost identical, however. Seeds of this plant are small and ovoid with three or four ridges on their sides.

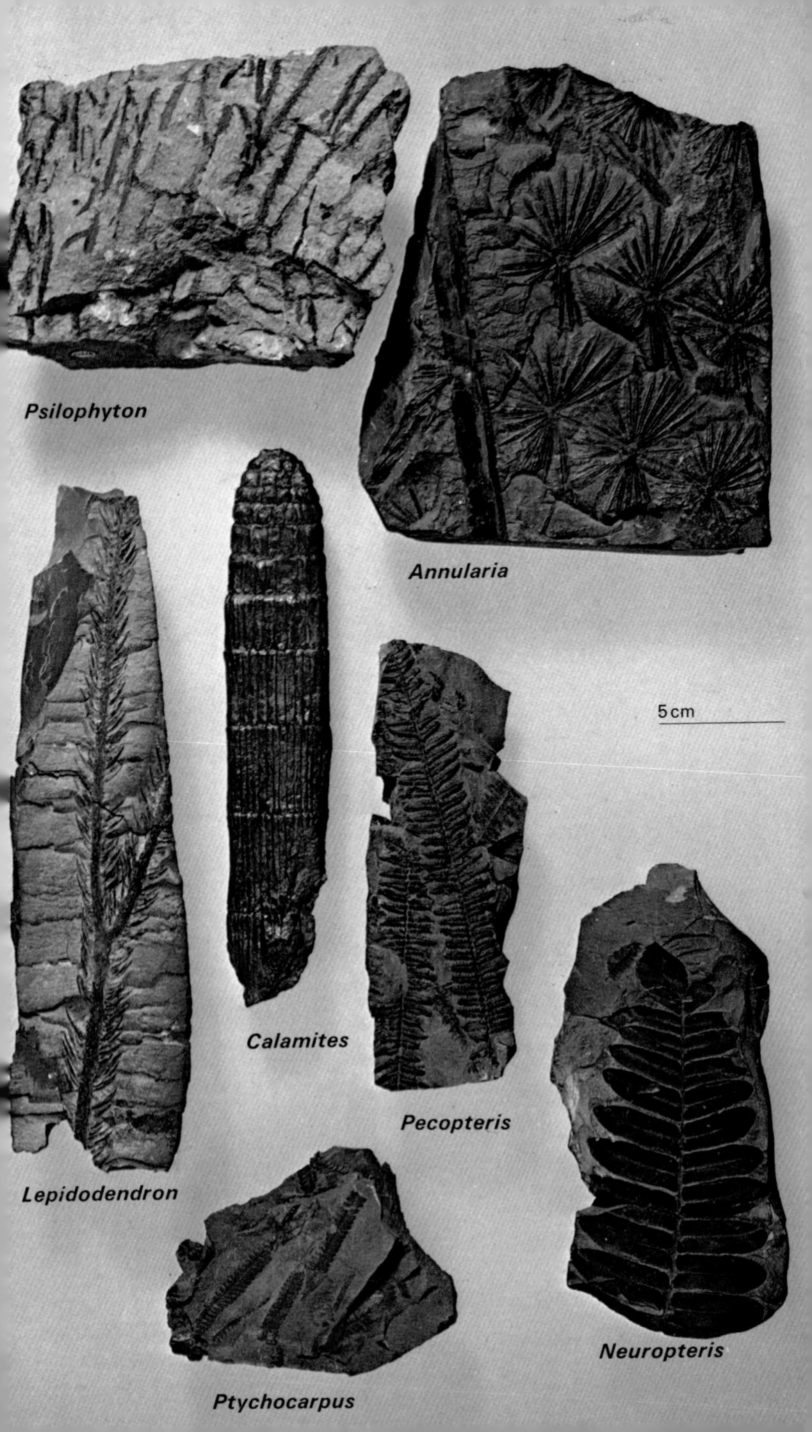

Psilophyton
Annularia
Lepidodendron
Calamites
5 cm
Pecopteris
Ptychocarpus
Neuropteris

Cordaitales Devonian–Triassic: Worldwide This extinct group includes the ancestors of the living conifers. In the Coal Measures it is represented by large trees having long, ribbon-like leaves and loosely constructed cones.

Cordaites A leaf fragment is shown. Note the narrow, strap-like form with almost parallel sides. Veins parallel to the long axis of the leaf.

Cordaianthus This is a loosely constructed cone and represents the female fruiting body of these early plants. Compare its structure with the much more compact cone of *Araucaria*.

Mesozoic and Tertiary plants

Ginkgoales Devonian–Recent: Worldwide Represented by a single living species *Ginkgo biloba*, but an important group in the Mesozoic. Trees similar to conifers but having deciduous leaves of characteristic shape and venation.

Ginkgo Jurassic Leaves are shown.

Coniferales Carboniferous–Recent: Worldwide An important living group of trees, including pines and redwoods. Leaves usually long and narrow, seeds contained in cones.

Araucaria Cretaceous: SA A compact globular cone with scales having spiral arrangement. The compact nature of the cone is shown here in the polished longitudinal section on the right.

Sequoiadendron Oligocene Closely related to the living redwood *Sequoia* of California. Cone small with relatively few scales.

Bennettitales Carboniferous–Cretaceous: Worldwide Important in most Jurassic floras. Possessing flower-like reproductive structures.

Williamsonia Triassic–Cretaceous The 'flower' consists of a discoid base from which are produced large, petal-like stamens that curve inwards and upwards.

Pterophyllum Triassic–Jurassic The leaves of the Bennettitales are fern-like and in *Pterophyllum* the leaflets are attached to a wide main stem and have parallel venation.

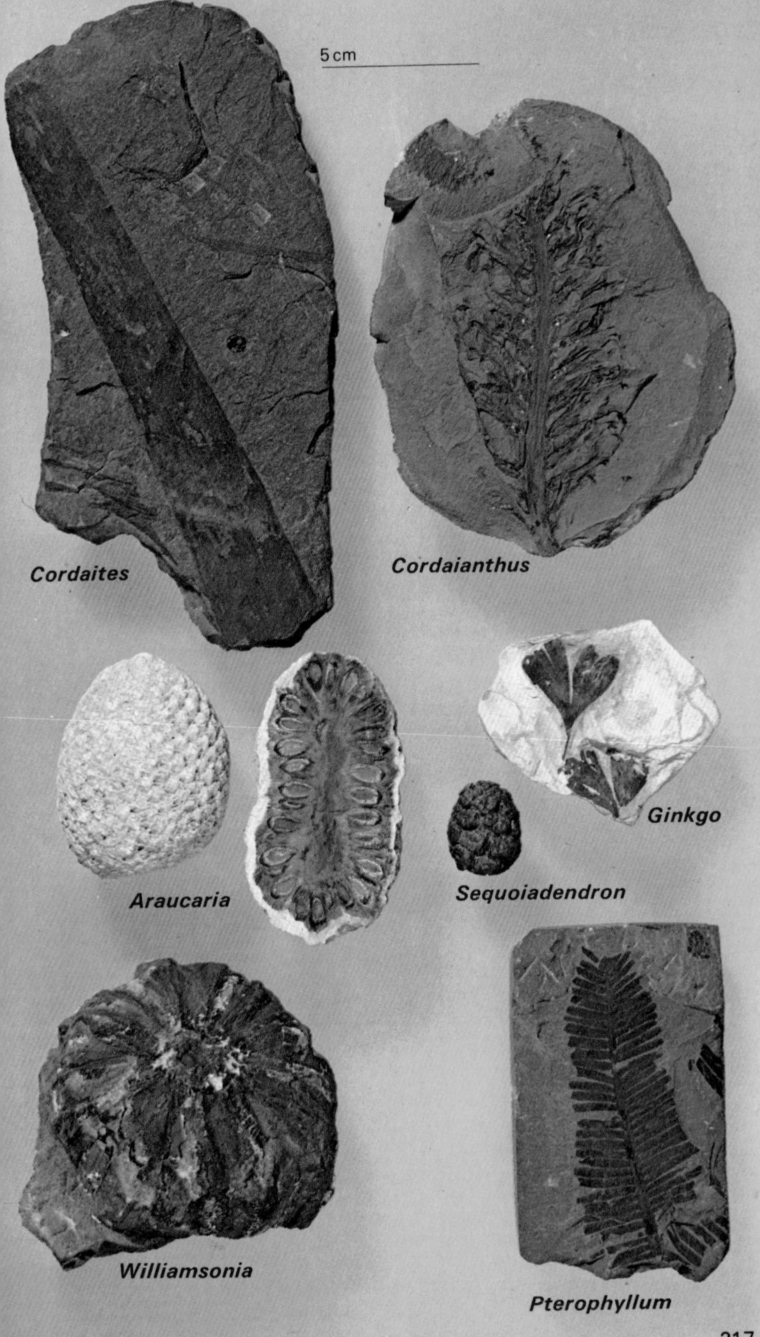
5 cm
Cordaites
Cordaianthus
Ginkgo
Araucaria
Sequoiadendron
Williamsonia
Pterophyllum

Nilssoniales Triassic–Cretaceous: NA E Asia A small group, closely related to the Bennettitales. Nilssoniales reproduced by seeds which were small and were carried in tightly bunched leaves, but cones were not developed.

Nilssonia Jurassic The small leaves are arranged along a central stem; they have fine parallel veins and each is attached along its complete base.

Angiosperms Cretaceous–Recent: Worldwide Angiosperms are by far the most important and best known plants. The group includes all the flowering plants and is divided into the dicotyledons and monocotyledons. Cotyledons are specialized leaves used for food storage in the seed, and they form the fleshy bulk of most seeds. In dicotyledons two cotyledons are present in the seed which can usually be easily split as the cotyledons are not firmly joined, for example pea and bean. The leaves of dicotyledons have a network pattern of veins and the vascular bundles of the stem are limited to a single ring around the outer part of the stem. In monocotyledons a single cotyledon is present in the seed which will not split easily, for example maize or wheat. The leaves of monocotyledons have parallel veins and the vascular bundles are scattered throughout the stem.

Dicotyledons

The angiosperms rose to importance during the Mesozoic and several living genera of dicotyledons are known from the Cretaceous.

Laurus (laurel) Representative of the family Lauraceae which was a very important family in the Cretaceous. The leaf edges are undivided and the veins are well shown here with the secondary veins diverging from a main central vein.

Platanus (plane) Cretaceous–Recent Representative of the family Platanaceae which was a very abundant and diverse group in the Northern Hemisphere during the Cretaceous. The single surviving genus *Platanus* was also common through Caenozoic and Recent times. The leaf edge is divided and the network venation is clearly shown here.

Tertiary dicotyledons

During the Tertiary the angiosperms were the dominant land plants, and by mid-Tertiary times floras of the Northern Hemisphere were probably very modern in appearance. Four fossil leaves are shown here. The last three are from Recent genera and the majority of Tertiary fossil plants may be identified by reference to modern floras.

Planera Miocene Closely related to the elm *Ulmus*.
Rhus Palaeocene–Recent Living forms of genus include poison ivy and sumach.
Acer Palaeocene–Recent Living forms of genus are sycamore and maple.
Populus Cretaceous–Recent Living form is poplar.

Nilssonia
Laurus
Platanus
cer
Planera
Populus
Rhus

Fossil wood Fossil wood is very common in many parts of the world and is frequently encountered in arid or desert areas. Heavily silicified wood may be cut and polished but special equipment is required for this. Polished sections of wood, such as the one shown, are frequently offered for sale in antique or curio shops.

Quercus **(oak) Eocene–Recent** Note the occurrence of growth rings with vascular bundles (shown as small black spots) arranged along them. This is a typical cross-section of wood from a dicotyledon; compare it with *Palmoxylon*.

Fossil fruits Fossil fruits occur frequently and may be as important as leaves for the identification of a plant. The legumes (family Leguminosae) were important throughout Caenozoic and Recent times, and the family includes the peas and beans.

Prosopsis **Oligocene** Seed case with six seeds ranged in a row.

Ficus **Cretaceous–Recent** Three fossil figs.

Monocotyledons

The commonest living monocotyledons are the grasses, but with the exception of their pollen, these are very poorly represented in fossil floras. The palms are by far the commonest fossil monocotyledons. These occur in fossil floras throughout the world and in many cases suggest the existence of tropical conditions in areas that are now cold or temperate. For example, the palm *Nipa* is common in the London Clay (Eocene) and the same genus now occurs in the tropical forests of Malaysia.

Palm leaf **Cretaceous–Recent** This fragment shows the parallel veins typical of monocotyledons. The veins are visible along the raised portions of the leaf.

Nipadites **Eocene** Palm fruits are also common as fossils, particularly those of palms which grow along shores or the banks of rivers or in shallow water.

Palmoxylon **Cretaceous–Recent** The stem or wood of the palm is typical of monocotyledons. In the polished cross-section shown here the vascular bundles are shown as small, black spots which are scattered throughout the cross-section, and growth rings are not developed, in contrast to *Quercus*.

5 cm
Prosopsis
Ficus
Quercus
Nipadites
Palm leaf
Palmoxylon

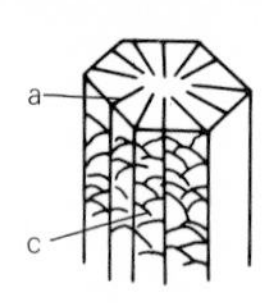

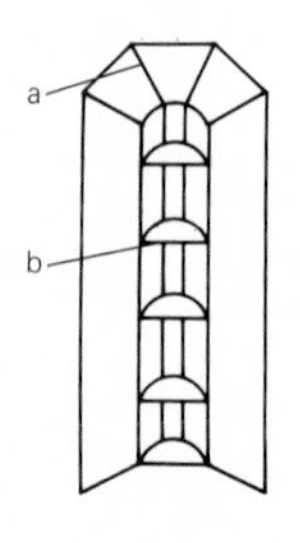

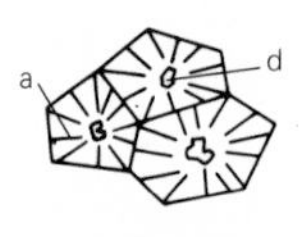

Typical structure of a corallite: side view (top) and two sectioned views

Corals

Simple animals having skeletons composed of calcite and frequently preserved as fossils. Much of their classification is based on microscopic details and is outside the scope of this book. Many forms can be identified from details of their gross anatomy, however, and with knowledge of the age of a deposit it is usually possible to place a specimen with its major group.

Corals may be *colonial*, that is consisting of many individuals (for example *Hexagonaria*), or *solitary* (for example *Caninia*). Shape is important and the growth form of colonial corals aids identification; they may be *massive*, forming clumps (for example *Hexagonaria*), or *branching* (for example *Coenites*), or *encrusting* (for example some species of *Echinopora*). An individual coral is termed a *corallite* and a complete group is a *corallum*. Important internal features are *septa* **(a)**, that is radiating vertical divisions, *tabulae* **(b)**, that is major horizontal divisions near the centre of the corallite, and *dissepiments* **(c)**, that is minor horizontal and slanting divisions near the walls. The *axial structure* **(d)** may be a rod-like process or a diffuse vertical structure in the centre of the corallite.

Scleractinia

Post-Palaeozoic, solitary or colonial, septa important, tabulae and dissepiments also present.

Parasmilia **Cretaceous–Recent: Worldwide** Usually small. Solitary, cylindrical or colonial. Cross-section circular. Septa numerous, of differing lengths and having a granular surface. Axial structure large and spongy. Outer surface with vertical ridges but lacking transverse ridges. Attachment visible on base.

Favia **Cretaceous–Recent: Worldwide** Colonial. Corallites small to medium-sized. Corallites separated by walls but septa join across walls. Corallum massive, encrusting and columnar. Branching not developed. Septa numerous, of variable length, with serrated edges. Axial structure large and spongy. Dissepiments well developed inside and outside walls of corallites. *Favia, Porites* and *Acropora* are very important Miocene to Recent reef builders.

Porites: septa

Porites **Eocene–Recent: Worldwide** Common from Miocene to Recent. Colonial, branching, massive or encrusting. Corallites small, lacking walls and having few septa. The septa are usually beaded and discontinuous in appearance. Axial structure may be present and the inner parts of the septa may be separated off as rods.

Acropora **Eocene–Recent: Worldwide** Colonial, usually branching. Corallites small, produced from surface and having ridged walls. Material between corallites spongy, granular and spiny. Septa short, axial structure and dissepiments absent. The commonest Recent coral.

Stylophora **Eocene–Recent: NA SA E Asia** Colonial, branching. Corallites very small, separated by thick walls. Septa few, major ones joining axial structure. Dissepiments present.

5 cm

Parasmilia

Favia

Porites

Acropora

Stylophora

Thamnasteria: septa

Thamnasteria Triassic–Cretaceous: NA SA E Asia Colonial, branching, massive or encrusting (shown here). Corallites medium-sized, lacking clearly defined walls. Centres of corallites joined by septa which fuse across walls. Axial structure slender. Relatively common and probably a reef builder.

Isastrea Jurassic–Cretaceous: NA E Af Colonial, massive. Corallites large, usually with five or six sides. Septa numerous and of varying lengths. Upper edges of septa beaded. Dissepiments present.

Cyclolites Cretaceous–Eocene: E Asia Af West Indies Solitary, usually from 2–10 cm in diameter, disc-shaped with deep central groove and very numerous radiating septa; these have serrated edges and carry small perforations which may be visible if there is a break in the specimen. The lower surface of *Cyclolithes* carries a pattern of concentric ridges and grooves which is characteristic of the genus.

Montlivaltia Triassic–Cretaceous: Worldwide Solitary and large, cylindrical, elongate conical or short conical in shape and having a circular cross-section. Septa long and very numerous, radiating from elongate central pit on upper surface and having serrated upper edges. Axial structure weak or absent. Dissepiments numerous (not shown here). On this specimen the outer surface carries numerous vert cal ridges indicating the septa. Septa may also be produced upwards from the upper surface and appear to overlap on to the outer face (clearly shown in *Thecosmilia*).

Thecosmilia Triassic–Cretaceous: Worldwide Colonial with large corallites similar to *Montlivaltia*. The body of each corallite is usually thick with a circular cross-section. Septa very numerous and, as in *Montlivaltia*, with serrated upper edges. Axial structure weak or absent, septa clearly visible here on the outer surface of the corallites. Transverse grooves usually absent.

Placosmilia Cretaceous–Eocene: E Similar to *Montlivaltia* but having a flattened cross-section. Body short and conical. Outer surface with strong transverse grooves and vertical ribs marking septa. Septa numerous and relatively thick. Axial structure long and flattened. This genus may possibly be confused with *Caryophyllia* (Jurassic–Recent: Worldwide). *Caryophyllia* has a more circular cross-section, however, and a rounded axial structure.

'hamnasteria
5 cm
Isastrea
Cyclolites
Cyclolites
Montlivaltia
Thecosmilia
Placosmilia

Echinopora **Miocene–Recent: Af Pacific** Colonial, leaf-like or encrusting. Corallites about 5 mm in diameter, separated by walls. Edges of corallites raised and spiny. Wall between corallites carrying small spines or granules. Axial structure large and spongy. Dissepiments numerous. A very similar form *Montastrea* (Jurassic–Recent) is common in the Caribbean region and also occurs in Europe. The walls of the corallites are less clearly defined than those of *Echinopora*.

Meandrina **Eocene–Recent: SA E West Indies** Representative of the group of brain corals. Colonial, massive, consisting of numerous elongate corallites in which the many short septa slope inwards to the elongate axial structure, thus forming numerous meandering valleys which are usually separated by high walls but may join.

Rugosa

Palaeozoic, solitary or colonial, septa important, tabulae and dissepiments also present. Very similar to *Scleractinia* and most easily distinguished on an age basis.

Palaeosmilia **Carboniferous: Worldwide** Usually solitary, medium-sized to large with very numerous septa of which the major ones reach the centre where they may form an axial structure. Outer surface with heavy transverse ridges. This specimen also carries vertical striations marking the septa.

Caninia **Carboniferous–Permian: NA E Asia Aust** Solitary or sometimes colonial; large, cylindrical or elongate conical. Septa short. Central part of corallite with numerous tabulae which are usually flat. Dissepiments around outer region. This specimen is weathered and the tabulae are clearly visible. The honeycomb effect on the vertical face is produced by the weathering which exposes the dissepiments.

Lithostrotion **Carboniferous: Worldwide** Colonial; corallites up to about 8 mm in diameter. Growth form variable, may be root-like (shown here) with adjacent corallites joined by connecting processes, but some species have a form similar to *Lonsdaleia* and in others the walls separating the corallites break down. Axial structure large, septa short. The tabulae are characteristic as they are conical and are arranged along the axial structure.

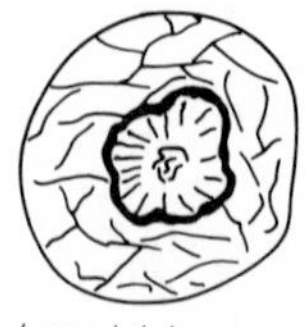

Lonsdaleia: cross-section of corallite

Lonsdaleia **Carboniferous: NA E Asia Aust** Colonial, massive. Corallites closely in contact as shown or slightly separated; delimited by strong walls. Septa long with large axial structure and central pit on upper surface. *Lonsdaleia* is similar to *Hexagonaria* (page 228) in general appearance, but may be easily distinguished by comparing cross-sections of corallites.

5 cm

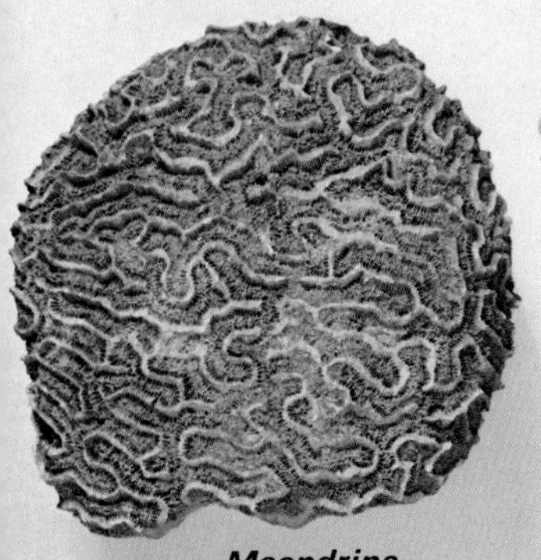

Meandrina

Echinopora

Palaeosmilia

Caninia

Lithostrotion

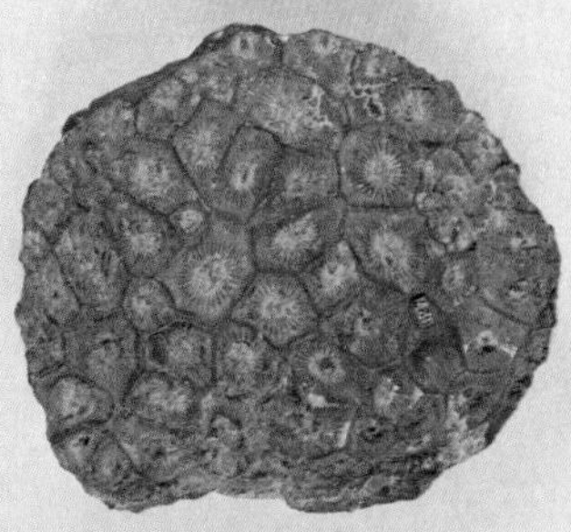

Lonsdaleia

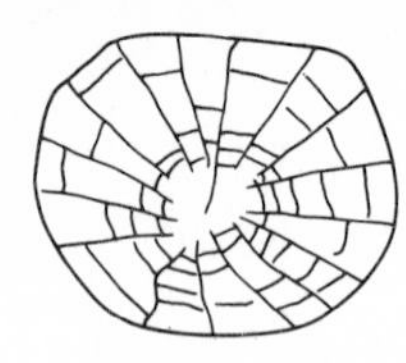
Hexagonaria: cross-section of corallite

Hexagonaria **Devonian: Worldwide** Colonial, massive with conical corallum. Corallites separated by strong walls. Axial structure absent; septa long and central pit on upper surface. Outer surface of corallum carrying strong horizontal wrinkles and vertical grooves which delimit corallites. Finer features include vertical striations indicating septa and fine horizontal lines indicating dissepiments. Corallite cross-section differs markedly from that of *Lonsdaleia* (page 226).

Tabulata

Extinct, almost entirely Palaeozoic corals in which the tabulae (horizontal partitions) are important and the septa are small or absent. Always colonial.

Favosites **Silurian–Devonian: Worldwide** Massive Corallites prismatic and in close contact; separated by thin walls. Septa small, represented as ridges or short spines. Tabulae numerous, flattened or slightly convex and extending across the corallite. Walls perforated by small holes (mural pores) which are visible as small black spots (shown here).

Syringopora **Silurian–Carboniferous: Worldwide** Colonies large; consisting of cylindrical, root-like corallites having numerous connecting processes. Septa small, spines or ridges when visible. Tabulae numerous (not shown here).

Coenites **Silurian–Devonian: NA E Aust** Leaf-like; massive or branching (shown here). Walls thick and openings of corallites small and crescentic as the corallites open obliquely to the surface. Mural pores usually absent.

Halysites: cross-section of corallites

Halysites **Ordovician–Silurian: Worldwide** Corallites elongate, more or less parallel and united along whole adjoining edges; arranged in rows one corallite wide, thus producing chain-like effect in cross-section. Corallite walls lacking mural pores. Septa sometimes present. The corallite cross-section is circular and a smaller corallite is often visible between each pair of larger ones. Only two other forms are grouped in the family Halysitidae and both are from the Silurian of North America. Both form similar chain-like patterns but *Labyrinthites* has more angular corallites and lacks the intervening smaller corallites, while in *Arcturia* the corallites are connected by tubes rather than fusion along their edges.

Labyrinthites: cross-section of corallites

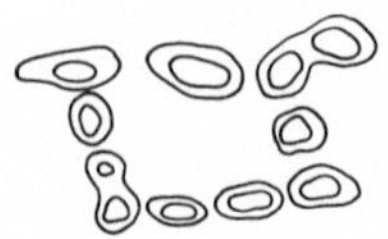
Arcturia: cross-section of corallites

Aulopora **Devonian: Worldwide** Corallum consisting of a network of fine tubes lying flat on the surface of attachment. Small vertical corallites are produced from these tubes, they are conical or trumpet-shaped and have strong transverse wrinkles.

Hexagonaria

Favosites

Favosites

Syringopora

Coenites

Halysites

Aulopora

Sponges

The simplest, multicellular animals. Usually having a radial structure with a central cloaca and surfaces covered with pores. Several forms are shown here that have characteristic shapes but the detailed identification of many sponges relies on the study of thin sections.

Chenendopora Cretaceous: E Medium-sized to large, usually 5–10 cm high. A vase-shaped sponge with a large, wide cloaca. Pores on outer and inner faces more clearly visible inside cloaca. Attachment stem shown at base.

Siphonia Cretaceous: E Globular, widening downwards. Cloaca narrow, less than 1 cm. Surface generally smooth with small pores. Stalk long and slender.

Ventriculites Cretaceous: E Thin walled, vase-shaped, high to flattened and saucer-shaped (both shown here). With strong vertical grooves on the outer surface marking the course of canals and large pores on the upper face. Cloaca varying in width with shape of whole animal.

Peronidella Triassic–Cretaceous: E A medium-sized form consisting of numerous cylindrical units each less than 1 cm in diameter and radiating from a common base. Each has a small cloaca at its tip.

Doryderma Carboniferous–Cretaceous: E A relatively large, branching, plant-like sponge. Branches at least 1 cm in diameter.

Hydnoceras Devonian–Carboniferous: NA E Small to large (shown here). Vase-shaped. Surface with a network pattern formed by large vertical and transverse ridges with finer ridges between them. Regularly arranged swellings are present, usually at the intersection of large ridges; these swellings delimit the eight faces of the sponge. This genus represents a group that is particularly common in the Devonian of New York.

Cliona Devonian–Recent: Worldwide A small burrowing sponge that forms nodular swellings in shells or on rock surfaces. These swellings are joined by slender connecting rods.

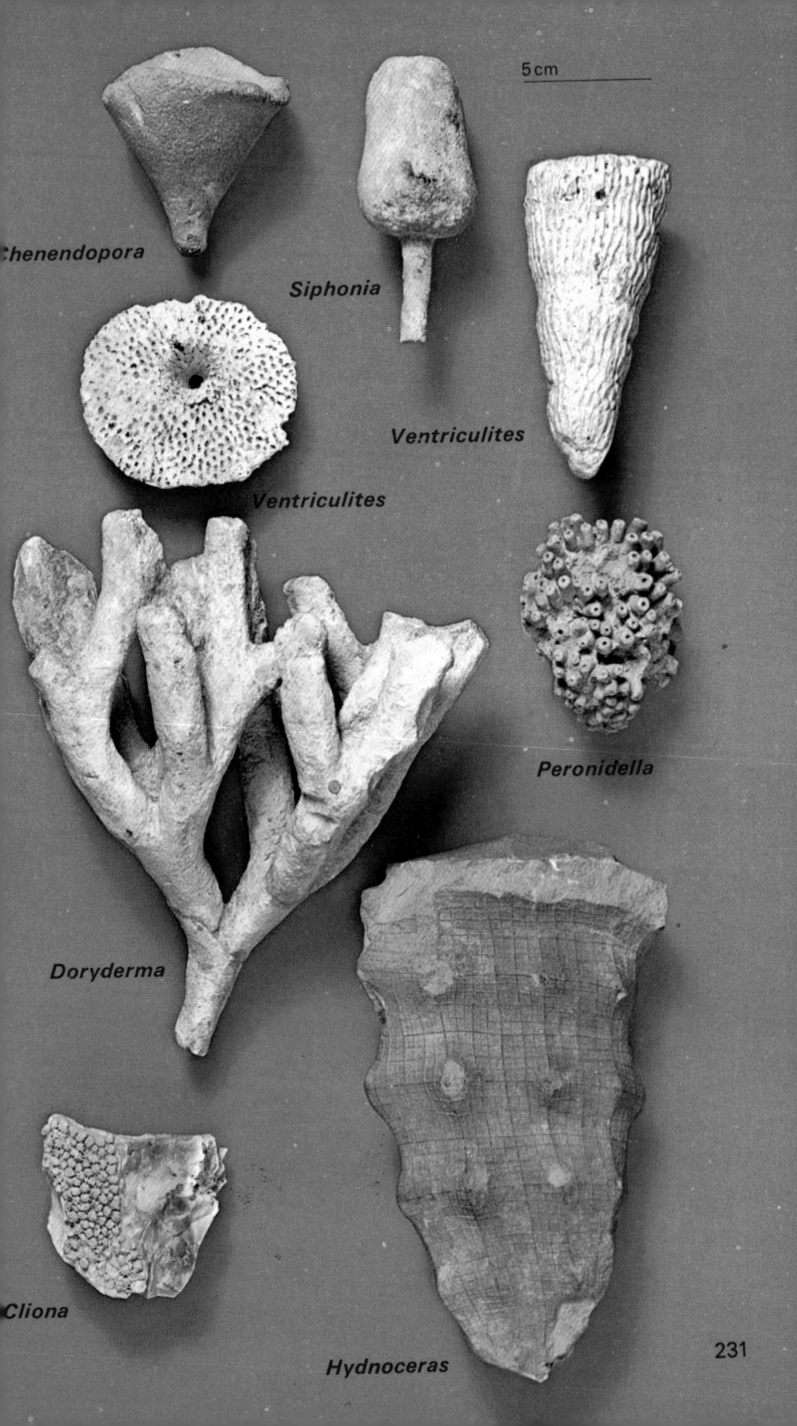
5 cm
Chenendopora
Siphonia
Ventriculites
Ventriculites
Peronidella
Doryderma
Cliona
Hydnoceras

Bryozoans

Moss-like, colonial animals, almost exclusively marine and important as fossils in most limestone deposits of Ordovician and later age. These delicate fossils may be collected from weathered surfaces or washed out of clays. Treatment of limestones with a weak solution (3 per cent) of hydrochloric acid allows good quality specimens to be recovered.

Each individual of a colony is housed in a tube known as a *zooecium* and the colony is a *zoarium*. The opening of each zooecium is an *aperture*.

Unfortunately most bryozoans can be identified and grouped only on the basis of their microscopic anatomy which can be seen only in thin sections. Where relevant these features are mentioned here but are not usually shown. Important microscopic features are the thickness of the walls of the zooecium and the presence or absence of cross-partitions or *diaphragms*. Although the more delicate bryozoans are easily distinguished from corals, the massive or encrusting forms are very coral-like, and some groups were treated as corals for many years. The naming of growth forms is the same as that given for corals (page 222). Four subgroups of bryozoans are mentioned here: Cryptostomata, Trepostomata, Cyclostomata and Cheilostomata. The first two are exclusively Palaeozoic, the third occurs from the Ordovician to Recent, and the Cheilostomata is exclusively Mesozoic and Caenozoic. Knowledge of the age of a deposit may therefore be an important aid to identification.

Cryptostomata

Aperture round, zooecial tubes very short. Diaphragms present. Outer and inner regions distinguished by differences in wall thickness. Outer region has thick walls and is known as *mature* region while inner walls are thin and known as the *immature* region.

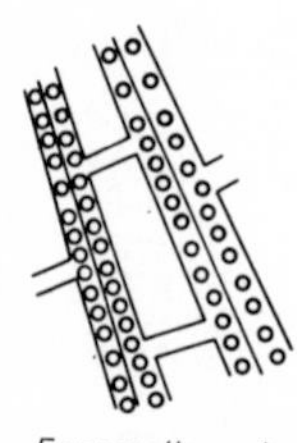

Fenestella: paired rows of pores

Fenestella Ordovician–Permian: E Net or lace-like, radiating elements much thicker than cross-bars. Zoarium fan-shaped or forming a funnel. Pores arranged in paired rows along radiating elements, divided by central keel. Pores present on upper surface only.

Archimedes Carboniferous–Permian: NA E Asia The most easily identified bryozoan. Consists of a central axis with strong spiral crest (shown here). Usually only axis preserved and this lacks pores. The axis carries a lace-like zoarium which is virtually indistinguishable from *Fenestella* when dissociated from the axis.

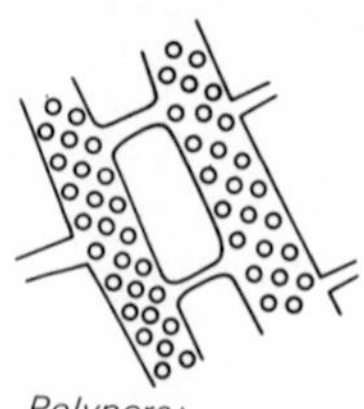

Polypora: rows of pores

Polypora Ordovician–Permian: NA E Like *Fenestella* but distinguished by pores which form two to eight rows. Central ridge not developed but a row of tubercles or lumps may be present.

Ptylodictya Ordovician–Devonian: NA E Fronds sickle-shaped, simple and narrow (shown here) or broad. Cross-section of fronds flattened. Fronds ribbon-like. Apertures oblong and arranged in lines.

5 cm
Fenestella
Polypora
Archimedes
Ptylodictya

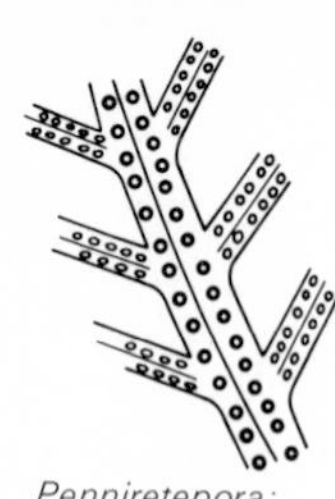

Penniretepora: surface features

Penniretepora Devonian–Permian: NA E Delicate and fern-like with a thin stem and short, regularly spaced side branches. Apertures restricted to one side of each frond and arranged in two rows separated by a median keel.

Trepostomata

A major group; growth form massive, branching or forming plates. Immature and mature regions developed.

Monticulipora Ordovician: NA Growth form massive (shown here); less commonly branching or leaf-like. Surface covered with small swellings known as *monticules* and formed from zooecia having apertures smaller than usual and surrounded by larger apertures. Walls of zooecia fused and thin. Representatives of the monticuliporids were for a long time regarded as corals.

Constellaria Ordovician: NA Growth form of flattened branches or leaf-like. Representative of a group—the family Constellaridae—which is characterized by the presence of star-shaped depressions containing zooecia having apertures smaller than usual. Spaces between rays of stars, raised and carrying apertures larger than usual.

Cyclostomata

Zooecia simple calcareous tubes, single or together. Diaphragms usually absent. Apertures not contracted. Walls thin, growth form delicate, thread-like (shown in *Stomatopora* on page 236) to large, massive (shown here in *Alveolaria*).

Meliceritites Cretaceous: E Consisting of large, branching stems carrying very numerous facets with small, triangular openings.

Fistulipora Silurian–Permian: NA E Growth form variable; encrusting, branching or forming large sheets up to 30 cm across. (A piece of such a sheet is shown.) Lower surface wrinkled. Large apertures rounded and surrounded by smaller pores. One of the commonest Palaeozoic cyclostomates.

Meandropora Pliocene: E Massive; consisting of cylindrical tubes joined by plate-like growths around each tube (shown here). Apertures circular and small, limited to the ends of the tubes. Sides of tubes ridged.

Alveolaria Oligocene–Pliocene: NA E Asia Similar in gross features to *Meandropora*, but zoaria are conical to pyramidal rather than tubular. Growth form lamellar, but forming nodules as lamellae grow over each other (shown here in broken side view). Surface pattern complex of triangles. Apertures very small and on side faces and ends of zoaria.

5 cm

Penniretepora

Monticulipora

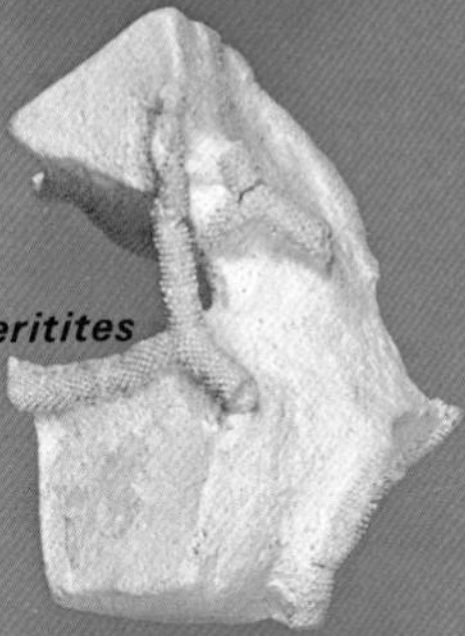

Meliceritites

Constellaria

Meandropora

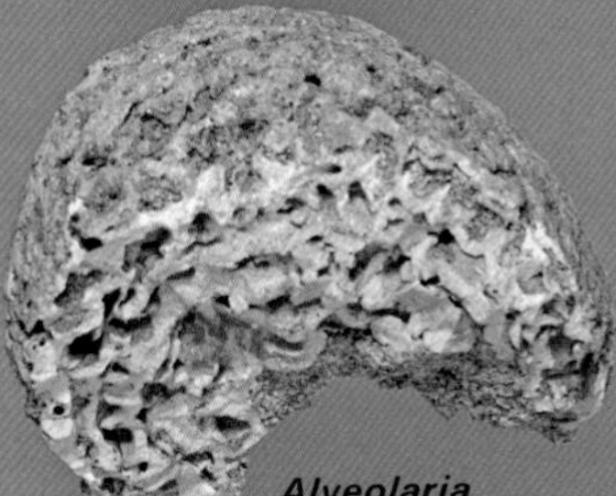

Alveolaria

Fistulipora

Stomatopora Ordovician–Recent: NA E Very small, less than 1 mm diameter but up to several centimetres long. Encrusting, thread-like, branching zoaria consisting of a single zooecium thickness with circular apertures at fairly regular intervals.

Berenicea Ordovician–Recent: NA E Af Small, usually less than 1 cm in diameter. Consisting of thin, encrusting sheet, a single zooecium thick, and usually almost circular. Zooecia with circular apertures arranged in irregular lines radiating from centre of zoarium. The specimen shown here is on the plate of a sea urchin.

Reticrisina Cretaceous: E Encrusting zoarium consisting of a network of compressed ribbon-like branches. Apertures circular and arranged in raised rows on the faces of the branches.

Cheilostomata

Post-palaeozoic. This group includes the commonest living bryozoans. Growth forms variable, usually delicate branching, as a network, flattened sheets or encrusting. Zooecia short and apertures constricted having a smaller diameter than the zooecium. Apertures usually not circular and in living forms each aperture is closed by a small cap *(operculum)*. In specialized forms this operculum is worked by a specialized, muscular zooid or *avicularium*, which has a small aperture.

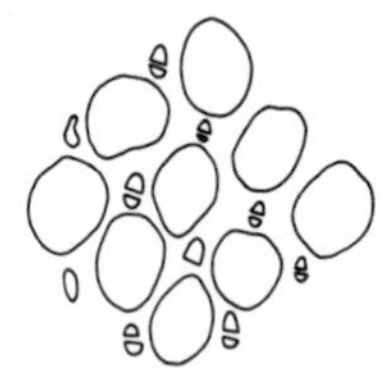

Membranipora: apertures and avicularia

Membranipora Miocene–Recent: NA E Encrusting zooaria of irregular shape but zooecia arranged in regular rows. Apertures of variable shape. Avicularia present, shown as small diamond-shaped apertures with cross-bars in well-preserved specimens. Specimen shown here is attached to a sea urchin plate. This is one of the commonest chalk bryozoans.

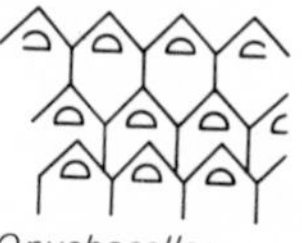

Onychocella: apertures and zooecia

Onychocella Cretaceous–Recent: NA E Encrusting or erect, forming sheets of variable size. Apertures small at centre of shallow depression and having straight edge and other edges rounded. Zooecia appear to be overlapping each other. Specimen shown here is attached to a bivalve mollusc shell.

Lunulites Cretaceous–Eocene: NA E Small zoarium usually less than 1 cm diameter. Zoarium flattened disc or conical. Apertures arranged in regular radial rows separated by channels. Aperture shape similar to that of *Onychocella* but rather more elongate. Avicularia present as small apertures between row of larger zooecial apertures.

5 cm

Stomatopora

Berenicea

Membranipora

Reticrisina

Lunulites

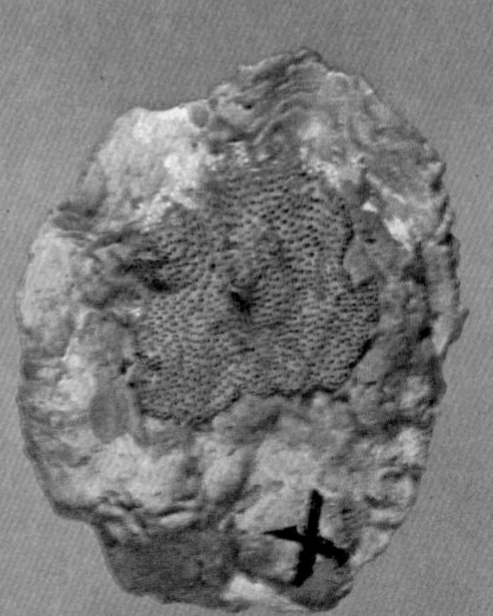
Onychocella

Molluscs

The most important class of fossil animals and including three major groups: Gastropoda (snails), Cephalopoda (squids) and Bivalvia (bivalves or clams).

Gastropods

The gastropod shell may be coiled (snails), uncoiled (limpets) or reduced (slugs). The only group with which confusion may occur is the ammonites or shelled cephalopods. Important features of gastropods are related to the coiling, aperture, columella and shell sculpture.

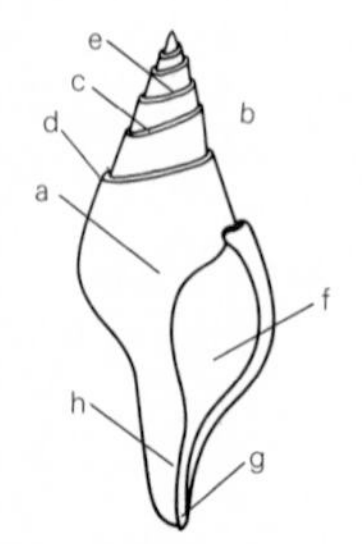

Typical structure of a gastropod as shown by *Clavilithes* (page 246)

A *whorl* is a complete coil of the shell; the *last whorl* **(a)** is the largest and the *spire* **(b)** is all the shell except the last whorl. The *suture* **(c)** is the line along which the whorls meet. If the whorls are angular, then the main angle, where the shell turns inwards towards the suture, is known as the *shoulder* **(d)**, and the part above the shoulder is known as the *ramp* **(e)**.

The *aperture* **(f)** is the opening to the outside. Its shape and features of the lips are important. Sometimes the aperture is rounded but in other cases it may be produced below and folded over, forming an *anterior canal* **(g)**. Less usually a *posterior canal* may be developed.

The *columella* **(h)** is the central column of the shell (clearly shown in *Clavilithes* on page 246). It sometimes bears ridges known as *columellar plications*. The columella may have a hollow centre known as the *umbilicus*. A pad of *callus* is often developed in the columellar area.

The sculpture of a gastropod shell may be *spiral*, that is following the line of coiling of the shell, or *axial*, that is parallel to the growth lines.

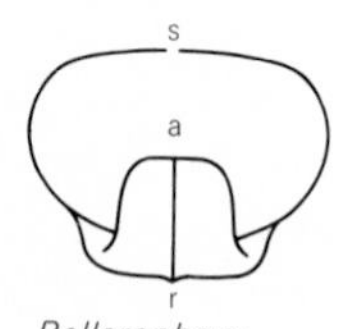

Bellerophon: aperture

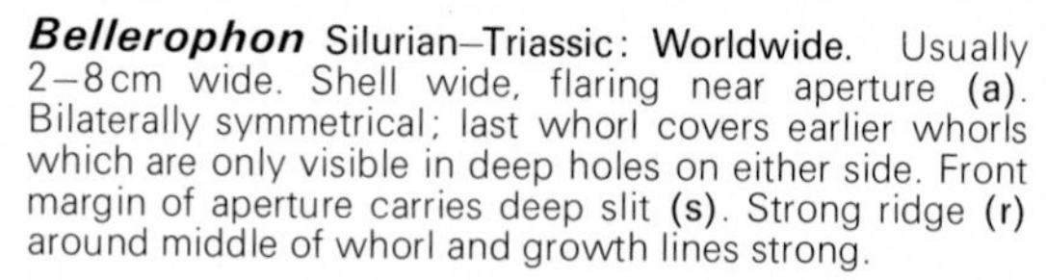

Bellerophon Silurian–Triassic: Worldwide. Usually 2–8 cm wide. Shell wide, flaring near aperture **(a)**. Bilaterally symmetrical; last whorl covers earlier whorls which are only visible in deep holes on either side. Front margin of aperture carries deep slit **(s)**. Strong ridge **(r)** around middle of whorl and growth lines strong.

Poleumita: aperture

Poleumita Silurian: NA E Usually 5–9 cm wide. Upper surface flattened. Ornament of fine lamellae and slightly raised spines on shoulder. A similar form *Straparollus* (Silurian–Permian: Worldwide) has an aperture of different shape.

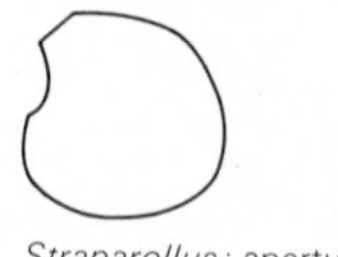
Straparollus: aperture

Trepospira Devonian–Permian: NA SA E Af Usually 2–4 cm long. Conical. Deep slit **(s)** on front edge of aperture. Faces of whorls flat; outer edge of whorl sharp. Aperture as shown. Surface smooth with row of tubercles just below suture; these distinguish *Trepospira* from *Liospira* (Ordovician–Silurian: NA E Asia) which has a completely smooth surface.

Trepospira: aperture

Mourlonia Ordovician–Permian: NA E Asia Aust Usually 3–7 cm long. Conical; sutures more deeply impressed than in *Trepospira*. Two to three ridges along shoulder and just above suture on earlier whorls. Strong slit on front edge of aperture (not shown here).

5 cm

Mourlonia

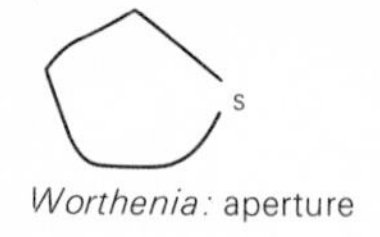

Worthenia: aperture

Worthenia Carboniferous–Triassic: Worldwide Medium-sized, usually about 3–5 cm high. Shell relatively higher than in *Mourlonia*. Whorls angular with flattened faces and strongly ridged shoulder bearing small tubercles. Under surface of first whorl with spiral ridges crossed by strong growth lines, thus forming a network pattern. Aperture almost square with thickened back edge and small slit **(s)** on front margin. Umbilicus absent.

Pleurotomaria Triassic–Cretaceous: Worldwide Usually up to 9 cm long and/or 7 cm high. Coiling low (as shown here) to high as in *Bathrotomaria*. Umbilicus present. Aperture rounded with long slit on upper front edge (shown here just below green spot). Heavy ornament of large swellings on shoulder of whorls and near suture. A spiral band of different sculpturing lies between the two rows of tubercles. Spiral grooves and strong growth lines also present.

Bathrotomaria Jurassic–Cretaceous: Worldwide Medium-sized to large, up to 7 cm high. Closely related to *Pleurotomaria* and similar to flattened or high forms of that genus. Deep slit on front margin of aperture (shown here) followed by strong spiral ridge which is visible as far as the apex. Ornament also includes numerous spiral ridges and grooves and weaker growth lines.

Platyceras Silurian–Permian: Worldwide Representative of a group (Platyceracea) in which the last whorl is very large and the other whorls are much smaller. The specimen shown is extreme and other members of the group may be more similar in general form to *Mourlonia*. Border of aperture may be wavy or straight. Ornament of growth lines. No slit on front margin of aperture.

Calliostoma: aperture

Calliostoma Cretaceous–Recent: Worldwide Medium-sized, usually 1–4 cm long. Conical with pointed, straight-sided spire. Aperture as shown. Umbilicus absent. Inner shell layer commonly like mother-of-pearl (shown here). Sutures may or may not be deeply indented. Ornament of spiral ridges, varying in distribution from near sutures only to covering whole surface of whorls, also varying in strength. No slit on margin of aperture.

Cirrus Triassic–Jurassic: SA E Medium-sized to large, 2–6 cm wide or high. Flattened to high conical (shown here). Umbilicus large, varying with height of shell. No slit on margin of aperture. Ornamentation of strong vertical ridges and weaker spiral ridges. Sutures shallowly depressed. Aperture almost circular. Coiling from right to left *(sinistral)*, (unusual in gastropods).

5 cm

Pleurotomaria

Worthenia

Platyceras

Bathrotomaria

Calliostoma

Cirrus

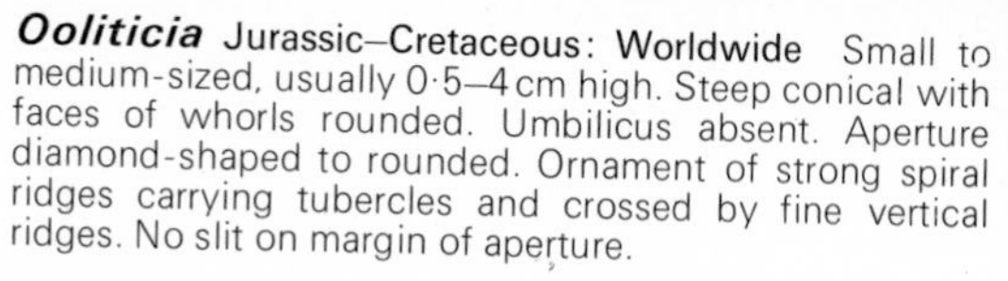

Ooliticia Jurassic–Cretaceous: **Worldwide** Small to medium-sized, usually 0·5–4 cm high. Steep conical with faces of whorls rounded. Umbilicus absent. Aperture diamond-shaped to rounded. Ornament of strong spiral ridges carrying tubercles and crossed by fine vertical ridges. No slit on margin of aperture.

Loxonema: aperture

Loxonema Silurian: **NA E** Medium-sized up to about 8 cm long. High, pointed spiral with whorls having rounded walls. Sutures deep. Umbilicus absent. No slit on margin of aperture but outer margin with a deep curved depression known as a *sinus*. Ornamentation absent.

Microptychia Carboniferous: **NA E** Medium-sized, up to 6 cm long. High pointed conical. Sutures deep with ornament of short vertical ridges, these increase in strength upwards and may completely cover the top whorls. Lower whorls smooth. Aperture almost circular and lacking sinus. Walls of whorls rounded but more convex near lower suture.

Natica Triassic–Recent: **Worldwide** Medium-sized, usually 1–5 cm high. Shape ranges from almost spherical (shown here) to conical. Walls of whorls rounded and sutures usually deep. Surface smooth and may be shiny with a few lamellar growth lines near aperture. Umbilicus usually present but columellar callus may cover it. Last whorl very large. Aperture oval to circular. Inner lip thickened, outer lip thin.

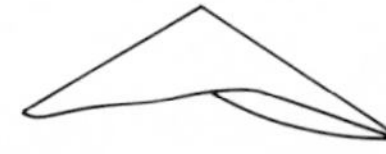

Xenophora: cross-section showing flattened base

Xenophora Cretaceous–Recent: **Worldwide** Medium-sized, up to 8 cm wide. Conical with flattened base. Last whorl with sharp outer margin. Wide umbilicus and characteristically shaped aperture (shown here). Inner margin thickened. Surface rough with depressions where shell fragments, and other foreign particles such as pebbles, were attached during life. Some shell fragments are still present on the right side of the specimen, and depressions on the upper side show the patterns of attached particles. Some species of *Xenophora* have lightly sculptured surfaces.

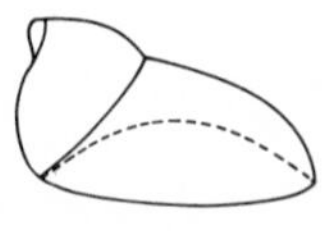

Calyptraea: cross-section showing concave lower surface

Calyptraea Cretaceous–Recent: **NA SA E** Medium-sized, up to 7 cm wide. Flattened to high conical shell consisting of a few wide whorls. Last whorl very large and lower surface deeply concave with a small internal shelf which has a twisted border (columella) and a small umbilicus at its highest point. Ornamentation of weak growth lines and occasional tubercles which are stronger near the lower edge.

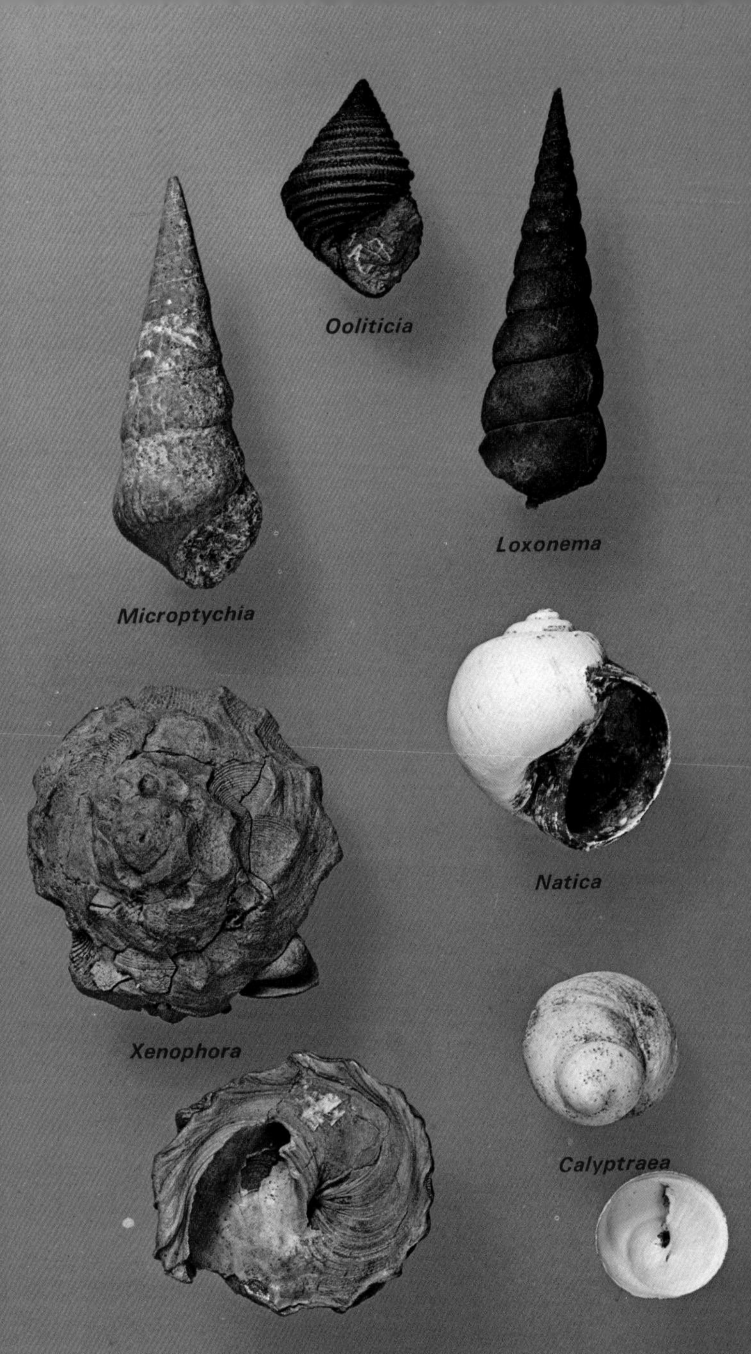
Oolíticia
Loxonema
Microptychia
Natica
Xenophora
Calyptraea
5 cm

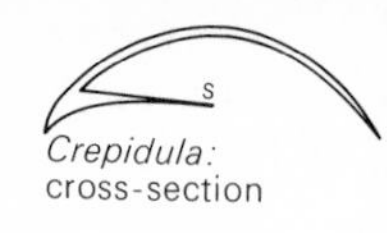

Crepidula: cross-section

Crepidula (slipper limpet) Cretaceous–Recent: NA E Medium-sized, usually 2–6 cm long. Flattened, convex and slipper-like. Whole shells consist of a single whorl. Under surface characteristic, having deep concavity as shown, and a wide concave shelf (s) which lacks the thickened columellar edge of *Calyptraea*. Ornamentation of ridges, spines and growth lines may be present on upper surface.

Architectonica Eocene–Recent: NA E Asia Small to medium-sized, up to 3 cm across. Flattened or slightly domed with a large, wide umbilicus which is strongly sculptured, often with prominent notches. Outer margin of last whorl sharp with a spiral ridge. Sculpture of a few spiral ridges above and below suture. Aperture sub-triangular and thickened at two outer angles.

Aporrhais Jurassic–Cretaceous: Worldwide Modern range is restricted to North Atlantic. Large to medium-sized, up to 12 cm high. Turretted with long anterior spine or elongate canal. Outer lip of aperture flared and notched with a variable number of spines developed. Sculpture of strong axial ridges and tubercles, spiral sculpture also developed with ribs often growing into the spines of the outer lip.

Cypraea (cowry) Cretaceous–Recent: Worldwide Small to large, 0·5–15 cm long. Highly characteristic egg-shaped with outer lip of last whorl greatly expanded and completely covering the rest of the shell. Aperture elongate, continuing for length of shell, with serrated margins and outer lip thickened. Surface smooth and usually shiny.

Ficus (fig shell) Eocene–Recent: NA E Asia Small to large, usually 1–12 cm long. Low spired, shell spindle-shaped (fusiform) with very large last whorl and lower region produced as broad, twisted canal. Whorls of spire rounded or with shoulders. Aperture large broad and elongate. Shell thin with little columellar callus. Sculpture of spiral and axial ribs.

Hippochrenes Eocene: E Asia Medium-sized to large with long spire, approximately equal to height of last whorl. Lower part of last whorl produced as elongate canal. Outer lip expanded as a large flare which is fused to the spire. The lower face of this carries a deep groove along the junction with the spire, and the shell below this groove may be expanded to cover it. The specimen shown has extreme development of the flare, but some forms may resemble *Aporrhais*.

5 cm

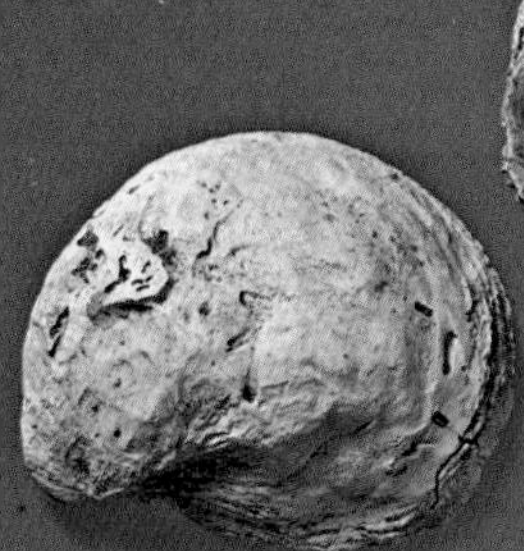

Crepidula

Architectonica

Aporrhais

Cypraea

Hippochrenes

Ficus

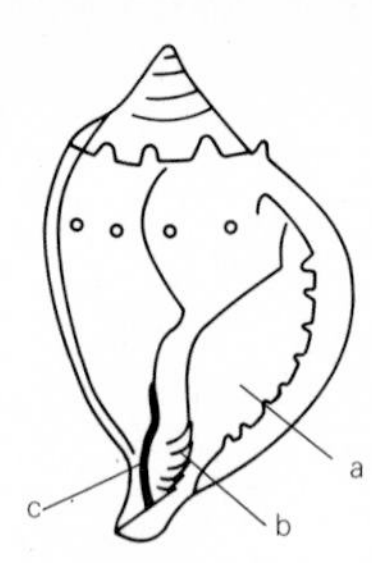

Galeodea:
showing features

Galeodea **Eocene: NA E Asia** There are surviving genera very like *Galeodea*, but this genus is extinct. Medium-sized, usually 2–8 cm long. Like *Ficus* but with higher, conical spire. Whorls angular with strong spiny projections at shoulder; spiral ridges and more swellings of variable strength on last whorl. Aperture **(a)** elongate with thickened outer margin carrying serrations. Strong columellar callus **(c)** with several strong ridges on inner margin **(b)**, especially at lower end.

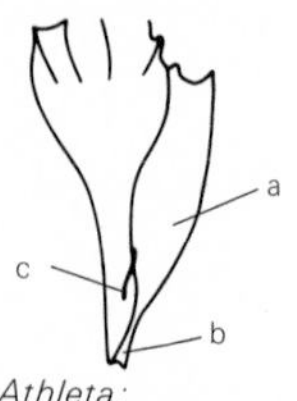

Athleta:
aperture region

Athleta **Cretaceous–Oligocene: NA E Af Asia** Medium-sized, usually 2–10 cm long. Representative of a group in which the spire is of intermediate height, the whorls are angular and usually carry ribs which bear spines at the shoulder. Aperture **(a)** narrow with short canal **(b)**. Columellar plications present **(c)**.

Marginella **Eocene–Recent: Worldwide** Small to medium-sized, less than 0·5 cm to 3 cm long. Often tapering equally at each end, oval or elongate. Surface smooth, unsculptured. Aperture **(a)** elongate (shown here), outer margin thickened and sometimes bearing teeth (not shown here). Several columellar plications present **(b, c, d)**. Sutures slightly impressed.

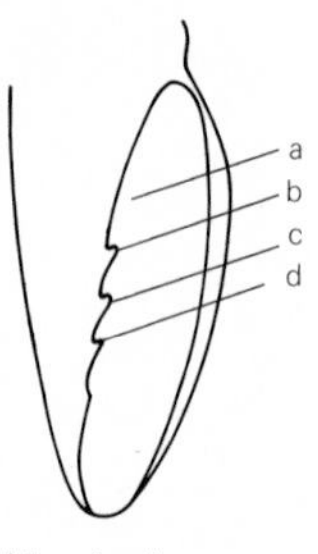

Marginella:
aperture region

Clavilithes **Eocene–Pliocene: NA E Asia** Medium-sized to large, usually 10–15 cm long. Shell elongate, conical with sutures deeply impressed. Spire short, pointed and often strongly sculptured at the apex, lower parts of shell smooth. Broad almost flat ramp above shoulder, rest of whorl almost vertical. Whorls increasing uniformly in size. Long canal. Aperture as shown. No columellar plications. A longitudinal section of the shell shows the ramp, whorl shape, canal, aperture and columella.

Murex **Cretaceous–Recent: Worldwide** Usually 3–8 cm long. Dominant ornament of three strong axial ribs per whorl, these usually have spines; spiral ridges also usually present. Aperture small. Outer lip expanded into rib as shown and having ridged inner margin. Inner lip thickened. Canal medium to long. Columellar plications absent.

Buccinum **(whelk) Pliocene–Recent: NA E** Medium-sized to large, usually 3–15 cm long. Fusiform shell with whorls increasing uniformly in size. Aperture wide and oval with short canal. Ornament of spiral and/or axial ribbing. Outer lip sharp, sometimes recurved. Columellar callus relatively weak.

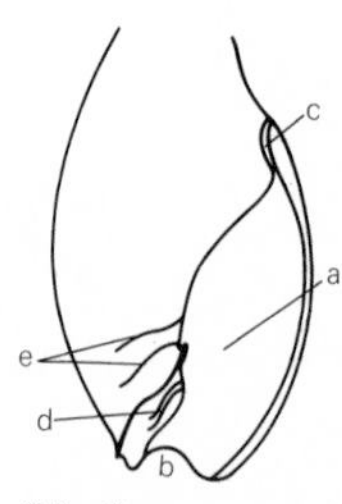

Olivella:
aperture region

Olivella **Cretaceous–Recent: Worldwide** Usually less than 4 cm long. Last whorl very high in relation to rest of shell. Height of spire variable. Aperture **(a)** elongate with short, wide canal **(b)** and notch at upper end **(c)**. Columellar plications present **(d)**. Outer lip thin and sharp. Very weak axial grooves may be present and several spiral grooves are usually developed near the lower end of the last whorl **(e)**.

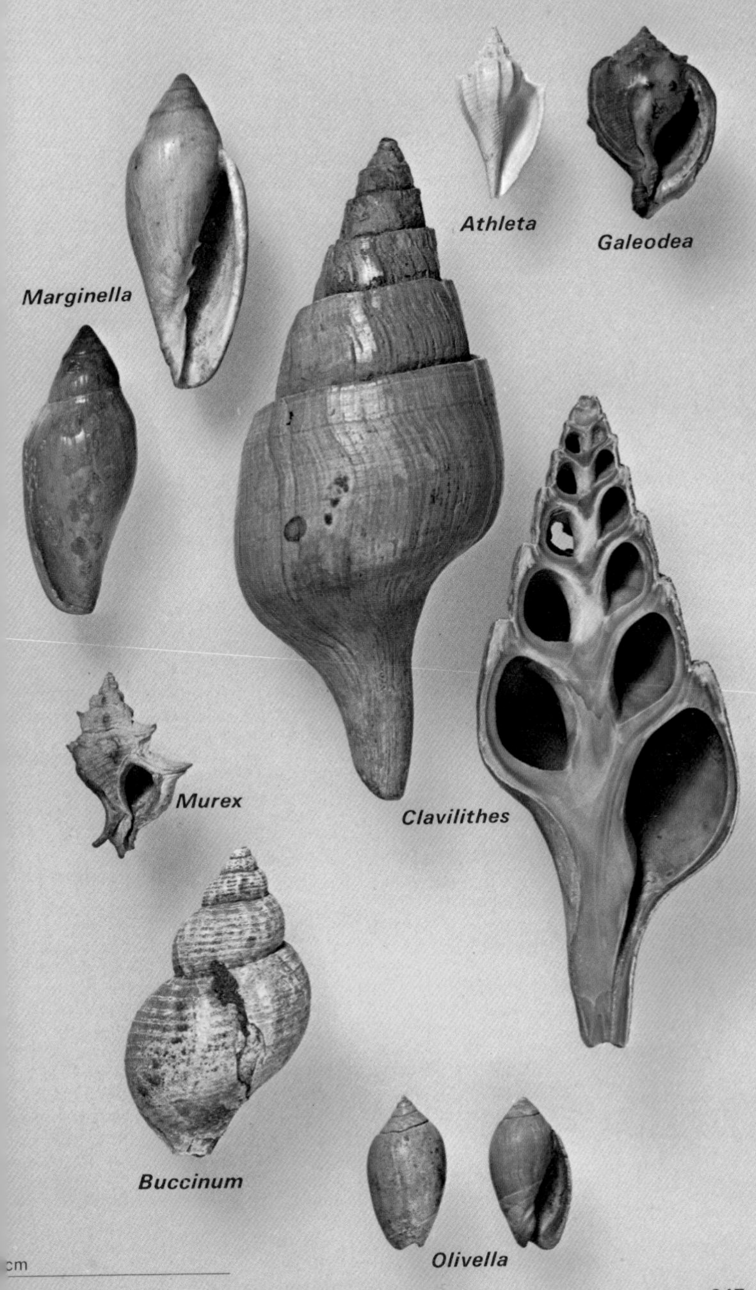
Athleta
Galeodea
Marginella
Murex
Clavilithes
Buccinum
Olivella
cm

Conus **Cretaceous–Recent: Worldwide** Small to large, usually 2–10 cm long. Steep upturned, conical with flat or shallow conical spire. Aperture parallel sided, long and narrow (shown here) with a notch at the upper end. Canal short and outer lip thin. Ornament of spiral grooves, ridges or túbercles. Spiral ridges of variable strength on spire.

Bathytoma **Cretaceous–Recent: Worldwide** Small to medium-sized, usually 1–8 cm long. Shell narrow, equally conical at both ends. Last whorl about half total length. Aperture elongate, almost parallel-sided. Columellar plications absent. Sutures deep. Growth lines flexed backwards at shoulder; with row of strong tubercles along shoulder. Spiral ribs present.

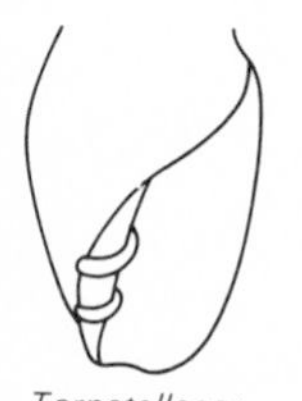

Tornatellaea: aperture region

Tornatellaea **Jurassic–Oligocene: Worldwide.** Small, usually less than 0·5–2 cm long. Spire whorls rounded. Aperture as shown, with two strong columellar plications. Outer lip with internal serrations. Ornament of numerous, strong spiral grooves.

Trochactaeon: aperture region

Trochactaeon **Cretaceous: Worldwide** Large to medium-sized, usually 3–8 cm long. Spire low, concavely pointed, body whorl large. Aperture elongate, parallel-sided. Columella with two or three strong folds at the lower end. Shell smooth and thick. In a closely related form *Actaeonella* (Cretaceous: NA) the last whorl is expanded and covers the spire, giving a superficial appearance similar to that of *Cypraea*.

Planorbis **Oligocene–Recent: E Af Asia** Small to medium-sized, usually less than 0·5–5 cm long. Flattened spiral (planispiral) shell with upper side flattened (shown here) and lower concave **(a)**, or with broad umbilicus and sutures deeply impressed (shown here). A low spire is sometimes present **(b)**, or both surfaces may be concave **(c)**. Aperture oval to wide crescentic. Outer margin sharp. Ornament of fine growth lines only.

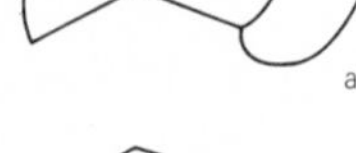

a

b

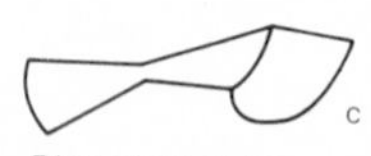

c

Planorbis: cross-sections showing grown forms

Scaphopods

A major group of molluscs of the same status as the gastropods, cephalopods or bivalves, but less common as fossils and in Recent faunas, and more uniform in appearance. Shells elongate, conical, open at both ends and usually slightly curved like an elephant's tusk. In life the concave side is upwards, the smaller opening is the top and the larger opening–the front–is deeply buried in the sand.

Fissidentalium **Cretaceous–Recent: Worldwide** Has the characteristic scaphopod shape and distinguished from the commoner *Dentalium* by the long slit at the upper end and by the nature of the ridges. In this genus they are asymmetrical with thick and thin ridges having no particular order, whereas in *Dentalium* the thin ridges have a balanced arrangement about the thick ones.

5 cm
Conus
Bathytoma
Tornatellaea
Trochactaeon
Planorbis
Fissidentalium

Cephalopods

Squid, octopus, cuttlefish and nautiloids are living cephalopods. Two important groups of extinct cephalopods—the ammonites and belemnites—are described here.

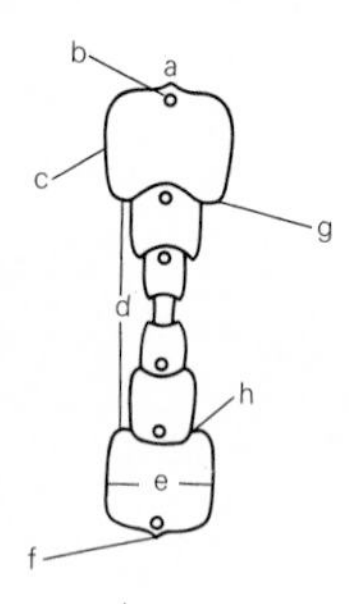

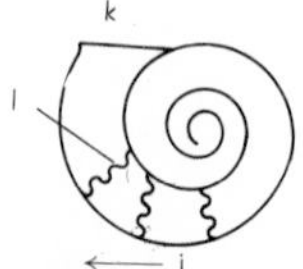

Typical structure of an ammonite: cross-section (top) and side view

Ammonites

Not known after the Cretaceous. Ammonites are one of the most important groups of fossils for dating Mesozoic rocks, as they changed very rapidly with time and had wide geographical distributions.

Ammonites are similar to flattened gastropods but distinguished by the *sutures* and *siphuncle*. Orientation and features are as shown. The *suture* is visible as a narrow, wavy line or depression on the face of the shell and is very important for identification. Suture diagrams trace the line of the suture from the *venter* (a) to the *seam* (h), which is where the whorls join. The arrow on the suture diagram points towards the *aperture* (k). *Lateral lobes* are backward swellings or troughs of the suture and *lateral saddles* are forward projections of the suture. A simple suture is shown here at the junction of the two colours on *Ceratites*, and a complex suture is shown on *Phylloceras* on page 252. The *siphuncle* (b) is a tube running through each chamber near the ventral face of each whorl. It is often indicated by a sharp flexion of the suture at the venter. The *umbilicus* (d) is the depression on the side of the shell produced by the coiling, and the *umbilical shoulder* is where the shell turns inwards (g) to the seam from the *lateral face* (c). The *keel* (f) is a ridge which may be present along the venter. The thickness is (e) and arrow (j) points backwards. Most specimens shown here are internal moulds but the shell or *test* is clearly shown on the lower part of *Goniatites*. The ammonites are grouped according to their time ranges.

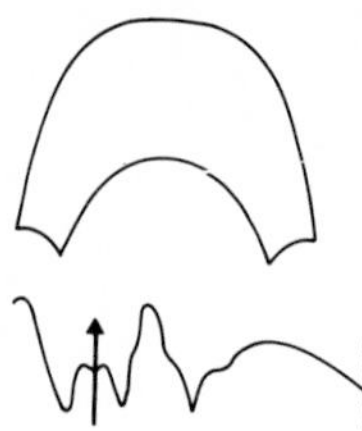

Goniatites: cross-section of whorl (top) and suture diagram

Palaeozoic ammonites

Goniatites Lower Carboniferous: Northern Hemisphere
Thick with narrow umbilicus. Whorl cross-section as shown. Suture is of the type characteristic of the major group (Goniatitacea) which ranges through the Carboniferous and Permian. In this type the ventral lobe is forked and the lateral lobe is smooth and convex. Part of the test is shown here and carries fine growth lines.

Gastrioceras: cross-section of whorl

Gastrioceras Upper Carboniferous: Worldwide
Globular (shown here) to flattened. Whorl cross-section as shown. Umbilicus narrow (shown here) to wide. Suture simple with smoothly curved saddles and pointed lobes. Ornament of strong ribs on umbilical face of lobes and very strong on shoulder; forking on ventral face but extending as fine ribs across venter, and flexed slightly backwards on venter.

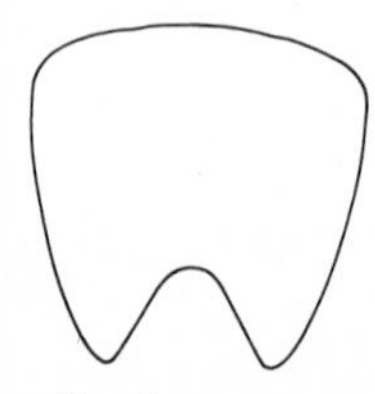

Ceratites: cross-section of whorl

Mesozoic ammonites

Ceratites Triassic: E Umbilicus wide and whorl cross-section as shown. Suture form known as *ceratitic* and characteristic of major group (Ceratitacea) which ranges through the Triassic. Has smooth, simple saddles and serrated lobes. Forms similar to *Ceratites* are known from the Triassic of the USA.

Goniatites

Gastrioceras

5 cm

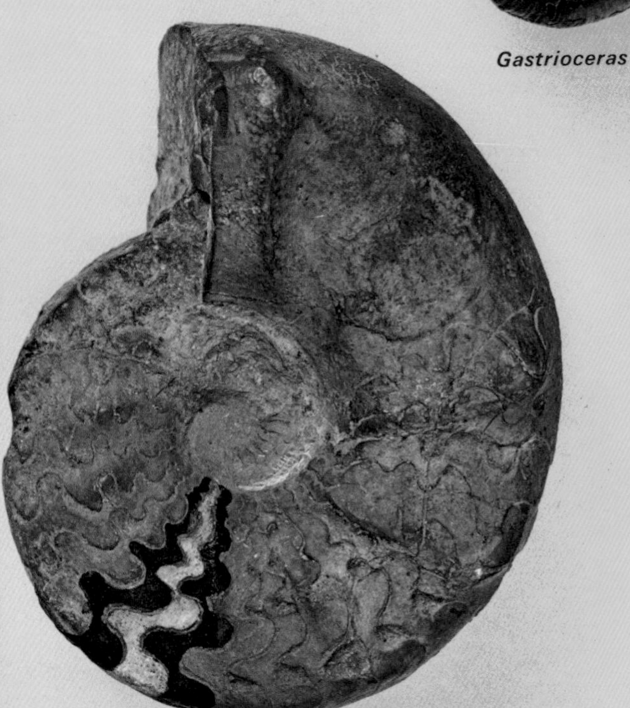

Ceratites

Lytoceras **Lower Jurassic–Upper Cretaceous: Worldwide** Umbilicus wide, whorl cross-section almost circular. Whorls increasing rapidly in diameter. Suture complex with ventral lobe divided, two lateral lobes and edges of all lobes extremely subdivided and fern-like. Ornament of fine ribs. If the test is preserved characteristic frills are present at intervals (not shown here).

Phylloceras **Lower Jurassic–Upper Cretaceous: Worldwide** Medium-sized, usually 10–15 cm across. Flattened with umbilicus very narrow or absent and last whorl covering earlier ones. Suture very complex and shown at junction of red and white paint on specimen; saddles with rounded, leaf-like projections and lobes with pointed projections. Surface smooth or with ornament of fine lines extending without a break across the venter. Whorl cross-section as shown.

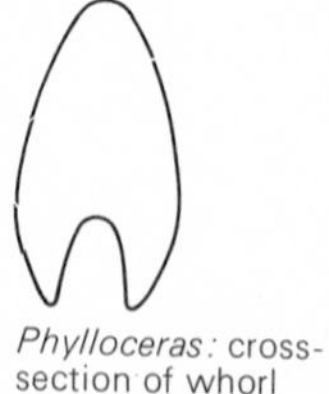
Phylloceras: cross-section of whorl

Jurassic ammonites Early Lower Jurassic

Arnioceras **NA SA E Af Asia** Flattened with very wide umbilicus. Whorl cross-section as shown. Ornament of strong ribs which bend forwards near ventral face. Keel strong and bordered by depressions on each side.

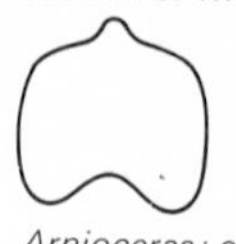
Arnioceras: cross-section of whorl

Asteroceras **NA E Asia** Medium-sized, usually about 10 cm diameter. Relatively thick with open umbilicus. Whorl cross-section similar to that of *Arnioceras*. Whorls increase in size rapidly. Seams deep. Ornament of strong ribs on lateral faces only, wider than in *Arnioceras*. Suture relatively simple (shown here). Keel strong with shallow depressions on either side.

Promicroceras **E** Small, usually 3–6 cm diameter. Whorl cross-section almost circular. Suture very complex. Ribs with sharp edges but flattened where they cross the venter.

Jurassic ammonites Middle Lower Jurassic

Amaltheus **Northern Hemisphere** Flattened with whorl cross-section as shown. Medium-sized to large, usually 7–15 cm diameter. Later whorls covering much of earlier whorls and umbilicus fairly narrow. Suture with complex, leaf-like saddles and simpler lobes. Ribs strong, present on lateral face only, fading outwards and flexing forwards. Keel very strong and characteristically serrated.

Amaltheus: cross-section of whorl

Dactylioceras **Worldwide** Medium-sized, usually 5–10 cm diameter. Umbilicus very wide and whorl cross-section almost circular. Ribs strong, forking on ventral face and joining across venter. Each rib slightly flattened on venter.

ytoceras
Phylloceras
rnioceras
Promicroceras
Asteroceras
Amaltheus
Dactylioceras
5cm

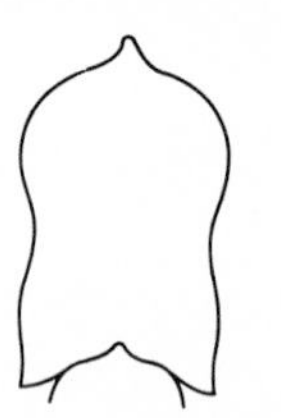

Harpoceras: cross-section of whorl

Jurassic ammonites Late Lower Jurassic

Harpoceras NA SA E Af Asia Medium-sized to large, up to 20 cm diameter. Umbilicus medium-wide. Whorl cross-section as shown. Sides flattened. Keel strong. Umbilical shoulder right-angled and sharp. Ribs fine, flexing forwards in middle of lateral face and near venter; increasing in strength outwards.

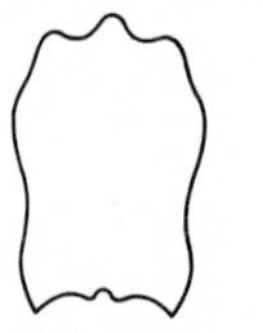

Hildoceras: cross-section of whorl

Hildoceras E Af Asia Flattened with wide umbilicus. Whorl cross-section as shown. Umbilical shoulder smooth, sloping inwards with tubercles near seam. Lateral face carrying central depression. Ribs absent on umbilical side of depression, but on outer side strong ribs curving backwards. Keel strong with characteristic wide channels on each side. Suture with wide lobes and saddles having serrated edges.

Jurassic ammonites Middle Jurassic

Parkinsonia E Af Asia Medium-sized, usually about 10–15 cm diameter. Umbilicus medium-wide. Ribs sharp-edged and numerous, bending forwards, forking with new ribs arising towards venter. Deep groove on venter, becoming shallower towards aperture.

Stephanoceras Worldwide Medium-sized. Whorl cross-section almost circular. Umbilicus wide. Row of tubercles near centre of lateral face. On the umbilical side of each tubercle there is a single strong rib, and on the other side there are usually three finer ribs.

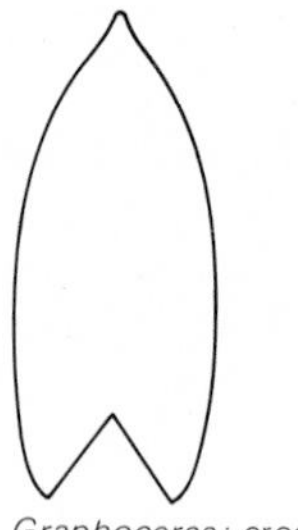

Graphoceras: cross-section of whorl

Graphoceras E Af Asia Small to medium-sized, usually about 5–10 cm diameter. Flattened with whorl cross-section as shown. Umbilicus medium-narrow. Shallow depression of lateral wall near umbilical shoulder. Ornament of wavy ribbing similar to that of *Harpoceras*, with ribs decreasing in strength towards aperture. Keel strong. Most of the face shown here is the test and growth lines are visible near the aperture; these follow the lines of the ribs.

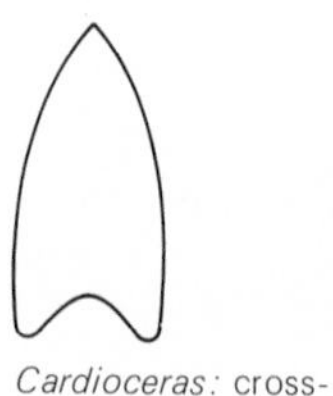

Cardioceras: cross-section of whorl

Cardioceras NA E Asia Umbilicus relatively narrow and deep. Whorl cross-section triangular as shown. Venter sharp with strong keel having a serrated edge. Umbilical shoulders sharp and overhanging seam. Strong ribs on early whorls reducing in strength and frequency towards aperture. Tubercles absent, but main ribs are very strong near umbilicus and decrease in strength abruptly near the middle of the lateral face. New ribs arise outwards and the main ribs fork.

Perisphinctes E Af Asia Large to very large, deep with almost square whorl cross-section and umbilical shoulders steep. Ribs very strong on side walls of later whorls, but forking and extending across venter away from the aperture. Ventral face smooth near aperture. Sutures very complex (shown here).

Harpoceras

Hildoceras

Parkinsonia

Stephanoceras

Graphoceras

Perisphinctes

Cardioceras

5 cm

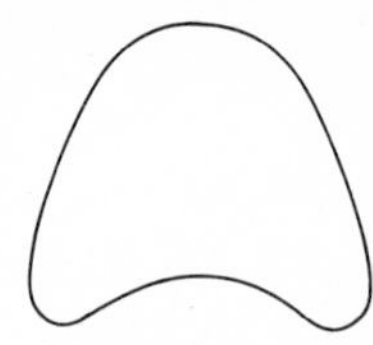

Pavlovia: cross-section of whorl

Pavlovia E Asia Medium-sized, usually about 10 cm across. Umbilicus wide and whorl cross-section as shown. Seam deep and umbilical shoulder reflected. Ribs very strong and sharp; forking once and extending across venter.

Cretaceous ammonites

Hamites: outline of complete shell

Hamites NA E Af Asia Shape characteristic as shown, shaded region only shown here. Very elongate open coiling with three or four straight regions joined by sharply curved regions. Ribs strong and entirely encircling shell which has a circular cross-section. Suture very complex.

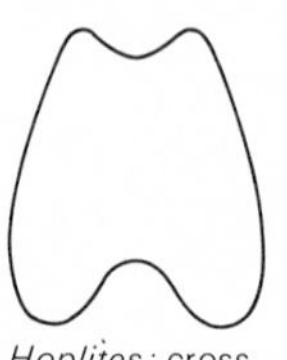

Hoplites: cross-section of whorl

Hoplites E Asia Medium-sized, 5–10 cm diameter. Umbilicus deep and relatively narrow. Whorl cross-section as shown. Strong swellings around umbilical shoulder give rise to strong forking ribs, all of which bend forwards at the venter. Venter depressed.

Mortoniceras NA SA E Af Asia Large and flattened, usually about 15–25 cm across. Umbilicus wide. Keel strong and bordered by furrows. Ribbing strong with tubercles near umbilical shoulder and new, smaller ribs developed near venter.

Cretaceous ammonites Lower Cretaceous

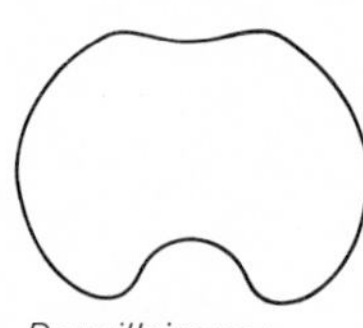

Douvilleiceras: cross-section of whorl

Douvilleiceras NA SA E Af Asia Umbilicus medium-narrow. Whorls rounded, cross-section as shown. Ribs with many tubercles increasing in numbers but decreasing in size forwards. All ribs depressed along venter.

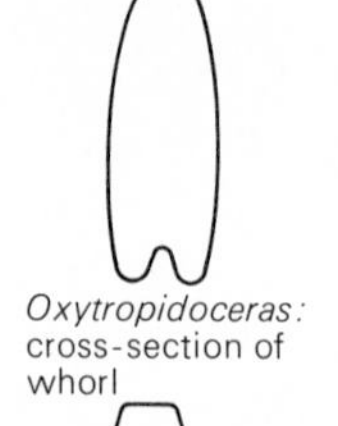

Oxytropidoceras: cross-section of whorl

Oxytropidoceras NA SA E Af Asia Medium-sized, usually 5–10 cm diameter. Flattened and whorl cross-section as shown. Umbilicus narrow and deep. Ribs numerous, forking, sharp and flexed forwards at ventral ends. Keel very strong with smooth outline.

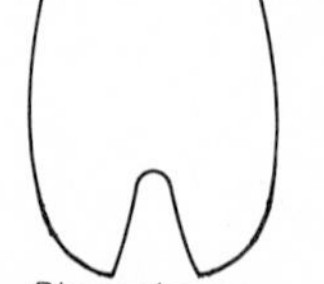

Placenticeras: cross-section of whorl

Placenticeras NA E Af Medium-sized to large. Flattened with narrow, deep umbilicus and rounded umbilical shoulders as shown. Sculpturing weak or absent, or with tubercles (shown here) bordering the umbilical shoulders, and another row on ventral face. Venter characteristic as a raised keel with a depressed outer face bordered by rows of numerous small tubercles.

5 cm
Pavlovia
Hamites
Hoplites
Mortoniceras
Douvilleiceras
Oxytropidoceras
Placenticeras

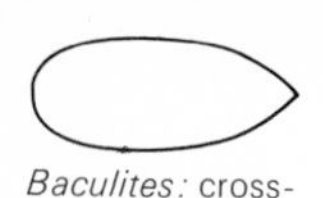
Baculites: cross-section of whorl

Baculites **Lower–Upper Cretaceous: Worldwide** Usually over 10 cm long. Consisting of one or two early whorls (not shown here) and a long straight section. Cross-section flattened as shown, and suture complex (shown here). Ornament weak or absent. Representative of a group of ammonites (Baculitidae from the Lower–Upper Cretaceous) all having straight or slightly curved shells.

Turrilites **Lower–Upper Cretaceous: NA E Af Asia** Large and coiling in a spiral, giving an appearance similar to some gastropods. The seam between the whorls is open, however. Ornament of tubercles with depression along venter. Suture complex (not shown here).

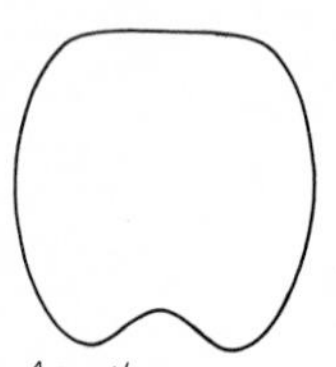
Acanthoceras: cross-section of whorl

Acanthoceras **Upper Cretaceous: NA E Af Asia** Medium-sized to large. Umbilicus wide and whorl cross-section as shown. Ribs straight with tubercles near ventral face. Venter with slight keel and bounded by rows of tubercles.

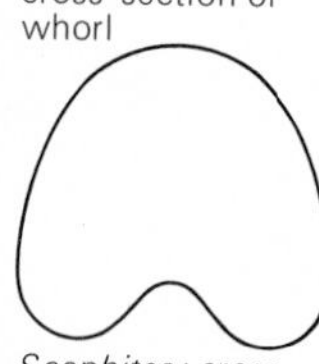
Scaphites: cross-section of whorl

Scaphites **Lower–Upper Cretaceous: Worldwide** Small to medium-sized. Umbilicus narrow and whorl cross-section as shown. Coiling abnormal with straight section and last part of whorl not in contact. Ornament of tubercles and many fine, branching ribs which extend across venter and curve forwards.

Belemnites An extinct cephalopod group which was particularly important during the Jurassic and Cretaceous. Usually only the back part of the shell or *guard* is found. This is bullet-shaped and was internal. In complete specimens, flattened regions are present at the front (that is opposite the point). Broken specimens show a structure of radiating calcite fibres and concentric growth lines. A hollow region at the front of the guard is the *alveolus*, and may house a chambered part of the shell (shown here in *Cylindroteuthis*) which is termed a *phragmocone*.

Neohibolites **Upper Cretaceous: E** Small, guard usually 5–10 cm long. Circular cross-section and widening forwards before narrowing rapidly to point. At the front of this specimen the crushed phragmocone is preserved. A short slit and groove is present on the guard at the back, near the alveolus, shown here.

Belemnitella **Cretaceous: E** Large, guard usually over 10 cm long. Cross-section almost circular but with upper face flattened and bounded by a pair of shallow, longitudinal depressions. There is a long slit on the ventral face of the alveolus near its edge.

Cylindroteuthis **Jurassic–Cretaceous: NA E** Large, guard usually about 15 cm long. Cross-section a wide oval with side faces slightly flattened, and lower face with a long groove becoming deeper backwards towards the point. The chambered phragmocone is clearly shown here at the front of the specimen.

Baculites

Turrilites

Acanthoceras

Scaphites

Neohibolites

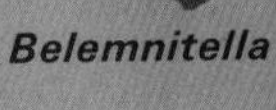

Belemnitella

Cylindroteuthis

Bivalves

Cockles, scallops, razor shells, oysters, mussels and clams are all bivalve molluscs. Bivalve molluscs resemble brachiopods (page 270) but closer inspection reveals important differences. In most bivalve molluscs each valve is asymmetrical *(inequilateral)* with the beak towards the front end; and the valves are mirror images of each other *(equivalve)*. Oysters are well known exceptions to this and they are *inequivalve*. In brachiopods each valve is usually symmetrical but the two valves differ in size and curvature.

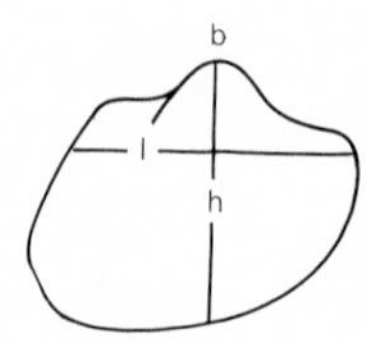

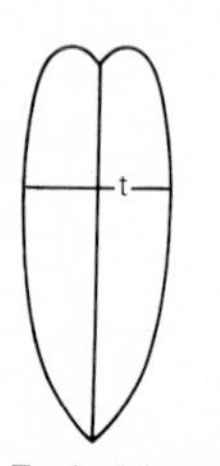

Typical features of a bivalve shell: side view (top) and cross-section

Important features of bivalve molluscs are height **(h)** length **(l)**, thickness **(t)**, the beak **(b)** and ornamentation. A flattened region between the beaks is an *area* (shown in *Arca* on page 264); a flattened depression in front of the beak is a *lunule* and behind the beak is an *escutcheon*. An opening or notch between or behind the beaks is a *ligamental notch*. Valves articulate at the *hinge line*. In some shells the valves do not meet at the front or back and a *gape* is left (shown in *Pholadomya* on page 262).

Projections of shell known as *hinge teeth* may be present below the beak and at either end of the hinge line; ridges or 'teeth' on the side and lower margins are termed *crenulations* (shown in *Glycymeris* on page 264). Front and back muscle scars may be present (shown here in *Mya*) or only one may be developed. The *pallial line* is a depression between the muscle scars marking the extent of attachment of the animal within the shell. The *pallial sinus* is an inflexion of this line near the back of the shell.

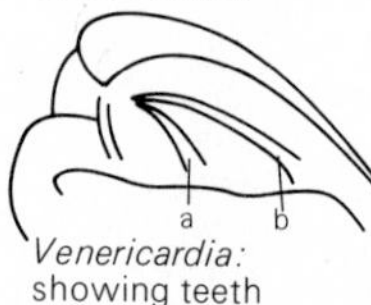

Venericardia: showing teeth

Venericardia Palaeocene–Eocene: NA E Af Ranging from 3–15 cm long. Equivalve and strongly convex. Beak points forwards. Ligamental notch behind beak. Two strong teeth **(a, b)** under beak on each valve as shown. Ornament of wide radiating ridges and concentric lamellae stronger near margin. Margins with small crenulations.

Arctica: beak showing teeth

Arctica Cretaceous–Recent: NA E Usually 3–10 cm long. Shape similar to *Venericardia*. Ligamental notch deep. Two or three teeth present as shown. Ornament of concentric ridges. Margins lacking crenulations.

Plagiocardium Palaeocene–Recent: E Af Aust Representative of group which includes cockles. Medium-sized. Valves almost symmetrical; beak points slightly forwards. Hinge line straight. Two central teeth on each valve, one side tooth at front and back on left valve; two front and one back on right valve. Ornament of strong ribs with beaded edges. Margins with strong crenulations.

Mya Oligocene–Recent: NA E Asia Usually 3–15 cm long. Elongate, flattened. Beak small pointing upwards. Ornament of concentric lamellae or smooth. Hinge line curved, lacking teeth but having a spoon-like process known as the *chondrophore*. Crenulations absent. Wide posterior gape. Front muscle scar high and curved, back muscle scar circular and deep. Deep pallial sinus.

5 cm

Arctica

Venericardia

Mya

Plagiocardium

Teredo Eocene–Recent: Worldwide Representative of a group of molluscs most commonly known from their borings in wood (shown here). Burrows are circular in cross-section and may have a calcareous lining. They may be filled with mud or contain remains of the shell. *Teredo* has a very small shell and the grouping of Recent genera in this group is based on the soft anatomy.

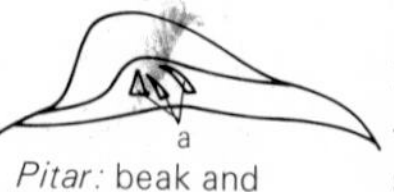

Pitar: beak and hinge showing teeth

Pitar Eocene–Recent: Worldwide Medium-sized. Valves very convex and similar in shape to *Arctica*. Beak points forwards, lunule shallow, escutcheon absent. Teeth as shown; front, side teeth well developed, usually three central teeth **(a)** in each valve. Ligamental notch behind beak, otherwise margins closed; lower margin smooth. Pallial sinus present. Ornament of concentric ridges.

Neocrassina Jurassic–Cretaceous: E Af Right valve shown. Medium-sized; shallowly convex to thick. Beak points forwards and front part of shell much smaller than back. Large lunule and escutcheon clearly defined. Ornament of concentric ridges. Two central teeth on each valve. Margins smooth or with small crenulations. Margins closed.

Pholadomya Triassic–Recent: Worldwide. Medium-sized to large, elongate. Valves very convex, shell thin. Beak near front end, not strong, rounded and pointing upwards. Ornament of radiating ridges over central region but with concentric ridges prominent at front and back ends. Teeth very weak. Valves with strong back and weak front gape. Pallial sinus present.

Sanguinolites Devonian–Permian: Worldwide Medium-sized. Elongate and curved with front end very reduced. Thick. Teeth absent from hinge line; escutcheon large and clearly defined; lunule less well defined. Ornament of concentric ribs. Margins smooth and leaving small gape at back end.

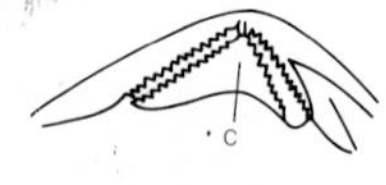

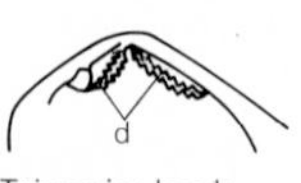

Trigonia: beak showing teeth on left (top) and right valves

Trigonia Triassic–Cretaceous: Worldwide Medium-sized to large. Almost triangular with front edge steeper than back. Beak pointing upwards or slightly backwards. Flattened face at back of shell delimited by high ridge and smooth channel. Ornament at front of strong concentric ridges and at back of weaker radiating ridges. Escutcheon large and defined by a high crest with a beaded edge. Large central tooth **(c)** on left valve, and two large teeth **(d)** on right valve, have strongly grooved surfaces. Margins closed and smooth.

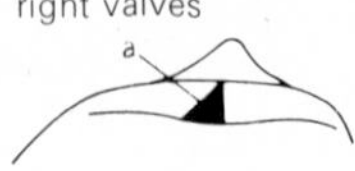

Schizodus: beak and hinge showing tooth

Schizodus Carboniferous–Permian: Worldwide Small to medium-sized. Thick with flattened margins. Beak strong, pointing upwards and front end reduced. Lunule and escutcheon absent. Single large tooth on each valve **(a)**, a few smaller teeth also present. Shell surface smooth or with weak concentric ripples. Margins smooth and closed.

Teredo

Pholadomya

Pitar

Schizodus

Neocrassina

Sanguinolites

Trigonia

5 cm

Anodonta (freshwater mussel) Cretaceous–Recent: NA SA E Af Asia 3–15 cm long. Shell elongate, beak well-formed pointing forwards or upwards. Shell flattened to thick. Surface smooth or with concentric rings. Hinge toothless or with small ridges. Ridge of variable strength runs backwards from the beak to the back margin. Back margin more pointed than front. Margins closed and lacking crenulations.

Carbonicola Carboniferous: E Fresh water. Medium-sized, flattened to thick. Elongate at back, shortened at front. Beak pointing upwards or forwards. Hinge line curved. Sometimes one or two tooth-like structures under beak on each valve. Margins smooth, closed. Front muscle scar circular and deep, back scar shallow and high. Ornament of concentric lines.

Modiomorpha Silurian–Permian: NA E Asia Medium-sized. Equivalve and valves expanded backwards. Beak low. Single tooth on left valve and socket on right. Margins smooth and closed. Ornament of concentric lines.

Arca Jurassic–Recent: Worldwide Medium-sized, usually 5–10 cm long. Elongate with beak well in front of mid-line and pointing slightly forwards. Valves very convex. Hinge line carrying very wide, flattened areas which separate the beaks. Hinge with long row of small, comb-like teeth. Lower margin with elongate gape, this is visible as the long dark region. Ornament of concentric and radial ribs.

a b

Parallelodon: beak and hinge showing teeth

Parallelodon Devonian–Jurassic: Worldwide Usually 5–15 cm long. Elongate with very long back region and shortened front end. Beak pointing forwards. Hinge line straight. Large flattened areas between beaks carry longitudinal ridges. Very few teeth near back of hinge **(a)**, and numerous shorter, curving teeth near front **(b)**. Elongate gape on lower margin. Margins smooth.

Glycymeris Cretaceous–Recent: Worldwide Small to medium-sized, almost circular. Beak almost centrally placed and pointing upwards (that is equilateral). Hinge teeth like *Arca* but arranged in gentle curve. Areas developed but smaller than in *Arca*. Crenulations on lower margin (shown here). Surface smooth or with radial ridges and concentric grooves.

Modiolus Devonian–Recent: Worldwide Medium-sized to large, up to 10 cm long. Generally similar to the common mussel but beak not at very front of shell. Hinge line without teeth. Shell surface smooth or with shallow concentric ridges. Equivalve with ligamental notch developed, otherwise margins smooth and closed.

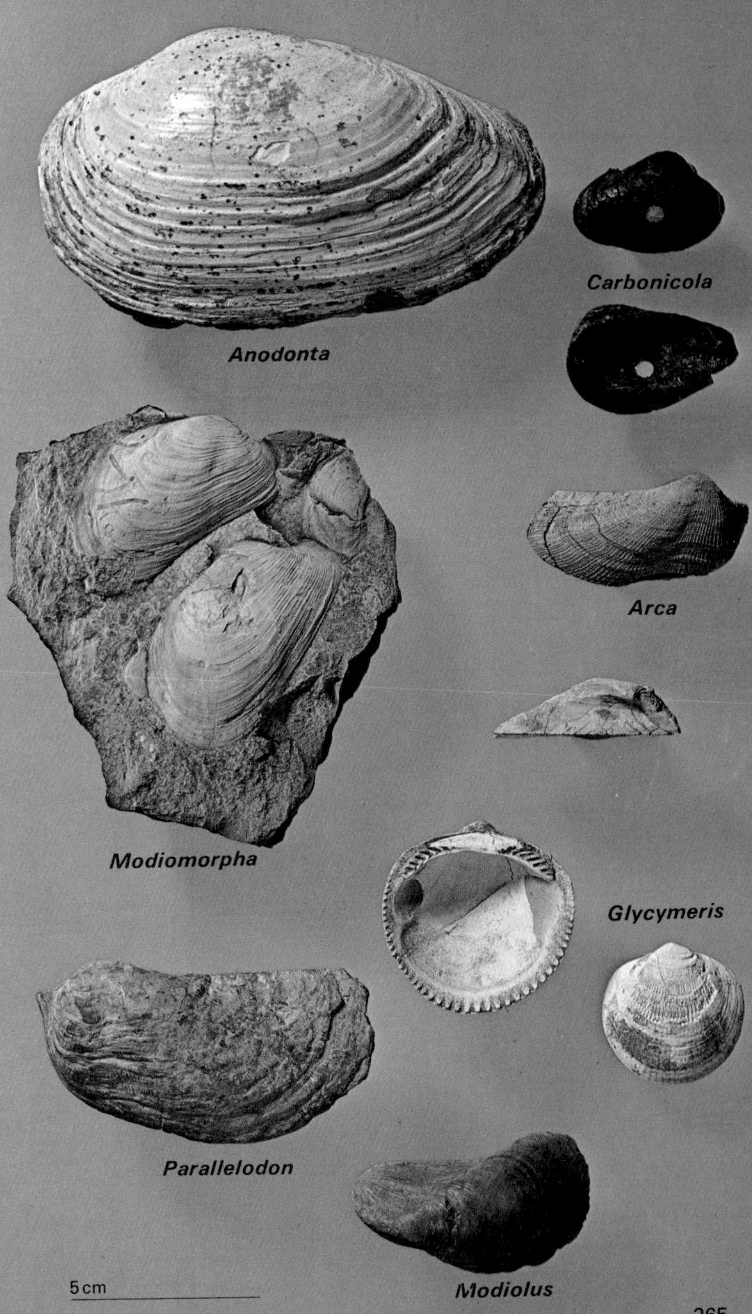
Anodonta
Carbonicola
Modiomorpha
Arca
Glycymeris
Parallelodon
Modiolus
5 cm

Pinna Carboniferous–Recent: **Worldwide** Medium-sized to large. Shaped like a half-closed fan, triangular and up to 25 cm long. Valves equal with beaks at anterior point. Ornament of wide ripples below and radiating ridges above. Shell surface often shiny (shown here). Lower margins with elongate gape near front and back margins wide open.

Gervillella Triassic–Cretaceous: **Worldwide** Medium-sized to large, up to 25 cm long. Very elongate with greatly lengthened back and reduced, sharply pointed front. Dentition of a few elongate teeth which are almost parallel to the long axis. Region above hinge line flattened with numerous (up to ten) vertical pits which hold the ligament. Ornament of concentric lamellae.

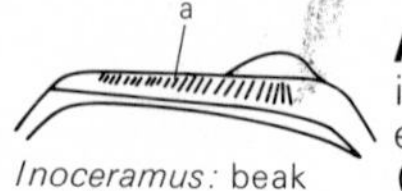

Inoceramus: beak and hinge showing ligamental pits

Inoceramus Jurassic–Cretaceous: **Worldwide** Medium-sized to large, usually 8–15 cm high. Back wing expanded as shown or reduced. Numerous ligamental pits (a) along upper edge of hinge line and wing. Hinge without teeth. Ornament of concentric, coarse ripples and fine grooves. Beak points upwards. Shell short and high, very convex.

Pterinopecten Silurian–Devonian: **Worldwide** Medium-sized. Beak pointing upwards. Hinge line straight with wings developed before and behind beak, back wing larger. Right valve usually less convex than left. Ornament of radial ridges of variable strength.

Oxytoma Triassic–Cretaceous: **Worldwide** Small to medium-sized. Beak pointing upwards with wings developed before and behind. Back wing usually longer and pointed. Right valve flattened, left valve convex. Hinge lacking teeth but with narrow areas, that of the left valve continues in plane of margin and that of the right valve is at about 90° to this. Ornament of coarse ridges and wide intervals. Ridges produced as spines around margin.

Meleagrinella Triassic–Jurassic: **Worldwide** Small to medium-sized. Small wings before and behind beak. Hinge lacking teeth. Left valve convex, right valve flattened. Left valve with radial ridges which have spiny edges; ridges weak or absent on right valve. A block with many small specimens is shown with mainly left valves visible.

Chlamys Triassic–Recent: **Worldwide** Medium-sized, rarely more than 15 cm high. Similar to living, common scallops. Equilateral, inequivalve, left valve more convex than right. Wings before and behind beak; back wing notched on left valve. Hinge teeth absent but triangular ligamental notch developed under centre of beak on both valves. Sculpture of strong ribs giving serrated edges at the margins. Concentric sculpture also usually developed.

5 cm
Pinna
Gervillella
Pterinopecten
Inoceramus
Meleagrinella
Chlamys
Oxytoma

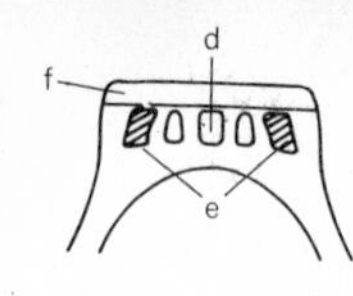

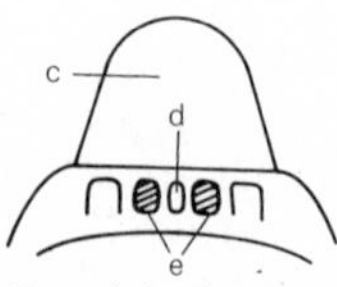

Spondylus: beak and hinge of left (top) and right valves

Spondylus **Jurassic–Recent: Worldwide** Medium-sized, up to 12 cm high. Nearly equilateral, strongly inequivalve. Valves high and right valve deeper than left. Hinge line straight. Beak of right valve with large area **(c)** which carries fine vertical and cross-striations. Area **(f)** of left valve low and sloping outwards. The specimen shown here is particularly spiny but in some forms the ridges predominate with only a few spines present. Two large teeth **(e)** are far apart on the left valve and close together on the right valve. Deep notch **(d)** below centre of beak on both valves.

Plagiostoma **Triassic–Cretaceous: Worldwide** Medium-sized to large, up to 15 cm long. Valves same size. Beak points backwards and the front edge is straight with an elongate, wide lunule. Margins usually closed. Teeth weak or absent. Surface smooth with fine concentric or radial striations.

Cardiola **Silurian–Devonian: NA E** Small, beak points upwards or forwards, equivalve. Hinge teeth absent. Triangular areas on both valves. Margins may have a gape. Strong radial ribs crossed by concentric grooves give a squared pattern. A block with external moulds and impressions is shown here.

Nucula **Cretaceous–Recent: Worldwide** Small, equivalve, beak points backwards. Comb-like teeth along margins before and behind beak (shown here). Internal ligamental process under beak. Lower margin has fine striations. Anterior and posterior muscle scars equal in size. Outer surface smooth with fine radial ridges and/or concentric rings. Inner surface often shiny (shown here).

Gryphaea **Triassic–Jurassic: Worldwide** Medium-sized to large, up to 15 cm long. Left valve much larger than right and very convex with beak rolled over onto right valve and displaced slightly backwards. Right valve flat or concave. Ornament of left valve numerous well defined lamellae. Right valve with smooth or rippled surface and lamellae near margin. Left valve with elongate curved swelling along back edge above margin.

Lopha **Triassic–Recent: Worldwide** Usually medium-sized. Valves convex, shape varying from similar to *Ostrea* to inequilateral (shown here). Almost equivalve. Radial ridges characteristic, varying from strong ripples to high ridges (shown here); these give lower margin zigzag contact. Inner faces with small tubercles near margins.

Ostrea **(common oyster) Cretaceous–Recent: Worldwide** Medium-sized to large, up to 20 cm long. Left valve convex; right valve smaller than left and flattened or concave. Shape varies from length equal to height, to length much less than height. Key feature is the radial ribbing of the left valve and the absence of ribbing on the right valve.

5 cm
Spondylus
Plagiostoma
Nucula
Cardiola
Gryphaea
Ostrea
Lopha

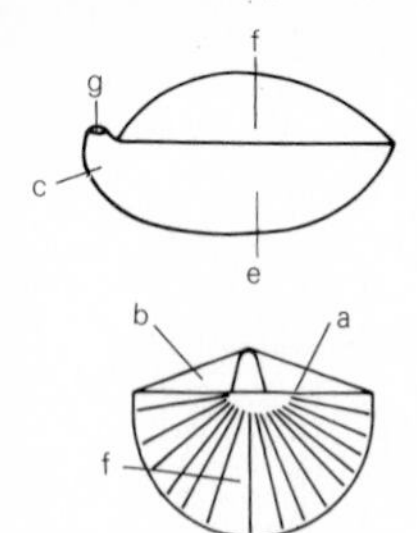

Typical features of a brachiopod: side view (top) and upper view (not same genus)

Brachiopods

Generally similar in appearance to bivalved molluscs as they consist of two shells. Distinguishing features of molluscs are given on page 238. Important features to study on brachiopods are the *hinge line* (**a**), *interarea* (**b**), that is flattened regions often present between hinge line and beak; *beak* (**c**); *front end* (**d**); the *fold* which is a long swelling (visible here on the brachial valve of *Spirifer*); and the *sulcus* which is a long channel (visible here on the pedicle valve of *Spirifer*). The fold and sulcus often occur together on opposite valves. The ornamentation usually consists of radiating ridges (as in *Spirifer*), but concentric growth lines may also be present (as in *Atrypa*). The two valves are termed the *pedicle* and *brachial* valve. The pedicle valve (**e**) always has the stronger beak, and is often larger than the brachial valve (**f**). Also, the beak of the pedicle valve often carries a small hole, the *foramen* (**g**), through which the attachment stalk emerges in the living animal.

Spiriferids

Spiriferids are defined by their internal structure and are very variable externally. Occasionally a spiral structure may be visible on a broken or weathered specimen. The other brachiopods may be placed with their major group on the basis of a few simple external features.

Spirifer Carboniferous: **Worldwide** Relatively wide and strongly biconvex; hinge line long. Wide, long interarea on pedicle valve only. Beak of pedicle valve strong. Strong sulcus on pedicle valve and fold on brachial valve. Ornamentation of strong ridges which fork and are present on the fold and sulcus. Growth lines may also be present. Foramen absent.

Eospirifer Silurian–Devonian: **Worldwide** Biconvex but pedicle valve not very deep. Beak strong and pedicle valve interarea almost horizontal. Hinge line long but less than maximum width of shell. Strong fold on brachial valve and sulcus on pedicle valve. Ornament of fine radiating ridges and concentric growth lines.

Atrypa Silurian–Devonian: **Worldwide** Medium-sized. Brachial valve very convex, pedicle valve flattened or shallowly convex flexing downwards at its edges. Interareas absent but hinge line long or short. Beak small and turned inwards. Ornamentation of ridges crossed by equally strong growth lines. Strong fold on brachial valve and sulcus on pedicle valve, particularly in old individuals.

5 cm

Spirifer (dorsal view)

Spirifer (ventral view)

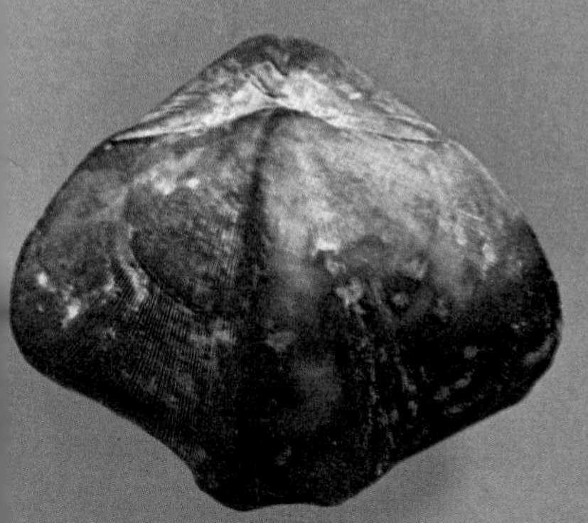

Eospirifer

Atrypa

Athyris **Devonian–Triassic: Worldwide** A small to medium-sized spiriferid with a smooth, biconvex shell. Interareas absent, hinge line short. Shape varying from wide to elongate. Fold on brachial valve and sulcus on pedicle valve, both single smooth curves of variable strength. Beak strong and foramen present. Ornament of growth lines which may have the form of thick lamellae.

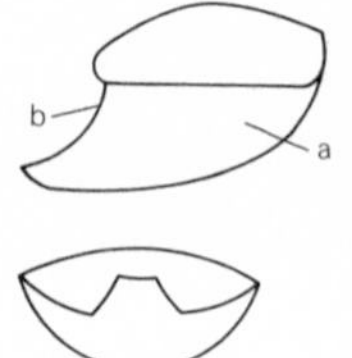

Cyrtia: side view (top) and front edge

Cyrtia **Silurian–Devonian: Worldwide** Medium-sized. Pedicle valve **(a)** convex and very deep. Fold on brachial valve and sulcus on pedicle valve. Interarea of pedicle valve very large **(b)** and almost vertical with high triangular projection in centre. Shell surface smooth or carrying fine ridges and grooves.

Orthids

Hinge line long and interareas present on both valves. Shells biconvex.

Orthis **Cambrian–Ordovician: Worldwide** Small to medium-sized. Pedicle valve convex, brachial valve shallowly convex or flattened. Hinge line equalling greatest width of shell. Interarea of pedicle valve large, interarea of brachial valve narrow. Interareas curve inwards and both have triangular swellings or depressions near the middle. Ornament of strong radiating ridges. Brachial valve usually with weak sulcus.

Platystrophia **Ordovician–Silurian: Worldwide** Large to medium-sized. Strongly biconvex. Hinge line may equal greatest width, produced as point or sharp corner at each end. Interareas large, almost equal in size. Beak curving inwards. Strong fold on brachial valve and sulcus on pedicle valve. Ornament of radiating ridges. Externally *Platystrophia* is indistinguishable from the spiriferids and is distinguished by its internal structure.

Schizophoria **Devonian–Permian: Worldwide** Medium-sized. Brachial valve more convex than pedicle valve. Interarea of pedicle valve larger than that of brachial valve; interareas shorter than hinge line which is less than greatest width. Low fold on brachial valve and sulcus in pedicle valve. Ornament of fine ridges and growth lines.

Dalmanella **Ordovician–Silurian: Worldwide** Medium-sized almost circular in outline. Brachial valve more convex than pedicle valve. Interarea of pedicle valve long with curved surface which slopes downwards. Interarea of brachial valve shorter and curving upwards. Weak sulcus sometimes on brachial valve. Ornamentation of fine ridges of variable thickness. Growth lines strong near edges of valves.

Dicoelosia **Ordovician–Devonian: Worldwide** Small to medium-sized. Strong sulci on both valves produce deep indentation on front edge. Hinge line shorter than greatest width. Interarea of pedicle valve longer than that of brachial valve. Ornamentation of ridges and growth lines.

5 cm
Athyris
Cyrtia
Orthis
Platystrophia
Dalmanella
Dicoelosia
Schizophoria

Strophomenids

Interareas present on both valves; one valve usually convex and other concave.

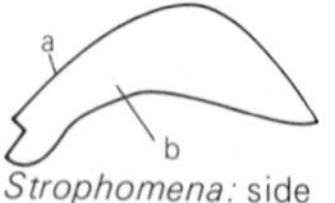

Strophomena: side view showing pedicle and brachial valve curvature

Strophomena **Ordovician: Worldwide** Brachial valve (**a**) convex, pedicle valve (**b**) concave. Hinge line long, corresponding to greatest width of shell. Interarea of pedicle valve wider than that of brachial valve. Triangular swellings in middle of upper and lower interareas. Ornament of fine radiating ridges and grooves.

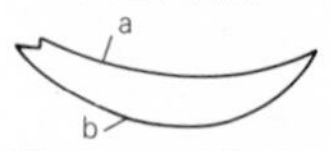

Chonetes: side view showing pedicle and brachial valve curvature

Chonetes **Devonian: Worldwide** Brachial valve (**a**) concave, pedicle valve (**b**) convex. Hinge line long but not always widest part of shell. Surface with fine radiating ridges and grooves. Interarea of brachial valve smaller than that of pedicle valve. A row of spines is present along the edge of the interarea on the pedicle valve; this feature is characteristic of the group to which *Chonetes* belongs.

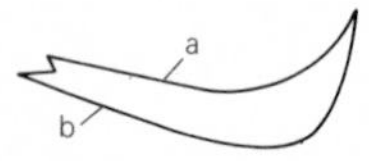

Rafinesquina: side view showing pedicle and brachial valve curvature

Rafinesquina **Ordovician: NA E Asia Af** Pedicle valve shown here. Large to medium-sized. This form is like *Strophomena* but with reversed convexity, that is the brachial valve (**a**) is concave and the pedicle valve (**b**) is convex. Hinge line long and small foramen on beak of pedicle valve. Ornament of radiating ridges of variable thickness with the stronger ridges reaching to the beak. Middle ridge of pedicle valve usually very strong (shown here).

Sowerbyella **Ordovician–Silurian: Worldwide** Small to medium-sized. Brachial valve concave, pedicle convex. Hinge line corresponds to greatest width of shell. Ornamentation of fine radiating grooves and ridges.

Leptaena **Ordovician–Devonian: Worldwide** Brachial valve concave, pedicle valve convex. Hinge line equals greatest width of shell and carries long, narrow interareas. Shells have very strong concentric ridges and grooves and finer radiating ridges and grooves.

Productella **Devonian NA E Asia** Small to medium-sized, hemispherical to almost square shell with deeply concave brachial valve (not shown here), and very convex pedicle valve. Interareas very narrow, straight and poorly developed. Small spines scattered over pedicle valve, but only rarely present on brachial valve.

Spinulicosta **Devonian: Worldwide** Small to medium-sized and similar to *Productella* to which it is very closely related. The shell is more elongate in *Spinulicosta* and carries an ornament of weak radiating ridges and grooves. Long slender spines may be present but are often not preserved. Interareas very narrow and straight as in *Productella*. Brachial valve (not shown here) is dimpled and may carry concentric grooves.

Strophomena

Chonetes

Rafinesquina

Sowerbyella

Productella

Leptaena

Spinulicosta

5 cm

Productus **Carboniferous: E Asia** Large. Pedicle valve (shown here) highly convex and overlapping the hinge line. Brachial valve flat. Ornament of radiating ridges. Spines may be scattered over the surface and rows of spines may be present on the pedicle valve near the hinge line.

Pentamerids

Interareas present on both valves; shells biconvex; hinge line short.

Sieberella **Silurian–Devonian: NA E Af Asia** Medium-sized and similar in general form to *Conchidium*, but with pedicle valve usually even more convex. Beak very strong. Sulcus on brachial valve and fold on pedicle valve strong and carrying an ornamentation of ridges, but the rest of the shell surface is smooth. Front edge with a single, strong, angular curve.

Conchidium **Silurian–Devonian: Worldwide** Large. Both valves very convex, pedicle valve more so than brachial valve. Beak of pedicle valve curves upwards and overlaps the beak of the brachial valve (shown here in side view). Interarea of pedicle valve small and interarea of brachial valve obscured by inwardly flexed beak. Ornament of strong ridges. Fold and sulcus not developed. Front edge straight or with shallow curve.

Terebratulids

Interareas on pedicle valves only, if visible. Shell surface usually smooth and foramen clearly visible on beak.

Dielasma: front edge

Dielasma **Carboniferous–Permian: Worldwide** Small to medium-sized. Biconvex, shell surface smooth. Shell elongate, tear-drop shaped. The front edge may show a single curve which may be only feebly developed as shown. Foramen open and beak pointing upwards and outwards.

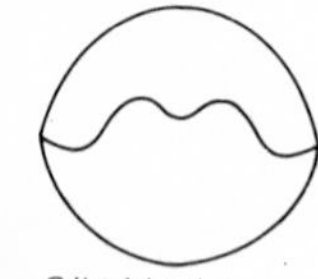

Gibbithyris: front edge

Gibbithyris **Cretaceous: E** Medium-sized, biconvex. Front edge showing double curve as shown. Foramen open and beak pointing upwards or upwards and inwards. Shell surface smooth and outline less elongate than that of *Ornithella*.

Ornithella **Jurassic: E** Small to medium-sized, biconvex, with a smooth surface and weak or strong growth lines. Outline an elongate oval and front edge having an upward curve which is depressed centrally. Foramen clearly visible and beak pointing upwards and outwards.

5 cm

Productus

Sieberella

Conchidium

Dielasma

Gibbithyris

Ornithella

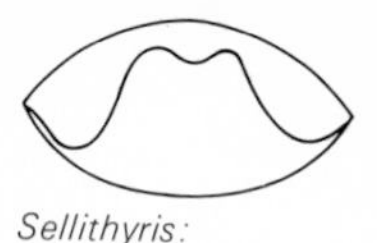
Sellithyris: front edge

Sellithyris Cretaceous: E Medium-sized. Body flattened and biconvex. Shell surface smooth with strong growth lines. Front edge complex as shown and similar to that of *Gibbithyris*. Foramen open and large. Beak pointing upwards or upwards and inwards.

Rhynchonellids

Interareas very small or not visible. Shell surface with strong ridges; usually angular. Beak usually strong.

Cyclothyris Cretaceous: NA E Relatively large rhynchonellid, similar in general form to *Goniorhynchia* but wider and more flattened. Upward fold of front edge weaker than in *Goniorhynchia*. Beak pointing upwards.

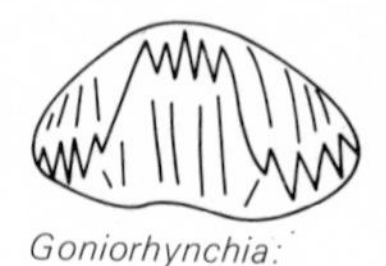
Goniorhynchia: front edge

Goniorhynchia Jurassic: E Medium-sized, biconvex, wider than long. Front edge as shown with single, strong, angular upward curve. Sulcus of brachial valve and fold of pedicle valve strongly developed. Line of contact between valves carrying strong interlocking teeth. Beak strong and pointing upwards and outwards. Ornamentation of strong, sharp-edged ridges.

Rhynchotrema Ordovician: NA Small, biconvex. Sulcus of brachial valve and fold of pedicle valve well developed. Front edge carrying interlocking teeth. Ornamentation of very strong ridges. Beak strong.

Hypothyridina Devonian: Worldwide Large to medium-sized. Shell very high and biconvex. The front edge is characteristic, as the pedicle valve is produced upwards as a strong process which meets the brachial valve near the top surface of the shell. The sulcus on the pedicle valve and fold on the brachial valve are well developed. Ornamentation smooth near beak but strong ridges near the front.

Inarticulate brachiopods

Valves not firmly joined. Interareas and hinge teeth never present.

Lingula Ordovician–Recent: Worldwide Elongate and nearly oval with small pointed hinge region. Shallowly biconvex and valves usually found separated. Ornament of fine ridges crossed by numerous growth lines. Shell very thin with slight thickening near hinge and may have appearance of mother-of-pearl.

Crania Cretaceous–Recent: Worldwide Small. Usually found attached to other fossils by all parts of pedicle valve, which is completely cemented to attachment surface. Shell conical or flattened and may carry radiating ridges and grooves as well as concentric growth lines. Four specimens of *Crania* are shown here arrowed on the surface of a strophomenid.

5 cm
Sellithyris
Cyclothyris
Goniorhynchia
Rhynchotrema
Hypothyridina (dorsal view)
Hypothyridina (anterior view)
Lingula

Crania

Graptolites

A group of colonial, usually planktonic animals. The class is extinct but its members were important and common from the Cambrian to the Carboniferous. Graptolites are very useful for dating Palaeozoic rocks as they changed very rapidly with time and many genera had worldwide distribution. Graptolites are common in shales and slates in which they are flattened along the bedding planes and are usually carbonized. They may be difficult to see on the rock surface, but by slanting the specimen to the light they are usually seen to have shiny surfaces.

Each graptolite colony is known as a *rhabdosome* and consists of a variable number of branches or *stipes* that diverge from the initial individual of the colony, which is known as the *sicula*. The *nema* is the thread-like process by which the rhabdosome may be attached. Each individual of the colony is housed in a cup-like structure known as the *theca*.

Diplograptus: biserial growth form

Dendrograptus Cambrian–Carboniferous: Worldwide An attached, plant-like form. Rhabdosome consisting of numerous stipes which give it a fern-like appearance.

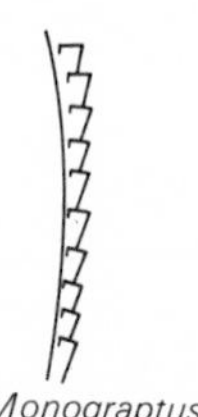

Monograptus: monoserial growth form

Diplograptus Ordovician–Silurian: Worldwide Member of the graptoloid group (Graptoloidea) of the graptolites, which also includes *Monograptus*, *Dicellograptus* and *Tetragraptus*. Members of this group were important planktonic forms in the Ordovician and Silurian. *Diplograptus* has thecae arranged on each side of the stipes as shown (that is it is *biserial*). This shows as double serrations on the ribbon-like specimens.

Monograptus Silurian: Worldwide Thecae arranged in a single row along the side of the stipe as shown (that is it is *monoserial*). Rhabdosomes may be coiled, spiral or straight, and fragmentary stipes of the other genera in the family (Monograptidae) also resemble *Monograptus*.

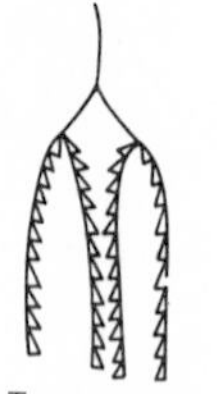

Tetragraptus: growth form

Tetragraptus Ordovician: NA E Asia Aust Rhabdosome consisting of four short, wide branches that diverge from the nema and then fork again as shown. Each branch has thecae on one side only. Serrated edges are clearly shown on this specimen.

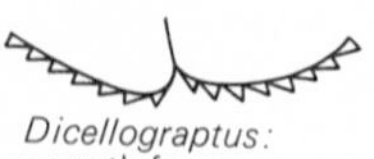

Dicellograptus: growth form

Dicellograptus Ordovician: Worldwide Consisting of two stipes that are characteristically flexed from the centre as shown, and carry thecae on one side only.

5 cm
Dendrograptus
Diplograptus
Monograptus
Tetragraptus
Dicellograptus

Echinoderms

Fossil echinoderms are easily identified as most of them are similar to living forms or have characteristic shapes. Five groups: sea lilies, starfishes, edrioasteroids, blastoids and sea urchins, are mentioned here. The sea lilies and sea urchins are most important as fossils.

Echinoderm groups

Plant-like; usually with stem, feathery arms and body consisting of large or small plates	sea lilies
Star shaped; arms long or short	starfishes
Discoid, carrying raised, sinuous star-shaped grooves	edrioasteroids
Bud-like with five radiating grooves	blastoids
Conical, flattened or lozenge-shaped; with anal and mouth apertures and radiating ambulacra raised or impressed on surface	sea urchins

Sea lilies (Crinoidea)

Consisting of a *theca* composed of *plates*, bearing five or more *arms* which are often branched and may carry small, hair-like processes known as *pinnules*. Theca usually supported by a *stem* (not in *Marsupites* on page 284 or *Uintacrinus* shown here). Features of *stem ossicles*, thecal plates, arms arising from *radial plates* and pinnules, are important. Often only stem ossicles are found (shown in *Cyathocrinites* on page 284).

Sagenocrinites Silurian: NA E A large crinoid. Theca consisting of numerous hexagonal plates. Arm bases not pronounced on surface of theca. Arms consisting of columns of single plates. Pinnules not present and stem with circular cross-section.

Taxocrinus Devonian–Carboniferous: NA E A large crinoid. Theca composed of many loosely joined plates, differing in shape from those of *Sagenocrinites*. Arm bases pronounced and ornament on radial plates may be easily traced to the base of the theca. Arms short, pinnules absent. Although not clearly shown by this specimen, the crown of *Taxocrinus* is usually higher and more elongate than that of *Sagenocrinites*.

Uintacrinus Cretaceous: NA E Large stemless form. Theca consisting of many small plates. Arms long and slender with pinnules. Arm bases strong and may be traced to radial plate near base of theca.

Pentacrinites Triassic–Cretaceous: NA E Large form. Theca small. Arms long and much branched. The stem of *Pentacrinites* is characteristic and isolated stem ossicles may be identified as they have a star-shaped cross-section. The stem may bear hair-like processes known as *cirri*, which have a diamond-shaped cross-section.

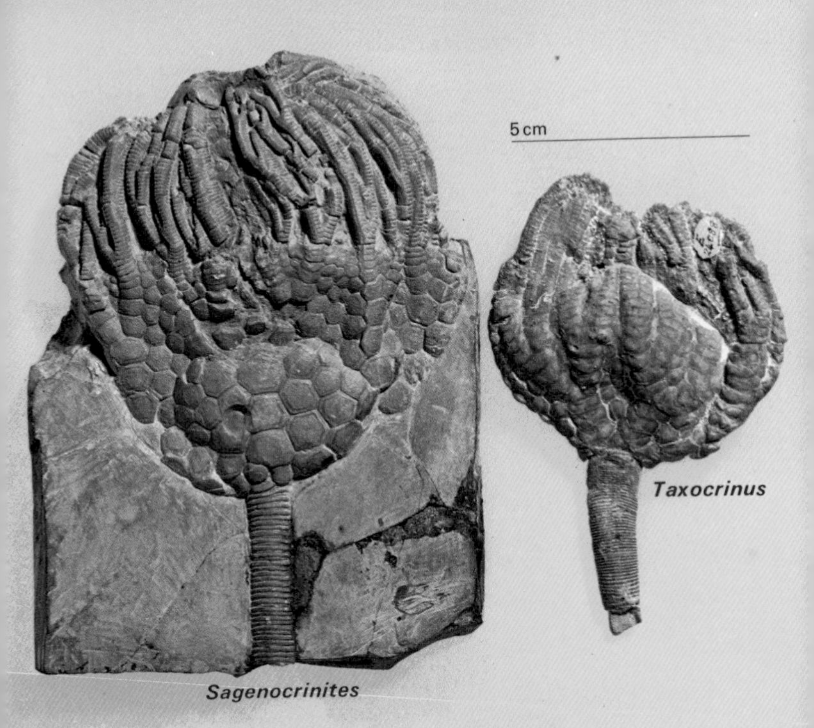
5 cm
Taxocrinus
Sagenocrinites

Pentacrinites
Uintacrinus

Marsupites Cretaceous: NA E A stemless form. Theca large consisting of very large plates which have ridged surfaces. Arms consist of columns of very large plates and bear pinnules.

Cyathocrinites Silurian–Permian: NA A medium-sized form with a relatively high theca consisting of few large plates arranged in three rows and firmly joined. Arms free from the top of the theca and branching by forking. Arms consist of columns of single plates and pinnules are not present. Several stem fragments are visible and scattered among the arms and to the top left of the specimen shown here. These have a wide canal with circular cross-section and a structure of radiating ridges and grooves.

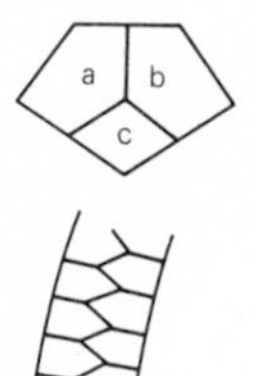

Phanocrinus Carboniferous: NA Smaller than *Cyathocrinites*, and having a low theca consisting of three rows of large, firmly joined plates. The base of the theca is concave with the stem-theca joint in the centre. Arms thick and long, consisting of columns of single plates. Ten arms are present in five pairs which divide at the top plate of the theca.

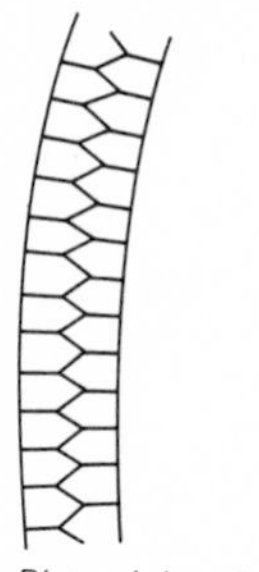

Platycrinites: base of theca (top) and biserial arm

Platycrinites Devonian–Carboniferous: NA E Theca cup-shaped, consisting of firmly joined plates which are large and fewer in number than in *Cyathocrinites*. The base of the theca is formed from three plates of which two (**a** and **b**) are large and one (**c**) small. Arms consisting of double columns of plates, a condition known as *biserial*. Arms with pinnules. The stem has a characteristic flattened cross-section and is twisted and ribbon-like (shown here).

Glyptocrinus Ordovician–Silurian: NA A small sea lily with a high theca consisting of many small plates which are fused in the regions between the arms. Arms fine with long feathery pinnules. The stem has a circular cross-section.

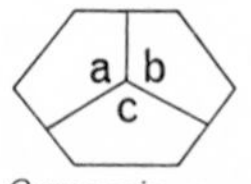

Carpocrinus: base of theca

Carpocrinus Silurian: NA Arms and general shape similar to *Phanocrinus* but the theca is very different in shape, consisting of numerous small plates. The base of the theca is formed from three equally sized plates (**a, b** and **c**).

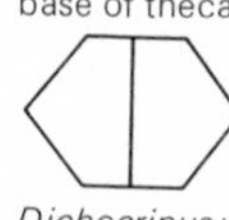

Dichocrinus: base of theca

Dichocrinus Carboniferous: NA Theca large and globular, consisting of a few large plates of which the upper ones are higher and narrower than in *Pentacrinites*. Two equally sized plates form the base of the theca and the stem has a circular cross-section. Arms long and feather-like, having large pinnules and consisting of double columns of plates. Here, *Dichocrinus* lies on top of a specimen of *Rhodocrinites*, which is similar in general appearance, but has a theca consisting of numerous small plates.

Marsupites

Cyathocrinites

Phanocrinus

Platycrinites

Dichocrinus

Glyptocrinus

Carpocrinus

5 cm

Starfishes (Asteroidea)

Star-shaped, usually with five arms which vary greatly in length from species to species and are not sharply marked off from the central disc. Skeletal plates along the edges of the arms are known as *marginal plates*. Complete starfishes are rare as fossils but when they occur they may be abundant.

Pentasteria Jurassic–Eocene: E Arms usually straight-sided, long and pointed. Disc relatively small. Marginal plates large. Inner plates less important, giving granular effect. Stout spines sometimes present on marginal plates. *Pentasteria* is very similar to the living *Astropecten*, from which it may be distinguished by the size of the contact facets between the marginal plates; these are small in *Astropecten* but large in *Pentasteria*.

Mesopalaeaster Ordovician: NA E Disc relatively larger than in *Pentasteria*. Arms narrow and straight-sided. Small spines sometimes present on marginal plates, raised on swellings or tubercles. Inner rows of plates reach ends of arms; these plates are relatively large and distinguish this form from *Pentasteria*. A fairly typical starfish, difficult to distinguish from several closely related forms.

Calliderma Cretaceous–Oligocene: E A fairly common European representative of a starfish group in which the marginal plates are clearly defined, relatively large and longer than wide, while the inner are small and irregularly arranged. The arms are short and the disc is large.

Palaeocoma Jurassic: E Medium-sized form belonging to the living group of brittlestars (Ophiuroidea) and characterized by the possession of long thin arms and a small disc. Ophiuroids are difficult to identify within their group but *Palaeocoma* is a typical fossil form, and differs only slightly from *Ophiura* (Cretaceous–Recent: Worldwide), which is the commonest living brittlestar.

Edrioasteroids (Edrioasteroidea)

Edrioaster Ordovician: NA E Fewer than thirty genera of edrioasteroids are known. Members of the group have the appearance of a starfish wrapped around a ball or disc. The rays consist of elongate cross-plates and the body between them is composed of numerous larger plates of varying size.

Blastoids (Blastoidea)

Pentremites Carboniferous: NA SA A bud-like echinoderm of the blastoid group. Rays have the form of elongate depressions and contain numerous cross-grooves marking small cover plates. The body consists of few plates. Side plates are 'V' shaped to accommodate the rays. Basal region consisting of two large and one small plate. Blastoids are locally abundant and there are many different genera. They are commonest in Upper Carboniferous deposits and *Pentremites* is the most abundant form.

Pentasteria

Mesopalaeaster

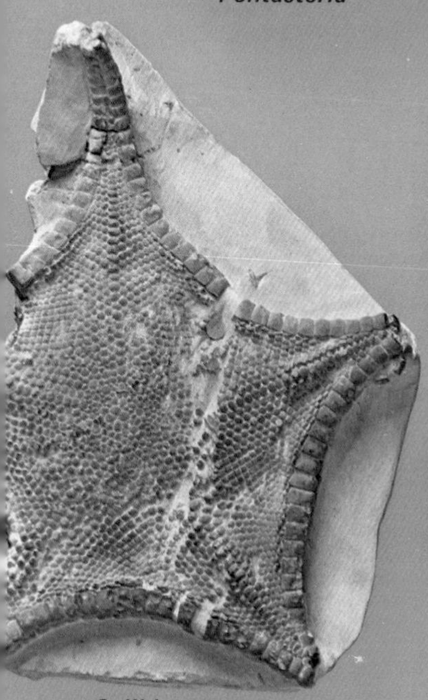

Calliderma

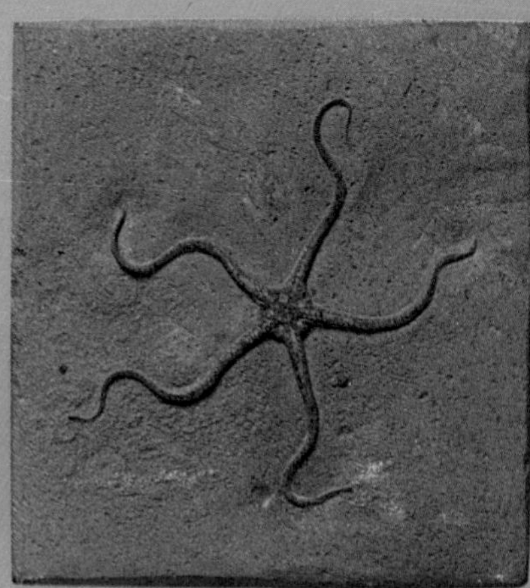

Palaeocoma

Pentremites

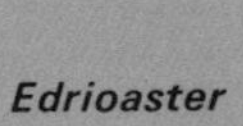

Edrioaster

Sea urchins (Echinoidea)

Round, pentagonal or heart-shaped animals with a *test* (body) bearing spines in well preserved fossils. Usually only the test is preserved. Radiating tracts delimited by rows of pores are termed *ambulacra*, and close inspection reveals the pores to be double. Spaces between the ambulacra are termed *interambulacra*. If the test is circular with an aperture at the apex and another vertically below it on the lower surface, then the specimen belongs with the *regular echinoids*. If the apex does not carry an aperture then the specimen belongs with the *irregular echinoids*, and in this group the outline is not usually circular.

Regular echinoids The lower opening is the *mouth* and the lower surface is known as the *oral surface*. The upper opening is the *anus*.

Pedina Jurassic: E Af Miocene: SA Test flattened. Anus enclosed by a ridge of plates. Small tubercles scattered over surface and larger tubercles in rows in ambulacra.

Pygaster Jurassic–Cretaceous: E Test flattened. Distinguished from *Pedina* by the anus which spreads from the apex as a large, oval aperture in the rear interambulacrum. Test covered with small flattened tubercles which in life would have supported short spines.

Psammechinus Pliocene–Recent: NA E Af Closely related to the common sea urchin *Echinus*. Test hemispherical or flattened, lacking deep sculpturing. Tubercles arranged in regular rows over surface. Tubercles lack holes at tips and their surfaces are smooth.

Acrosalenia Cretaceous: E Af Test flattened. Large tubercles in interambulacra; each with a small hole at its tip and a circlet of grooves just below the tip. Anus displaced slightly from centre of upper surface and surrounded by raised plates. Ambulacra narrow, carrying paired rows of small tubercles.

Hemicidaris Jurassic–Cretaceous: NA E Af Asia Anus surrounded by raised plates. Large tubercles in interambulacra. Tubercles perforated each with a circle of grooves just below the tip. Large tubercles also in lower parts of ambulacra; these decrease in size just above the junction of the upper and lower surfaces. Mouth large with several deep notches around its edge.

Coelopleurus Eocene–Recent: Worldwide Large tubercles in ambulacra, not perforate, one feature which helps distinguish this form from *Acrosalenia* and *Hemicidaris*. Interambulacra very depressed. Large tubercles in lower part of each interambulacrum, decreasing in size abruptly above junction of upper and lower surfaces. A clear space is present between the ambulacra near the apex.

Pygaster

Pedina

Psammechinus

Acrosalenia

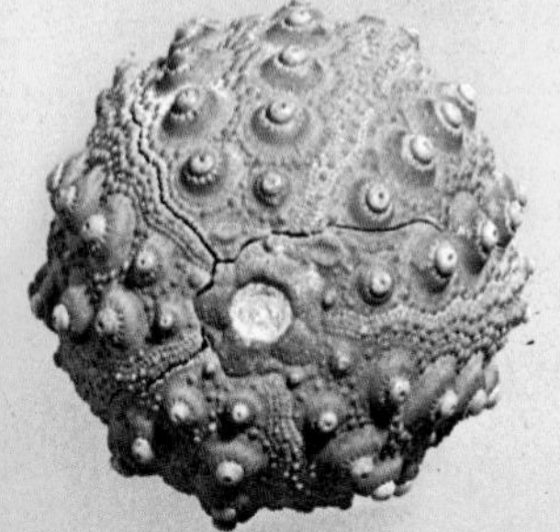
Hemicidaris

Coelopleurus

Irregular echinoids The anus is shifted away from the apex and lies at the back of the animal. Outline usually not circular.

Micraster Cretaceous–Palaeocene: Worldwide A heart-shaped form, very similar to the living heart urchin *Spatangus*, or *Echinocardium*. Back interambulacrum projects as a strong ridge with the anus at its end. Test swollen between anus and oral surface. A large area, composed almost entirely of two plates covered with large tubercles, occupies the middle of the oral surface. Mouth at front of test and a deep depression runs from the mouth to the apex, forming the notch of the heart-shape. Test swollen behind mouth. *Micraster* is very common in Europe.

Pygurus Cretaceous–Eocene: Worldwide Test flattened and pentagonal in outline with the anus near the point. Ambulacra petal-like and smooth. Outer pores of each ambulacrum elongate and slit-shaped, inner pores shorter. On oral surface an elongate swelling runs from the anus to the mouth, which lies in the middle of the oral surface below the apex.

Holaster Jurassic–Cretaceous: Worldwide Heart-shape less pronounced than in *Micraster*. Mouth in front ambulacral notch and anus near back point. Ambulacra not depressed and pores slit-shaped. Tubercles very small and thinly scattered over upper surface, but an area with many larger tubercles is present on the lower surface as in *Micraster*.

Echinolampas Eocene–Recent: Worldwide Generally similar to *Pygurus* but usually less flattened. Ambulacra petal-like and open as in *Pygurus* and the pores may be rounded or slit-like. Lines of pores usually unequal Ambulacra may be swollen above general surface of test.

Clypeaster Eocene–Recent: Worldwide Almost oval in outline with oral face deeply depressed with mouth in centre, vertically below apex. Upper surface elevated with ambulacra petal-shaped and wide with pore pairs joined by grooves. Anus on lower surface near flattened end of body.

Conulus Cretaceous: NA E Af Asia Round or pentagonal in outline with flattened lower surface and hemispherical to highly conical body. Anus at border of lower and upper surfaces and mouth central, lying immediately below apex. Ambulacra narrow and not petal-shaped. *Conulus* is very common in the chalk of Britain.

Micraster (dorsal view)

Micraster (ventral view)

Clypeaster

Holaster

Conulus

Pygurus

Echinolampas

Arthropods

The largest phylum of animals; includes insects, spiders, scorpions, crustaceans, millipedes, centipedes and several extinct groups, of which the trilobites (pages 294 to 301) are the most important. The group was well established by the start of the Cambrian. The most characteristic feature of the group is the hard outer coating, which is slightly flexible in most arthropods and provides attachment for the muscles. In most arthropods the body is divided into a head, thorax and abdomen, with the jointed legs attached to the thorax. With the exception of trilobites, the arthropods are relatively uncommon as fossils, though insects or crustaceans may be locally abundant. The arthropod groups are extremely large and it is possible to show only a few representatives of the phylum here.

Crustaceans

Lobsters, crabs, crayfish, shrimps, prawns, barnacles. One of the most important and diverse groups of marine invertebrates.

Hoploparia Cretaceous–Eocene: Worldwide A small lobster. Note the jointed legs, large *chelipeds* (pincers) and long, segmented abdomen.

Balanus (barnacles) Eocene–Recent: Worldwide Highly specialized crustaceans, sedentary as adults, with rigid plates. Opercular valves, across the opening, are retained in this specimen.

Insects

Body clearly divided into three parts; thorax carries three pairs of legs. Wings usually present.

Marquetia Oligocene Similar in general appearance to a small dragonfly but distinguished by features of the wings. Belongs to the family Nemopteridae which is now almost entirely restricted to the warmer parts of the world.

Insect in amber (*Leptis* Oligocene: E) Amber is fossilized resin and in some deposits numerous insects and spiders are preserved, complete, inside pieces of amber. Amber is rare but insects such as this are occasionally seen for sale as jewellery. *Leptis* is a member of the Diptera which includes the true flies.

Eurypterids

Eurypterus Ordovician–Carboniferous: NA E Asia An extinct group closely related to the scorpions and important during the Palaeozoic. Some eurypterids attained great size, being well over a metre long. Complete specimens are rare but fragments may be locally common. Eurypterids are popularly known as giant water scorpions and the largest, *Pterygotus*, was about 3m long, and is also the largest known arthropod.

Insect in amber

Eurypterus

Hoploparia

Marquetia

Balanus

Trilobites

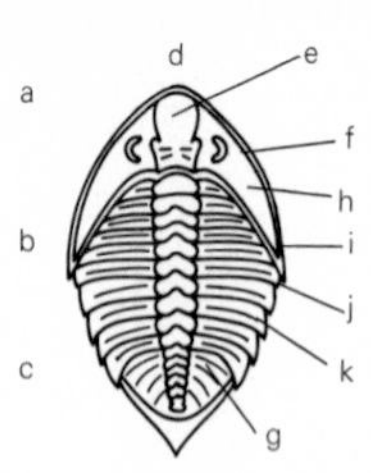

Typical structure of a trilobite as shown by *Dalmanites*

The most common fossil arthropods. The body is divided transversely into the *head* (a), *thorax* (b) and *tail* (c). It is divided along its length by two furrows delimiting the central *axis* (d) from the side regions. On the head the axial region is termed the *glabella* (e), and the sides are *genae* or *genal regions* (f). On the thorax and tail the sides are termed *pleural lobes* (g).

Eyes may be present on either side of the glabella (shown in *Phacops* on this page). The back outer corner of each genal region is termed the *genal angle* (h) and may be produced as a *genal spine* (i) (shown in *Dalmanites* on this page). A *front border* may be present; this is a raised rim around the front of the glabella and genae.

The thorax consists of *segments* defined by *thoracic grooves* (j) and the number of these is important. The side region of each segment is a *pleuron*. A *pleural furrow* (k) is a groove sometimes present on the upper face of each pleuron.

The tail also shows segmentation and transverse furrows may be present on the axis. The tail is known technically as the *pygidium* but this term is not used here.

The undersurface of the trilobite is only rarely exposed but a large plate, the *hypostome*, from the underside of the mouth may be locally very common.

Dalmanites Silurian–Devonian: Worldwide Medium-sized. Tail about same size as head. Glabella with deep grooves, widening forwards; eyes prominent; front border wide; genal spines long. Thorax about eleven segments; pleural furrows marked. Tail about eleven segments; back border smooth, carrying a spine. Ornament of small tubercles.

Phacops Silurian–Devonian: Worldwide Head larger than tail. Glabella wide and widening forwards; eyes large, lenses visible here; front border convex and bounded by deep groove; genal angles rounded. Thorax about eleven segments. Back edge rounded and smooth. The specimen shown here is rolled up.

Ogygopsis Cambrian: NA Medium-sized to large. Elongate; tail larger than head. Glabella parallel-sided with faint cross-grooves; eyes long and narrow; front border wide and flattened; genal spines short (not shown here). Thorax about eight segments; axis strong and wide; pleurae with deep wide pleural furrows. Tail about ten segments; axis tapering; tail pleurae with deep segmental grooves and furrows; back edge with convex border and smooth outline.

Calymene Silurian–Devonian: Worldwide Medium-sized. Tail smaller than head. Glabella very convex and sloping steeply at the front, narrowing forwards and carrying three pairs of swellings; eyes large; front border convex and separated from glabella by deep groove; genal angle rounded. Thorax about thirteen segments. Tail about six segments.

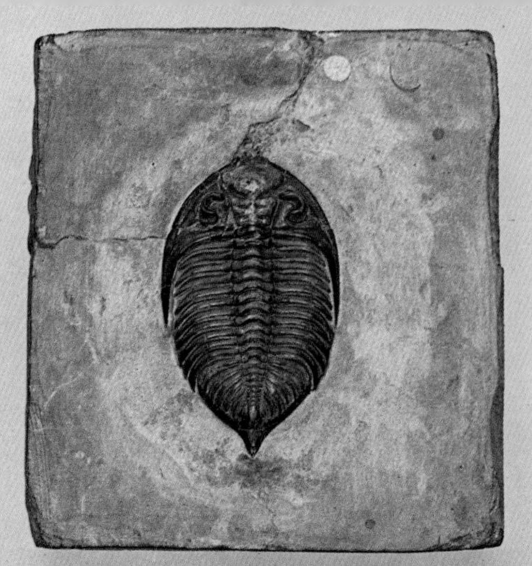

Dalmanites

5 cm

Phacops

Ogygopsis

Calymene

Paradoxides Cambrian: NA E Af Aust Head much larger than tail. Glabella expanding forwards, carrying about three pairs of cross-furrows; eyes large; genal spines about half body length. Thorax of about eighteen segments; pleural furrows strong and diagonal; pleurae produced as spines at sides, which increase in size backwards. Tail small with straight back edge.

Paedeumias Cambrian: NA E Asia Head large with flattened cheeks. Glabella deeply furrowed. Rounded swelling at front of glabella is connected to front border by a ridge; genal spines long. Thorax about fourteen segments, decreasing in size backwards from the second; first segment with short spine; second large with spine extending back beyond tail region; other pleurae with long spines; pleural furrows deep. Tail small, carrying long spine (twisted to left in the specimen shown here).

Olenoides Cambrian: NA SA Asia Head and tail about same size. Glabella with several furrows, expanding slightly forwards and reaching front border; eyes medium-sized; front border convex and wide; genal spines short. Thorax about seven segments; axis wide and tapering with cross-furrows and tubercles or spines on each segment; pleurae produced as short spines. Tail of at least five segments with axis tapering backwards; back edge with several pairs of spines.

Oryctocephalus Cambrian: NA SA E Asia Tail and head almost equal in size. Glabella parallel-sided with three or four pairs of cross-furrows which have deep pits at each end; eyes small; genal spines long (not clearly shown here). Thorax about seven segments; pleurae produced as spines; pleural furrows deep and diagonal. Tail axis with six cross-grooves; sides and back of tail produced as long spines (not clearly shown here)

Encrinurus Ordovician–Silurian: Worldwide Head larger than tail. Glabella widening forwards; eyes pronounced; genal spines small, directed outwards. Thorax of eleven or twelve segments. Tail of five to ten pleural segments; back edge serrated. Ornament of strong tubercles.

Cheirurus Ordovician–Devonian: Worldwide Tail smaller than head. Glabella produced forwards to overhang front border; eyes medium-sized; genal spines small. Thorax about eleven segments; pleural furrows short and diagonal. Tail with well-defined, deeply grooved axis; back edge with three pairs of spines separated by small central spine.

Leonaspis Silurian–Devonian: NA SA E Head very wide; eyes large; front border with strong spines; genal spine large (broken off in the specimen shown here). Thorax about eleven segments; pleurae produced backwards as spines. Tail small; back edge with one pair of large spines and two pairs of smaller spines.

5 cm

Paradoxides

Paedeumias

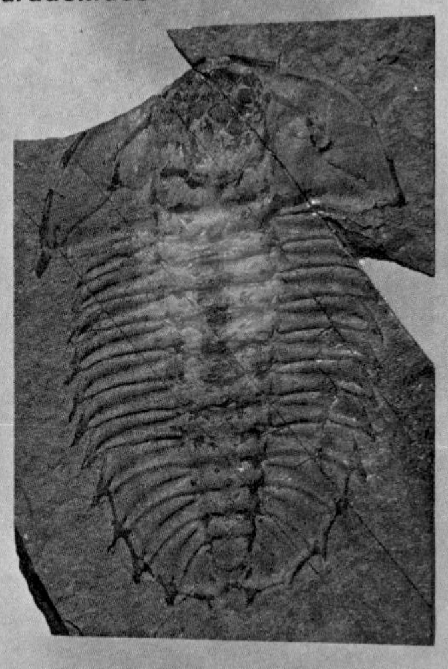

Olenoides

Oryctocephalus

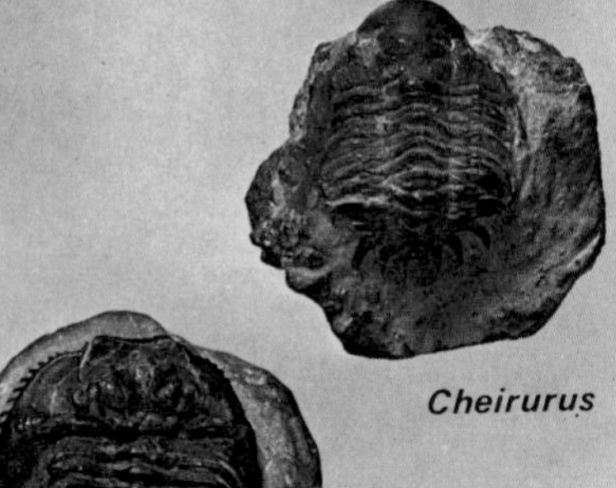

Cheirurus

Encrinurus

Leonaspis

Triplagnostus Cambrian: NA E Asia Aust Small, less than 1 cm. Head and tail same size. Glabella divided into triangular front and elongate hind lobes, less convex than in *Eodiscus;* cheeks curved and divided at front by a groove; front border strong and convex; eyes absent; genal angle rounded or with small genal spine. Thorax of two segments. Tail very similar to head; axis of tail slightly wider than glabella, divided into larger triangular back region and a shorter front region which may carry a strong swelling. A groove at the back separates the two curved side regions of the tail. Back border similar to front border.

Eodiscus Cambrian: NA E Very small, less than 0·5 cm. Head and tail same size. Head consists of a short, very convex glabella carrying a single pair of indistinct furrows, and curved cheek regions which are divided at the front by a deep groove extending to the narrow front border; eyes absent; genal angle sharp or strong genal spines may be present. Thorax of two or three segments. Tail axis pronounced and carrying many strong cross-grooves; tail pleurae swollen and curved. Head and tail regions are here shown separated, and *Eodiscus* is often found in this condition.

Cedaria Cambrian: NA Head and tail almost equal in size. Glabella lacking furrows, having rounded front end and terminating well behind border; eyes medium-sized; front border strong and convex; genal spines (not shown here) fairly long. Thorax about seven segments; axis well defined by furrows; pleural furrows long. Tail with strong axis and four or five furrows; back edge rounded.

Ctenocephalus Cambrian: NA E Af Asia Only head region shown. Glabella very convex, tapering forwards, carrying three pairs of strong furrows; cheeks swollen, convex; eyes absent; front border very convex; genal spines long, extending over half the length of the thorax (not shown here). Body similar in shape to that of *Elrathia* with small tail and about fifteen thoracic segments. Ornament of fine tubercles covering head region.

Bonnaspis Cambrian: NA Head slightly larger than tail. Glabella very convex, expanding strongly forwards to front edge; furrows not present on glabella; eyes small; genal spines short (not shown here). Thorax about seven segments; pleurae with deep furrows. Tail up to five segments, poorly defined; back edge rounded.

Elrathia Cambrian: NA Medium-sized. Head much larger than tail. Glabella tapering forwards with rounded front end well behind front border; glabella surface carrying several pairs of weak furrows; eye ridges strong; front border wide; genal spines short. Thorax about thirteen segments; pleural furrows long and deep. Shallow furrows on tail indicate about five segments; back edge smoothly rounded. The most frequently encountered North American trilobite.

Triplagnostus

Eodiscus head

tail

Ctenocephalus

Cedaria

Bonnaspis

Elrathia

5 cm

Cryptolithus Ordovician: NA E Head much larger than tail. Glabella narrow and very convex, widening forwards and carrying a single pair of furrows; eyes not visible. The most characteristic feature is the wide front border which slopes downwards and outwards, and carries radiating rows of deep pits. Genal spines long Thorax about six segments. Tail smooth with raised central region and smooth back edge.

Bumastus Ordovician–Silurian: Worldwide Elongate with head and tail regions equal in size. Glabella not clearly defined but head carries large swellings on either side; genal angles rounded. Thorax of eight to ten segments; axis not clearly defined. Tail convex with steep back border and smooth outline. Surface ornament very weak.

Trinucleus Ordovician E Similar in general shape to *Cryptolithus*. Glabella convex and carrying three pairs of deep furrows; front border wide, carrying radiating grooves; genal spines long (not shown here). Thorax of six segments; axis strong. Tail much wider than long; back edge smooth.

Harpes Devonian: E Af The head region only is shown here. Glabella very convex with lobes at sides; eyes strong; genal spines almost as long as body and very wide; front border wide, carrying many fine pits and tubercles. Thorax about twenty-nine segments. Tail small.

Griffithides Carboniferous: NA E Medium-sized; elongate. Head and tail almost equal in size. Glabella wide and expanding slightly forwards; eyes small; front border narrow; genal angle rounded. Thorax about nine segments; axis strong. Tail of numerous segments.

Isotelus Ordovician: NA E Asia Head and tail equal in size. Glabella not clearly defined; eyes medium-sized, produced as conical swellings; genal angles rounded. Thorax of eight segments; axis very wide and defined by shallow furrows. Pleurae short; pleural furrows short, deep and diagonal. Tail region pointed with weakly defined axial region and weak furrows in the pleural areas.

5cm

Cryptolithus

Trinucleus

Harpes

Isotelus

Bumastus

Griffithides

Vertebrates

Fishes, amphibians, reptiles, birds and mammals. Vertebrates have internal skeletons of cartilage or bone. Complete fossil skeletons are rare and identification is often based on isolated bones or teeth.

Fishes

The largest group of living vertebrates with over 20,000 species and a huge number of fossils forms.

Armoured fishes Many Palaeozoic fishes had a heavy external armour of bony plates. These are usually found isolated. Especially of Silurian and Devonian age.

Cephalaspis Silurian–Devonian: NA E Asia One of the best known armoured fishes. A complete specimen is shown. Note the wide head and slender tail region which are covered with bony plates.

Sharks and rays The skeleton is composed of cartilage and rarely fossilizes. Teeth are the commonest remains in the Carboniferous and later, becoming common in the Cretaceous and Tertiary.

Hybodus Triassic–Cretaceous: Worldwide Teeth low and wide, high central point and numerous side points. Spine long and pointed with grooved sides. These spines support the fins in members of the hybodont group of sharks which were common during the Mesozoic.

Carcharodon Palaeocene–Pleistocene: Worldwide Very large teeth with a single point and serrated edges.

Lamna Cretaceous–Pliocene: Worldwide A medium-sized shark. Tooth with a large point and a pair of side points. No serrated edges.

Orodus Carboniferous–Permian: NA E Long, wide teeth with a single point and rippled sides.

Ptychodus Cretaceous: NA E Af Asia Flattened teeth suitable for crushing mollusc shells. This is a hybodont shark, but its teeth are similar to those of many rays.

Myliobatis Cretaceous–Pliocene: Worldwide A ray. Flattened crushing teeth indicating a shellfish diet.

Bony fishes Most living fishes such as the salmon, cod and herring. The group was important in fresh water by the end of the Palaeozoic, and has since become important in marine conditions. Identification is very difficult.

Ceratodus Triassic–Palaeocene: Worldwide Lung fish generally known as fossils from teeth only. Shape and ridges are characteristic. Surface with many small pores.

Perleidus Triassic: E Af Asia Complete bony fishes may be found in nodules and the presence of the dead fish sometimes appears to have caused the formation of the nodule.

Brookvalia Triassic: Aust Fossil fishes may be found flattened out along bedding planes and are discovered when the rock is split.

5 cm

Cephalaspis

Hybodus teeth

Hybodus spine

Lamna tooth

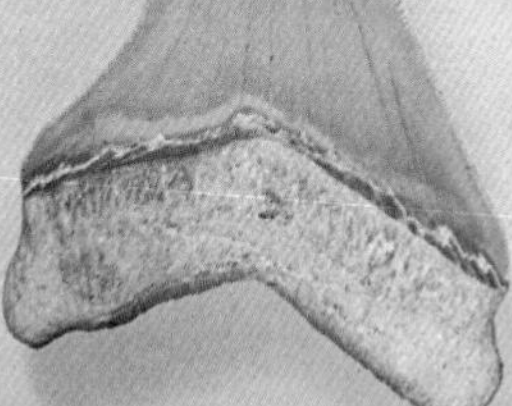

Carcharodon teeth

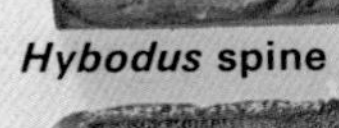

Ptychodus teeth

Myliobatis teeth

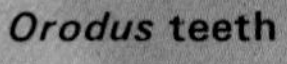

Orodus teeth

Ceratodus tooth

Perleidus in a nodule

Brookvalia on a bedding plane

Reptiles

Dinosaurs, crocodiles, turtles, ichthyosaurs, lizards and snakes. Reptiles were very important on land and in the sea from the Permian to the end of the Cretaceous. Their remains are relatively rare, but may be locally abundant especially in North America and Africa.

Crocodiles Triassic–Recent: Worldwide Crocodiles are among the commonest fossil reptiles, but they are very difficult to identify generically. A bony plate and two teeth are shown here. Plates are arranged in rows along the back of the animal and always have heavily pitted upper surfaces. Crocodile teeth vary greatly along the jaw of the same individual. They usually have short, sharply pointed crowns (the black upper part), and long roots.

Trionyx Jurassic–Recent: NA E Af Asia Pieces or scutes of the turtle shell or carapace are the commonest parts found. A scute from the upper part of the shell of *Trionyx*, a freshwater turtle, is shown here. In freshwater turtles the scutes have patterns on their upper surfaces but in marine turtles the surfaces of the scutes are smooth.

Ichthyosaurs Jurassic–Cretaceous: NA SA E Asia Aust Rarer than crocodiles or turtles but important marine reptiles in the Mesozoic. The most frequently found parts are the centra of the vertebrae. The two swellings at the top indicate where the neural arch was broken off. A fragment consisting of part of the upper and lower jaws and teeth is also shown here. The crowns of the teeth carry deep vertical grooves.

Dinosaurs Triassic–Cretaceous: Worldwide Regions where dinosaur remains have been found are usually well known.

Aublysodon Cretaceous: NA Flesh-eating dinosaurs have high sharp teeth. A single tooth of *Aublysodon* is shown here.

Iguanodon Jurassic–Cretaceous: E Af Asia Plant-eating dinosaurs have square crowned teeth with flat upper surfaces and ridged sides. A single tooth of *Iguanodon* is shown here.

Hypsilophodon Cretaceous: E Not all dinosaurs were large and a femur of *Hypsilophodon* is shown here. This dinosaur was about one metre tall.

Birds

Birds originated in the Jurassic and have been common since the Palaeocene. Their bones are very fragile as they have thin walls and an empty internal cavity; as a result they are only rarely preserved as fossils. You are most likely to find them in Pleistocene deposits and they can usually be identified by comparison with living bird bones. Shown here is the metatarsus (lower leg) of a dodo (Pleistocene: Mauritius) which has a form characteristic of birds, as there are three articulating surfaces for the toes at the lower end; a feature not found in mammals or reptiles. Other bird bones may sometimes be confused with bones from members of these two groups.

5 cm

Crocodile bony plate

Crocodile teeth

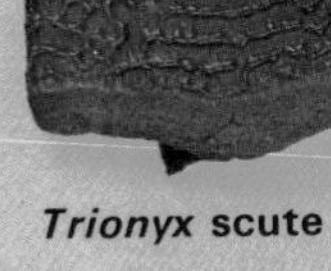

Trionyx scute

Ichthyosaur jaws and teeth

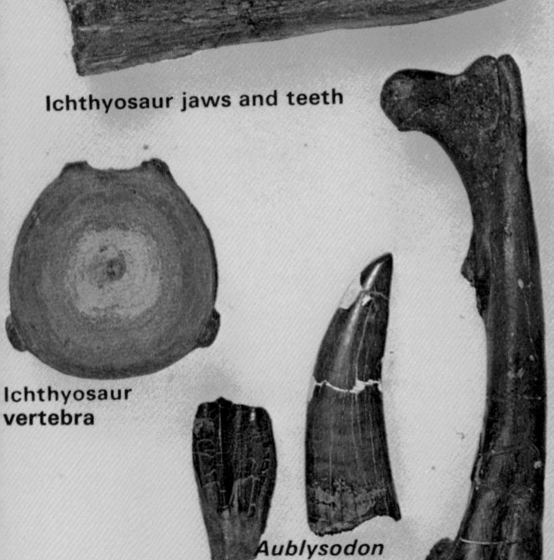

Ichthyosaur vertebra

Iguanodon tooth

Aublysodon tooth

Hypsilophodon femur

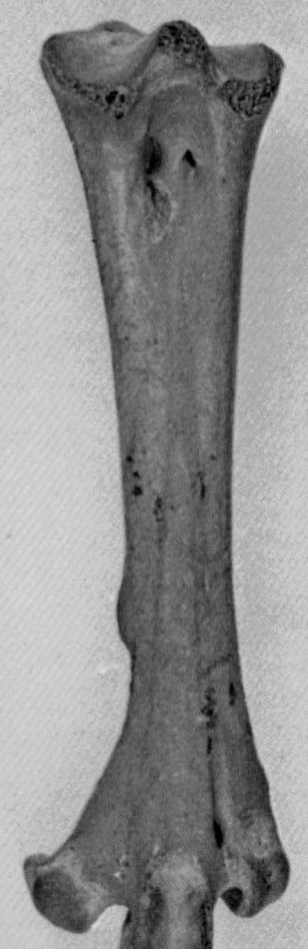

Dodo metatarsus

Mammals

Man, horse, elephants, whales, bats and dog. Important since the end of the Cretaceous. Remains common in the Pleistocene and may be locally abundant in earlier deposits, for example Oligocene: South Dakota. Identification of most mammals is based on features of the cheek teeth (the back three cheek teeth are molars). The part of the tooth above the gum is the *crown* and larger swellings on the chewing surface are *cusps*. The pattern made by the cusps is important for identification.

Plant-eating mammals Cheek teeth with high crowns which are usually square or rectangular with flat chewing surfaces. Crests often developed

Rhinoceroses Oligocene–Recent: NA E Af Asia Upper teeth with continuous outer walls (upper in plate) and two inner crests. Lower teeth consisting of two crescentic ridges.

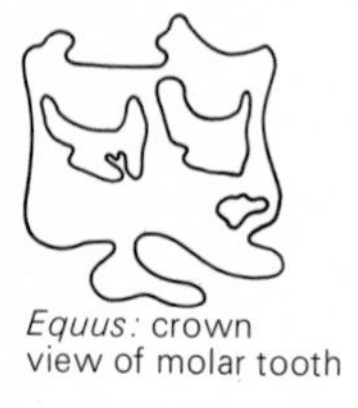

Equus: crown view of molar tooth

Equus Pleistocene–Recent: NA SA E Af Asia Horse, donkey, zebra. Teeth very high with square crowns (upper) and rectangular crowns (lower). Pattern complex. Eocene (for example *Hyracotherium* shown on page 308), Oligocene and early Miocene horses have low crowned teeth similar to those of small rhinoceroses.

Bos Pleistocene–Recent: Alaska E Af Asia Cattle, includes the domestic cow. Upper teeth with four crescentic cusps forming square crown. Lower molars rectangular with an extra cusp at the back of the last molar. Bison (NA), antelopes and gazelles (E Af Asia), deer and giraffes (E Af Asia) have cheek teeth with similar patterns.

Hippopotamus Pleistocene: E Af Asia A lower molar. Four cusps arranged in a rectangle; similar in general pattern to *Bos* but cusps less crescentic. Some pigs have similar teeth, as do members of an extinct group, the anthracotheres.

Ursus Pliocene–Recent: NA E Asia Includes grizzly bear and brown bear. Cheek teeth with low crowns, low rounded cusps and many additional small swellings and grooves. Some pigs have similar cheek teeth. Fangs (canines) large with swollen root and sharply pointed crown.

Castor Pliocene–Recent: NA E Asia Beaver. Complete lower jaw shown. Front tooth extremely long with almost triangular cross-section and enamel on front face only. Cheek teeth few in number, separated from front tooth by a space, very high crowned, flat-topped with several cross-crests.

Flesh-eating mammals Cheek teeth with low narrow crowns, usually with sharp edges or points. Cross crests rarely present.

Canis Pliocene–Recent: Worldwide Wolf, domestic dog, dingo. One of the cheek teeth is large and elongate with a sharp chewing edge that is used for slicing flesh. Cats, hyenas, weasels and civets have slicing teeth generally similar to these.

Rhinoceros upper molar

Equus upper molar

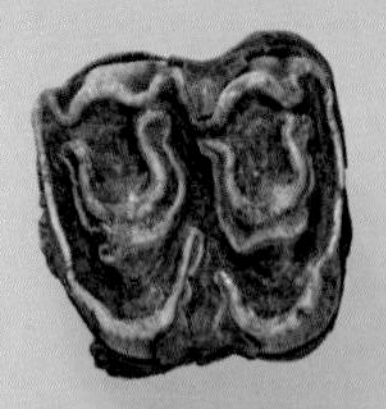

Bos upper molar

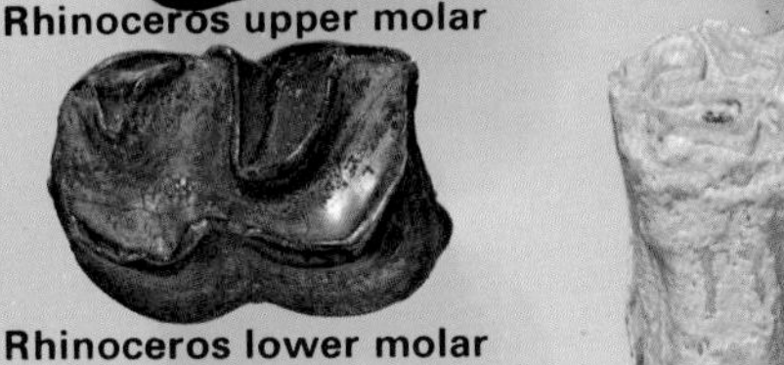

Rhinoceros lower molar

Bos lower molar

Equus lower molar

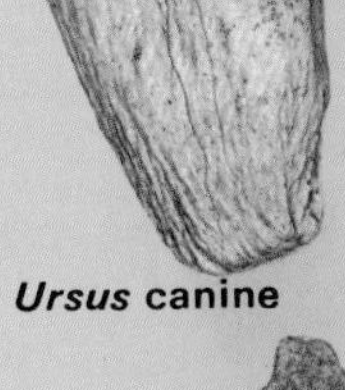

Ursus canine

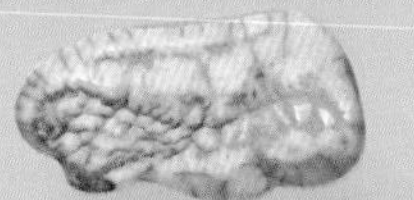

Ursus lower molar

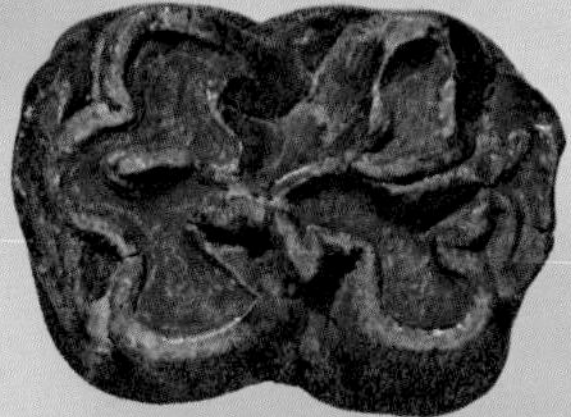

Hippopotamus lower molar

Castor lower jaw

Canis lower jaw fragment

Elephas **Pleistocene: E Asia Recent: Asia** Living representative is the Indian elephant. Very large cheek teeth consisting of wide, almost parallel-sided platelets forming ridges on chewing surface. Mammoths (*Mammuthus* Pleistocene: NA E Af Asia) have similar teeth but with many more platelets per unit length. In African elephant cheek teeth the platelets have an almost diamond-shaped cross-section.

Mammut **Miocene–Pleistocene: NA E Af Asia** Remains relatively common in North American Pleistocene, known as American mastodon. Cheek teeth large with several cross-crests but these are much lower and more triangular than in the elephantids. The enamel on this type of molar is very thick.

Merycoidodon **Oligocene: NA** Also known as *Oreodon* and very common in Oligocene of Mid-West where beds are known as 'Oreodon beds'. Skull relatively short and deep. A plant-eating mammal having upper molars similar in general crown pattern to *Bos;* consisting of four crescents but crowns much lower. Upper canine relatively large, giving skull a pig-like appearance.

Hyracotherium **Palaeocene–Eocene: NA E** Also known as *Eohippus:* the first horse. Skull long and low. Cheek teeth, low crowned with four rounded cusps on the upper molars. You are unlikely to find remains of this animal but it is displayed in most museums.

Diprotodon **Pleistocene: Aust** Upper teeth shown. These each have a pair of low sharp-edged cross-crests. *Diprotodon* is a marsupial and is therefore related to the kangaroo, koala, wombat and opossum. Remains of marsupials are the commonest mammalian remains in Australia and also occur in South America (mainly Eocene–Pliocene), but are very rare in North America and Europe and unknown from Asia and Africa. In the Miocene to Pleistocene of Europe, Africa and Asia large teeth similar to *Diprotodon* are from *Deinotherium*, while smaller ones are from pigs or tapirs (not Africa). *Pyrotherium* from the Oligocene of South America also had similar teeth.

Hyaena **Pliocene–Recent: E Af Asia** Hyaena. This is a skull of a young individual, but it shows the long slicing cheek tooth and the relatively small number of teeth. The upper fangs are not erupted, but the point is visible near the front of the jaw. The arch of the jaw is wide to accommodate large jaw muscles, and the face is relatively short. These are features of most flesh-eating mammals, but are highly developed in the hyaenas which are adapted for crushing bones.

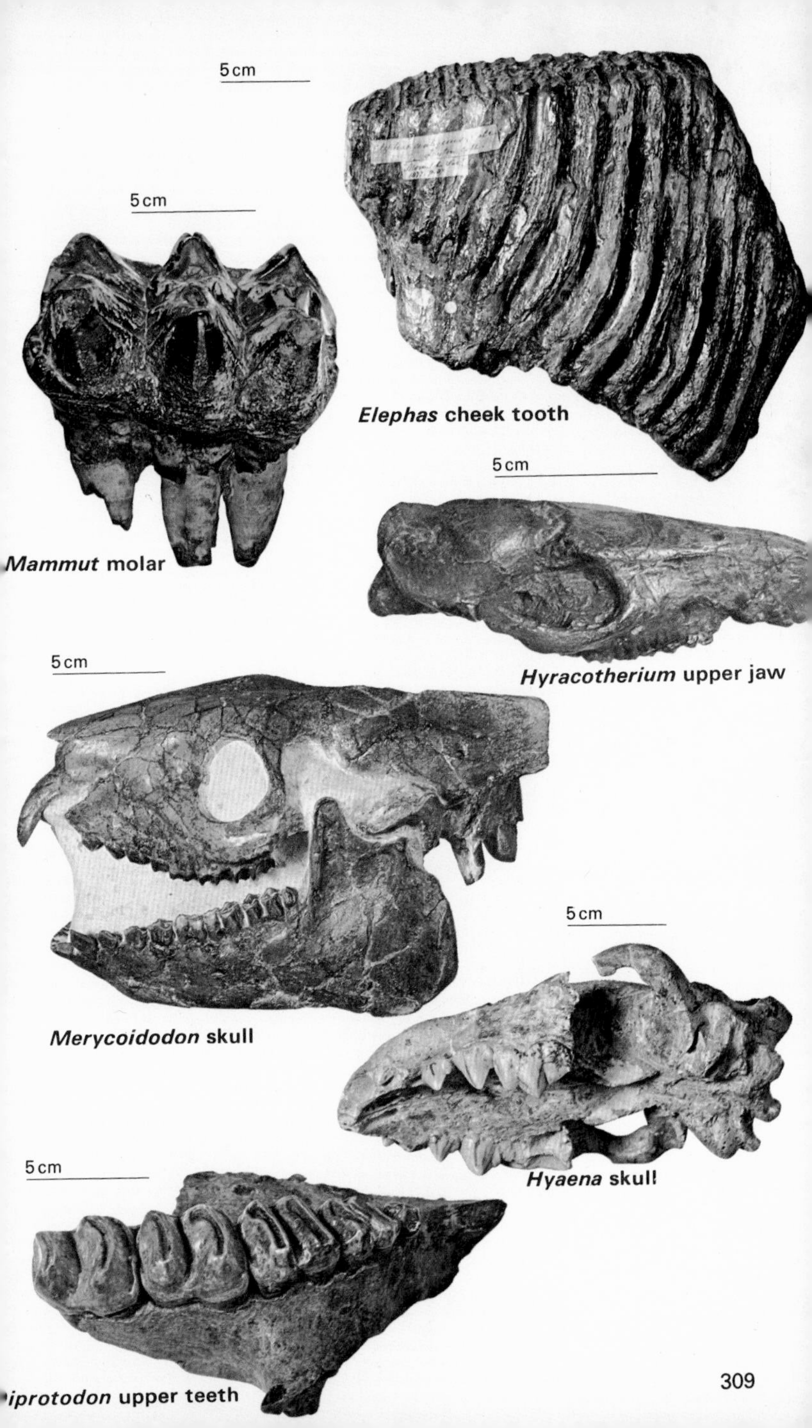

Elephas cheek tooth

Mammut molar

Hyracotherium upper jaw

Merycoidodon skull

Hyaena skull

iprotodon upper teeth

Geological time-scale

This stratigraphical column gives the time span of each geological period (Oligocene, Pliocene etc) and its absolute age. Periods are grouped into larger time units called eras (Caenozoic, Mesozoic etc).

Time in millions of years	Era	Period
2	Caenozoic	Quaternary
7	Caenozoic	Pliocene
25	Caenozoic	Miocene
40	Caenozoic	Oligocene
55	Caenozoic	Eocene
70	Caenozoic	Palaeocene
135	Mesozoic	Cretaceous
195	Mesozoic	Jurassic
225	Mesozoic	Triassic
280	Palaeozoic	Permian
345	Palaeozoic	Carboniferous
395	Palaeozoic	Devonian
440	Palaeozoic	Silurian
500	Palaeozoic	Ordovician
600	Palaeozoic	Cambrian
		Pre-cambrian

Further reading

General geology

Gilluly, J, Waters, AC and Woodford, AO **Principles of Geology**, second edition. WH Freeman and Company, San Francisco and Reading, 1959.

Holmes, A **Principles of Physical Geology**, new and fully revised edition. Nelson, London, 1965.

Read, HH and Watson, J **Introduction to Geology, volume I, principles**. MacMillan, London and New York, 1962.

Robson, DA **The Science of Geology**. Blandford Press, London, 1968.

Minerals

Bishop, AC **An Outline of Crystal Morphology**. Hutchinson, London, 1967.

Correns, CW **Introduction to Mineralogy, Crystallography and Petrology**, translated by Johns, WD. George Allen and Unwin, London, 1969.

Dana, ES **A Textbook of Mineralogy**, fourth edition, revised and enlarged by Ford, WE. John Wiley and Sons, New York, London and Sydney, 1932.

Dana, JD **A System of Mineralogy**, seventh edition. John Wiley and Sons, New York, volume I, 1944; volume II, 1951; volume III, 1962.

Deer, WA, Howie, RA and Zussman, J **An Introduction to Rock-forming Minerals**. Longmans, London, 1966.

Desautels, PE **The Mineral Kingdom**. Hamlyn, London, 1969.

Hurlbut, CS **Dana's Manual of Mineralogy**, seventeenth edition. John Wiley and Sons, New York, 1959.

Hurlbut, CS **Minerals and Man**. Thames and Hudson, London, 1969.

Jones, WR **Minerals in Industry**. Penguin Books, Harmondsworth, 1963.

Kostov, I **Mineralogy**. Oliver and Boyd, Edinburgh and London, 1968.

Kraus, EJ, Hunt, WF and Ramsdell, LS **Mineralogy: an Introduction to the Study of Minerals and Crystals**, fifth edition. McGraw Hill, New York, 1959.

Mason, B and Berry, LG **Elements of Mineralogy**. WH Freeman and Company, San Francisco, 1968.

Pough, FH **A Field Guide to Rocks and Minerals**. Houghton Mifflin Company, Boston, 1960.

Read, HH **Rutley's Elements of Mineralogy**, twenty-sixth edition. Thomas Murby and Company, London, 1970.

Rocks

Harker, A **Metamorphism**. Methuen, London, 1950.

Hatch, FH, Wells, AK and Wells, MK **Petrology of the Igneous Rocks**, twelfth edition. Thomas Murby, London, 1961.

Mason, B **Meteorites**. John Wiley, New York and London, 1962.

Pettijohn, FJ **Sedimentary Rocks**. Harper and Brothers, New York, 1949.

Fossils

Treatise on Invertebrate Paleontology. Edited by Moore, R C. Geological Society of America and University of Kansas Press, 1953 onwards. This is a reference book and can be consulted in a library. There are volumes dealing with all the phyla of fossil invertebrates; every known genus is described with figures, geological and geographical ranges.

Moore, R C, Lalicker, C G and Fischer, A G **Invertebrate Fossils.** McGraw Hill, New York, 1952.

Romer, A S **Vertebrate Paleontology,** third edition. University of Chicago Press, 1966.

British Palaeozoic Fossils (1969), **British Mesozoic Fossils** (1972), **British Caenozoic Fossils** (1971). British Museum (Natural History). With good illustrations of most species of fossils occurring in Britain.

Kummel, B and Raup, D **Handbook of Paleontological Techniques.** W H Freeman and Company, Reading, 1965.

Index

Page numbers in heavier type refer to main description and facing illustration.
Page numbers in italic refer to line drawings.